Evolution to the Fourth Enlightenment

Donald Pribor

University of Toledo

Kendall Hunt
publishing company

Kendall Hunt
publishing company

www.kendallhunt.com
Send all inquiries to:
4050 Westmark Drive
Dubuque, IA 52004-1840

This edition has been printed directly from print-ready copy.

Copyright © 2009 by Kendall Hunt Publishing Company

ISBN 978-0-7575-7144-2

Printed in the United States of America
10 9 8 7 6 5 4 3 2 1

DEDICATION

I dedicate this book to President Barack Obama who exemplifies the Fourth Enlightenment as a result of his commitment to: collaboration between rational, control objectivity and feeling, participatory subjectivity; transcending ideological conflicts by means of creative, mindful dialogue; non-ideological pragmatism; Faith-Hope as the basis for individual and cultural transformations, and to the ultimate SOURCE, who he calls God that generates Christ, that also may be understood as Allah, or non-theistically as Brahman of Hinduism or Emptiness of Buddhism.

EVOLUTION TO THE FOURTH ENLIGHTENMENT

CONTENTS

UNIT VII: EMERGENCE OF THE FOURTH ENLIGHTENMENT

UNIT VIII: THEMES OF THE FOURTH ENLIGHTENMENT

one."[2] Finally, though Murray is committed to the ideal of a liberal education, he presents reasons why this ideal no longer is attempted at virtually all institutions, even the very best ones, and where it is attempted, most students cannot achieve it.

However, "the times they are a changing." The University of Toledo is in the bottom of the fourth tear (in part because 35 to 40% of our students need remedial courses in math and English), but UT is edging toward major changes. In a draft (February 2009) of self-evaluation and planning – with the aid of consultants – a roundtable of 30 faculty members, administrators and other stakeholders in the Arts and Sciences College (A & S College) made four evaluations of the college. First, there is no shared vision. Second, the College as a whole has not yet realized its potential. Third, though there are many academic strengths and achievements, the focus is on individual departments (especially the biology department with millions of dollars of grant money) and scholars rather than on the College as a whole, and fourth, the College does not have "a language to describe itself in ways that are unique and that convey clearly and coherently its sense of purpose." Based on the above evaluations of the A & S College, the roundtable made four major overlapping recommendations. First, the College should "develop an identity as a hub of integrative learning … that encourage[s] faculty members and students to work across boundaries of knowledge in ways that allow for new insights and [non-ideological, pragmatic] solutions to complex societal problems." Second, students should be educated for "learning in creative ways." Third, the College should become a place of connections "in which students acquire the ability to see … across the boundaries of academic disciplines …. [so that it provides] a kind of integrative learning that extends beyond the realm of interdisciplinary studies *per se*." Fourth, the College must "foster modes of thinking and inquiry that build linkages across academic departments and disciplines … [The result is that students and some faculty will have] breath as well as depth … [and will have] the ability … to approach problems through the conceptual framework of more than one discipline." The roundtable concluded that these recommendations produce the chaos and challenge of how to construct a shared vision and a strategy for accomplishing these recommendations, especially in the context of the longstanding tradition of academic institutions resisting any profound transformation.

I suspect that many colleges and universities – even those in the first and second tear – are exploring ways to transform to better meet the needs of the millennials who come to them and the needs of a transforming American culture. The outstanding achievements of specialized undergraduate education and graduate education distracts us from acknowledging the many ways our educational system – like our health care system – is broken and unsustainable. I suggest that our higher education system is like GM before its downfall. If academia does not start to make profound changes, it may "self-destruct" the way GM did, and this may begin to occur as early as the 2010-2011 academic year. Profound changes in colleges and the way they educate teachers will lead to profound transformations in elementary and high school education.

Unit VIII of this book provides some major elements of the fourth enlightenment. These insights could provide guidelines for transforming higher education. Chapter 23 describes postmodern scientific constructivism that is centered on non-ideological pragmatism that may be expanded to all empirical problems. Scientific constructivism evolved to narrative, scientific constructivism (as implied by Thomas Kuhn's idea of scientific paradigms). In chapter 24 I describe how narrative, scientific knowing may evolve to narrative constructivism that, among other things, provides the language for reforming education. In particular, it leads to student-

---

[2] Charles Murray. *Real Education* (New York: Crown Forum ,2008), p. 104.

teacher constructivism that produces creative learning in which the learner constructs a subjective understanding of abstract ideas in terms of his/her personal experiences. While many people believe that science opposes spirituality or mysticism, in fact, the cutting edge of postmodern science (systems science of chaos and self-organization) may be shown to imply, as indicated in chapter 25, that there is an ultimate SOURCE generating all of reality. The SOURCE may be interpreted as creator God (Jewish-Christian-Muslim God) or non-theistically as Te of Taoism, Brahman of Hinduism or Emptiness of Buddhism or Other. Especially since the late 1960s, many thinkers proposed that humans are on the verge of transforming to a higher level of consciousness, which I call the move to the fourth enlightenment. Instead of science-dominated education opposing this transformation, it could, as described in chapter 26, guide it. Narrative constructivism applied to systems science is vision logic of postmodern science called (by Wilber) "constructivism postmodernism." In particular, narrative constructivism promotes sequential dialectical collaboration between subjective, (Eros-chaos), narratives and objective, (Eros-order), logical models. This chapter also describes narrative constructivism leading to generalizations about human creativity and individuation and to a psychic framework for human creativity. I further describe the idea of human individuation as moving toward the first stage of transpersonal, human consciousness. I call this stage, "subjective, narrative, constructivism mind-self." According to my personal experience of it, this stage is related to the experience of many in alcoholic anonymous (AA) and to some religious people who have experienced being "born again."

Universal creativity produces harmony in our biosphere wherein the ecology of this creativity exhibits seven characteristics, which chapter 27 describes. In this chapter I also describe aspects of my spiritual journey in relation to the human partnership with nature that generates the evolution of the knowledge universe and to the Inner Witness that generates mindfulness. I relate the knowledge universe to stages of individuation that include the "psychic self" (natural mysticism), openness to each of the three mind centers (knowing, acting, and feeling) to creative mindful dialogue, no knowing-no mind self, and no ego. All this provides the foundation for understanding how to engage in creative, mindful dialogue. Such dialogue occurs in the sequence of participatory dialogue and collaborative dialogue described in chapters 28 and 29 respectively.

The economic collapse in September 2008 represents one expression of radical, ego constructivism that intensified the drive toward radical, postmodern democratization. In everyday language people subcoming to this drive do whatever they want to do without any external constraints. As a result, many trends indicate that American society is degenerating toward self-extinction. But out of this chaos a new order may emerge, which is the fourth enlightenment. Chapter 30 describes spiritual constructivism that is a middle way between dogmatic hierarchies and radical, ego constructivism. If this were incorporated into the education system so as to transform it, American society would move to a new level of collective awareness and express elements of the fourth enlightenment. In particular, this final chapter describes how spiritual constructivism could reformulate the three core ideas of the Declaration of Independence to fit the needs of postmodern democracy.

UNIT I UNIVERSAL THEORY OF CREATIVITY
Chapter 1
FOUNDATION IDEAS IN PHYSICS

FIRST GLIMMER OF THE FOURTH ENLIGHTENMENT IN SCIENCE

The fourth enlightenment began as an expansion of the mystical vision of the third enlightenment, which produced modern science. Newtonian mechanics that represented the first emergence of modern science proclaimed that nature is not transparent; it does not produce immediately evident truths. Rather, one can discover empirical patterns and then construct mathematical models, such as Newton's law of gravity, to represent these patterns. However, the models do not explain why patterns occur, as medieval philosophy sought to do. Also, the scientific models do not give insight into paradoxes that sometimes manifest in these patterns such as Newton's idea of a force of gravity at a distance. Thus, the scientist mathematically describing aspects of Nature does not comprehend her. Nature remains a mystery and at times paradoxical. But she continually provides new patterns and thereby entices the scientist to engage with her by constructing new models to represent these patterns.

Western culture in the period of the last third of the 1800s and up to the beginning of World War I in 1914 burst forth in a flourish of creativity that produced radically new ideas. Maxwell's theory of waves was a validated way of describing patterns such as light but it contradicted aspects of the traditional Newtonian mechanics. This generated an intellectual-spiritual crisis: how can humans deal with the prospect of nature expressing contradictory patterns of behavior? The young Einstein presented a core idea of the fourth enlightenment, which is creativity that transcends rather than resolves paradoxes. Einstein's special and general theories of relativity put forth a mathematical perspective that incorporates rather than replacing modified versions of Maxwellian and Newtonian models. In Einstein's theories energy replaces force as the unifying idea of physics. However, a new perspective describing interrelations between motion and temperature contradicted aspects in both the old Newtonian and the new Einsteinian perspectives. In these mechanistic perspectives time is like a dimension of space; it is reversible. Furthermore, during any change in the universe, the loss of order in one place is equal to a gain of order elsewhere; order is conserved. But the new science of motion-temperature, called *thermodynamics*, claims that in any change there is a loss of order that never is made up elsewhere; that is, as the universe ages, chaos always increases, which implies that time is irreversible. Time only flows forward, never backward.

In the 20th century, the integration of the unifying idea of energy with the new science of motion-temperature (thermodynamics) produced a paradoxical view of energy. Energy represents both Order and Chaos. Energy has two aspects: 1) potential and 2) flux. *Potential*, when applied to a system or applied to a system's environment, represents Order and the possibility of generating a type of event. *Flux* is the way energy as potential manifests itself. Energy flux leads to one or both of two kinds of change: (1) motion, that is, change of location in space and (2) change in temperature. Energy flux leading to motion is called *work* and energy flux leading to change in temperature is called *heat*. Potential representing Order "points to" sequential, reversible changes of structure in which time is reversible and Order is conserved. That is, at any moment during these changes Order never is lost. Rather the disappearance of Order associated with one event is coupled with the emergence of a new and equal degree (quantity) of Order of the sequential next event. For example, in the ideal sea saw, as one person goes down, the other person goes up; the quantity of going down equals the quantity of going up. The conservation of Order means that there is no Chaos.

Flux may be metaphorically thought of as the transfer of a quantity of energy that leads to motion or as the transfer of *thermal energy* that leads to a change in temperature. When we imagine these transfers of energy to occur "infinitely slowly," then flux also involves reversible processes in which Order is conserved. However, energy flux of finite human experience implies: (1) time is irreversible, (2) time based events are irreversible, (3) Order is not conserved, which means there always is a loss of some degree (or quantity) of Order and a corresponding gain of degree (or quantity) of Chaos, and (4) flux may involve only heat or heat and work, but it never only involves work.

The fluxes associated with the operation of any machines we shall describe are continuous. There are fluxes in nature and in mechanical systems that are discontinuous. Descriptions of these systems lead to profound insights that will not be talked about in this text.

MECHANISTIC COSMOLOGY

Mechanical potential energy due to the force of gravity is a mystery. We cannot visualize how two masses such as the sun and the earth that are separated by empty space can effect one another's motion, but Newton's laws enable us to measure this interaction. Potential energy is not a "substance," but it is convenient to think of it as such; each body is thought of as possessing potential energy. In like manner how the temperature of the sun can increase the temperature of the earth is a mystery, but we can describe radiation of light from the sun through empty space to the earth. We also can measure the effects of this radiation in terms of increase in temperature and in terms of motion. Again it is convenient to think of this interaction as an energy "substance" traveling from the sun to the earth. Other types of radiations such as X rays and radio waves as well as light are thought of as "possessing" energy. Thus, a great diversity of events in the universe are metaphorically related; they all can be described as work or heat events which in turn may be measured in units of kinetic energy (Joules). Energy is the unifying idea amidst the diversity of observed events, and in particular it underlies both thermo (heat) and dynamic (motion) events. If we agree to think of it as a substance, then we can formulate a fundamental law of the universe, the so-called *first law of thermodynamics*: during any event in the universe, energy is neither created nor destroyed; it is transferred from one place to another, and it may exist in any one of many different forms. Starting with the metaphor of energy as a substance we can go on to think of any object in the universe as a system. Any system (or object) is embedded in a network of work and heat potentials. A system does not create or "cause" work or heat to be done; rather it merely transfers energy to and from its environment. These energy transfers are what we call work and heat.

A more precise formulation of the first law of thermodynamics depends on three ideas:
1. *Events* in nature may be understood in relation to a *system* interacting with its *environment*. In thermodynamics the boundary between the system and its environment must be clearly specified.
2. Any system interacts with its environment in *only* two ways:(1) heat process and (2) work process.
3. Energy is like a substance and so it can be transferred into or out of a system via heat or work or both.

Let U = total internal energy of a system where this energy may be present in many different forms; and let q = heat and w = work and ΔU = change of internal energy of a system. Then the first law of thermodynamics states:

$$\Delta U = q + w$$

that is, the change in internal energy of a system is equal to the amount of energy transferred into or out of the system via q plus the amount of energy transferred into or out of the system via w. The first law also allows that sometimes $\Delta U = q$ and $w = 0$. Other times $\Delta U = w$ and $q = 0$. In this kind of event the flux of a change in internal energy (ΔU) shows up totally as work. This means that the loss of Order in one place (decrease of a potential) shows up as the creation of an equal amount of Order (increase in potential) some place else.

BREAK FROM MECHANISTIC KNOWING

Entropy

Sadi Carnot (1796-1832) in 1824 "mentally constructed" a hypothetical engine that operates on a reversible cycle he called a Carnot cycle. This model of an *ideal Carnot machine* provided the core insights for formulating what came to be known as the second law of thermodynamics. Initially Carnot's model was ignored by scientists and even by those engineers working on steam engines. But a former classmate, Benoit Paul Emile Clapeyron, in 1834 revived it from obscurity, expanded and represented Carnot's model in mathematical terms that was more acceptable to the scientific community. This, in turn, led to Rudolf Clausius (1822-1888) in 1865 to invent a measure of the inability of any system in the universe to convert the energy "within it" or the energy "it receives" into work. Clausius called this measure, *entropy*; the higher the entropy, the greater the inability of the system to do work. The concept of entropy reminds us of the fallacy of thinking of energy as a substance in an autonomous system. If this metaphor were universally valid rather than merely a convenient way of thinking, then we could compare a machine doing work to filling up a glass with water from a pitcher of water. We simply empty the pitcher and fill up one or more glasses with water. All of the water, which is the metaphor for energy, in the pitcher, which is the metaphor for the machine, is transferred to the work of filling glasses with water. William Thomson's statement of the second law of thermodynamics and what may be described as the limitation of any machine is that even the best machine represented by the ideal Carnot heat engine is less than 100% efficient. The energy in it or the energy it receives cannot be totally converted into the work of some desired task.

According to Clausius, every system has a characteristic property called entropy which represents its inability to express spontaneous change -- the higher the entropy, the lower the ability to be spontaneous. In an isolated system, the entropy will increase to the maximum value the environment will allow. Another way to say this is that entropy is a measure of the degree of Chaos of a system. *The higher the entropy, the greater the degree of Chaos and the lower the degree of Order.* This means that the system is less able to do work, that is, is less able to spontaneously change. In an isolated system, the degree of Chaos will increase to the maximum value the environment will allow. That is, if no thermal energy is transferred into or out of the system and if the environment allows the system to do work on it due to a work potential equal to the possible change in internal energy of the system, then the entropy will increase to a maximum value. In the process of the system becoming more disorganized (increase in Chaos), it loses its potential to do work on the environment.

Second Law of Thermodynamics As the Entropy Principle

The most general statement of the second law is that in an open system, which is one that allows matter or energy into or out of the system, the net change of entropy of the system and its environment always increases. If the entropy of the system decreases (in an irreversible manner, which in real life always is the case), there always will be an even greater increase in the entropy of the environment or of some other system. Stated in this way the principle describes the only possible way in which new order is created anywhere in the universe. For example, any living system continually converts some of its internal energy into work. Much of this "organismal work" keeps the organism alive such as rebuilding proteins. Of course if nothing else happened, as the organism's internal energy decreased, it would no longer be able to maintain itself; it would die. But instead, the organism takes in food, which contains a high level of internal energy. The food is broken down in such a way that some of its internal energy is converted into the work of keeping the organism alive. The loss of Order during the process of living is renewed – the organismal work creates new order so that its entropy decreases. But, according to the entropy principle, this only can occur as a result of the food loosing Order thereby increasing its entropy. Overall the organism stays alive by the collaboration of food increasing its entropy with the organism decreasing its entropy thereby renewing itself. Furthermore, the increase in entropy of the food always is greater than the decrease in entropy of the organism renewing itself.

Mechanistic Science "Grows Up"

Primitive societies recognize that living and experiencing nature always involve Order and Chaos. They formulated this realization into participating in nature's cycles in which Order and Chaos alternate. For example, each day is a cycle in which daylight equal to Order gives way to darkness (night) equal to Chaos, which gives way to daylight that starts a new cycle, a new day. A year is a longer, more complex cycle involving four seasons: spring, summer, fall and winter. Summer is Order and winter is Chaos. Spring is rebirth to Order and fall is the dieing trend toward Chaos. Humans psychologically and ritually participate in Order/Chaos cycles. The Order of childhood is life; puberty is death of childhood, and emergence of adulthood is rebirth to new Order. People in modern societies often are disconnected from the cycles of nature and not mindful of the transformation that puberty usually produces. As a result, the realization that Chaos along with Order is a fundamental aspect of living is traumatic and slow in coming. It often may take several experiences of failure or unexpected and "undeserved" downturns such as death of a grandparent or parent or divorce of one's parents or an accident or illness before one finally "gets it." Eventually one must acknowledge the inevitability of Chaos and learn how to cope with it. This is the process of "growing up," though many people postpone it or avoid it altogether by means of addictions.

Mechanistic science is analogous to childhood in that it promotes one to believe in a mathematical universe consisting of concepts such as force, velocity, energy, etc. constructed by one's rational imagination (analogous to children believing in Santa Claus). It is not surprising that many scientists – and, in fact, many academicians – retreat to their ivory towers of the mind to avoid the experience of Chaos. And they may do this is a totally irrational way. For example, Einstein totally rejected the entropy principle except as a useful engineering idea and clung to the notion that time and events in nature are reversible. Thus, he lived in the world of science fiction involving time machines. All this in spite of experiencing himself as growing old and knowing that he like everyone else before and after him will die. Many scientists in the 19th century and early 20th century rejected the entropy principle as a fundamental aspect of nature. Thus, they refused to grow up while insisting that persons committed to spiritual traditions or religions are "childish." However, those scientists and engineers in the 19th century that did grow up produced a theory of machines wherein every cycle of a machine involves collaboration between Order (potential) and Chaos (flux). Scientists in the mid-20th century that matured to a greater extent than the 19th century engineers produced systems science of creativity and self-organization that also involve collaboration between Order (potential) and Chaos (flux).

THEORY OF MACHINES

Characteristics of Any Machine

Our ideas of force, especially action at a distance, and energy, especially potential energy, are rather nebulous. But in assigning measures to these notions we can treat them as if they really exist. In particular we can describe energy as like a substance that can be transformed and/or transferred from one place to another until it results in a particular type of motion, a task, which we desire to happen. We can devise formulas which enable us to create machines that take in energy and transform it into the type of work we wish to accomplish. The machine is our magic wand, for as Eliade notes:

> modern science [is] the secular version of the alchemist's dream,
> for latent within the dream is "the pathetic program of the
> industrial societies whose aim is the total transmutation of Nature,
> its transformation into `energy.'" The sacred aspect of the art
> became, for the dominant culture [of Europe], ineffective and
> ultimately meaningless. In other words, the domination of nature
> always lurked as a possibility within the Hermetic tradition, but
> was not seen as separable from its esoteric framework until the
> Renaissance. In that eventual separation lay the world view of
> modernity: the technological,..., as a logos.[1]

At the core of the operation of any machine is one or a sequence of energy couplers. An *energy coupler* is a part of a machine that goes through a two step cycle: (1) the coupler receives energy and goes to a particular higher energy state; (2) the coupler gives off energy and goes to a lower energy state wherein (a) some of the energy given off always accomplishes a particular type of work and (b) the lower energy state is ready or can be modified to repeat step 1. A human using a hammer to pound a nail into a block of wood is a "nail pounding machine," and the hammer is the energy coupler. The hammer receives energy from arm muscles and rises above the nail and then acquires kinetic energy in its downward movement toward the nail. The

---

[1] Morris Berman, *The Reenchantment of the World* (Ithaca, New York: Cornell Uni. Press, 1981), p. 99. [quote from Eliade, *Forge and Crucible*, pp. 172-173, Cf. Brown, *Life Against Death*, p.56.]

hammer is at its highest energy level just before hitting the nail. Upon hitting the nail some of the Kinetic Energy of the hammer accomplishes the task of moving the nail into the block of wood. The hammer then is ready to start a new cycle.

In general, machines exhibit three characteristics: (1) Energy flows from an energy source into the machine, through it, and out again as work. (2) Incoming energy causes an energy coupler to go from a low to a high energy level; then in going from a high to a low energy level, the coupler accomplishes some task. (3) A large task is accomplished by the sum of many small tasks each accomplished by one cycle of an energy coupler. For example, in a one-cylinder motorbike, the combination of gasoline and oxygen injected into the cylinder has chemical potential energy. Just after the spark plug initiates the chemical reaction between gasoline and oxygen, the products of this reaction (CO_2 and H_2O gases) have high kinetic energy (KE). As these expand, the K.E. decreases as the piston in the cylinder rises. When the valve opens allowing the gases to escape through the exhaust pipe, the piston falls and becomes ready to start a new cycle. The rise and fall of the piston is converted into one rotation of the wheel of the motorbike. Thus, each cycle of the piston accomplishes the small task of one rotation of the wheel. The addition of many of these cycles accomplishes the large task of the bike moving from one place to another.

Fundamental Energy Limitation of Any Machine

The most fundamental limitation of any machine is that it functions only in accord to the heat and work potentials it can express and continually re-create. The machine must be in a context in which it continually receives energy and therefore continually re-creates a potential which then is focused on accomplishing a specific task. The best, that is, the most efficient, machine imaginable is the ideal heat machine first described by Sadi Carnot in 1824.

Generalizations about the Ideal Heat Machine

The generalizations about the ideal heat machine apply to any machine no matter how complex it is and how far from ideal. In any real machine, that is, non-ideal machine, any energy transfer shows up as work *plus heat*. This is the "non-ideal" aspect of real machines. The energy transfers that show up as heat influences the efficiency of the machine. The more heat that is expressed, the less efficient is the machine.

Seven Generalizations from the Functioning of an Ideal Heat Machine

One: Machine efficiency. Even the ideal heat machine is not 100% efficient where efficiency is defined as task work divided by total energy input to the machine for a single cycle of the energy coupler. If energy input equals task work, the machine is 100% efficient. But some energy input always shows up as environment contraction potential which, of course, does not contribute to the task work. Thus, task work is less than the energy input; how much less this is determines the inefficiency of the machine. For any real machine the inefficiency is even greater because some of the energy input shows up as heat rather than as task work or environment contraction potential.

Two: Machine Taoism. An energy coupler cycle is seen to consist of two component processes. These two processes are radically different but mutually dependent on one another.

There is the *action phase* where the energy coupler accomplishes task work and creates an environment contraction potential. There is the *active-passive phase* where the environment contraction potential created in the action phase <u>prepares</u> the energy coupler to start a new cycle. Even though the two sub-processes of the cycle of an energy coupler are radically different, their mutual dependence is represented by the reversible changes that occur within them.

    Nature's cycles. At the core of machine function is the cyclic activity of an energy coupler. Taoism as expressed in the *I Ching* describes the same idea as the core of all activity in the universe.

> The Chinese philosophers saw reality, whose ultimate essence they called Tao, as a process of continual flow and change. In their view all phenomena we observe participate in this cosmic process and are thus intrinsically dynamic. The principal characteristic of the Tao is the cyclical nature of its ceaseless motion; all developments in nature... show cyclical patterns. The Chinese gave this idea of cyclical patterns a definite structure by introducing the polar opposites yin and yang, the two poles that set the limits for the cycles of change: "The yang having reached its climax retreats in favor of the yin; the yin having reached its climax retreats in favor of the yang."[2]

An energy coupler cycles between high and low energy levels. During phase 2 an energy coupler, the gas in an ideal heat machine, goes from a higher to a lower energy level and accomplishes work on the environment; $\Delta U = w$. In the Taoism perspective this is the yang process which having reached its climax now must give way to a yin process. In phases 4, the energy coupler goes from a lower to a higher energy level. This is the yin process which having reached its climax now must give way to a yang process.

    A yin-yang cycle as described... " by Manfred Parkert in his comprehensive study of Chinese medicine,"[3] is explicitly analogous to the cycle of the energy coupler of an ideal heat machine.

> According to Parkert, yin corresponds to all that is *contractive,* responsive, and conservative[phases 3 & 4], whereas yang implies all that is *expansive,* aggressive, and demanding[phases 1 & 2].[4]

---

2 Fritjof Capra, *The Turning Point* (New York: Bantam ed., 1983) p. 35; quote in passage taken from: Wang Ch'ung, quoted in F. Capra, *The Tao of Physics* (Berkeley: Shambhala, 1975) p. 106.

3 Fritjof Capra, *The Turning Point* (New York: Bantam ed., 1983) p. 35; quote of Manfred Porkert, The Theoretical Foundations of Chinese Medicine (Cambridge, Mass.: MIT press, 1974) p. 9.

4 Fritjof Capra, *The Turning Point* (New York: Bantan ed, 1983), p.36.

In the ideal heat machine the ideal gas is made ready to do work on the environment by contracting during phases 3 and 4, the yin process; then the gas does work on the environment by expanding during phases 1 and 2, the yang process.

The direction of harmony. The ideal heat machine only is able to function by allowing the flow of thermal energy from high to low. Using the metaphor of energy as a substance we say that thermal energy flows into the machine during phase 1, but some thermal energy must flow out of the machine in phase 3 in order for the cycle to start again. The machine functions only as a result of being interposed between two heat reservoirs, one that adds heat energy and the other that takes away heat energy. If the machine is not open to receive thermal energy, it will not do work. If the machine does not giveup thermal energy, it will not be prepared to receive thermal energy. The machine functions by being in harmony with a heat potential in nature. The machine merely couples the flow of thermal energy down a heat potential to accomplishing work.

The heat potential is an intrinsic aspect of the universe that becomes manifest in the functioning of a machine. This thermodynamic insight is in accord with an ancient Chinese recognition that:

> activity -- "the constant flow of transformation and change," as
> Chuang Tzu called it -- is an essential aspect of the universe.
> Change, in this view, does not occur as a consequence of some
> force but is a natural tendency [a potenial that spontaneously may
> disappear in producing work or heat], innate in all things and
> situations. The universe is engaged in ceaseless motion and
> activity, in a continual cosmic process that the Chinese called Tao -
> - the Way.[5]

The machine coupling to a heat potential in order to accomplish work is what Taoist philosophy means by *wu wei.*

> In the West the term is usually interpreted as referring to passivity.
> This is quite wrong. What the Chinese mean by *wu wei* is not
> abstaining from activity but abstaining from a certain kind of
> activity, activity that is out of harmony with the ongoing cosmic
> process.... [this is in accord with] a quotation from Chuang Tzu:
> "Nonaction does not mean doing nothing and keeping silent. Let
> everything be allowed to do what it naturally does, so that its
> nature will be satisfied." If one refrains from acting contrary to
> nature or as Needham says, from "going against the grain of
> things," one is in harmony with the Tao and thus one's actions will
> be successful [the task will be accomplished]. This is the meaning
> of Lao Tzu's seemingly puzzling statement: "By non-action
> everything can be done."[6]

---

5 Fritjof Capra, *The Turning Point* (New York: Bantan ed., 1983), p. 37. quote in passage taken from quote in Fritjof Capra. *The Tao of Physics* (Berkeley: Shambhala, 1974, p. 114.)

6 Fritjof Capra, *The Turning Point* (New York: Bantan ed., 1983, p. 37.

The flow of thermal energy down a heat potential has direction; it always is from high to low. The great heat potential of the universe is the measurable manifestation of time, and time is the existential unfolding of finite being. The universe is like a great water wheel coupled to the directional flow of water, which is time. All things are in harmony with the Tao when "they go with the flow."

Machine harmony. The Tao of an energy coupler. Let Y represent a type of energy coupler that only accomplishes a small task in going from high to a low energy level; then let Y^h = the energy coupler at the higher energy level and let Y_1 = the energy coupler at the lower energy level. Then the energy coupler cycle may be represented as follows:

$$Y^h$$

YIN   **YANG**

work done on Y work done by Y = task

$$Y_1$$

Wu Wei of an energy coupler. One way of defining *wu wei* operationally is to discover a particular work or heat potential in the universe and an energy coupler that can convert this potential into a desired task. Then one can connect a machine containing the energy coupler to the potential in such a way that: (1) the *YANG* process of the energy coupler cycle will accomplish the desired task and then immediately give way to the *YIN* process; (2) the *YIN* process of the cycle will re-create the machine potential for accomplishing the desired task and then immediately give way to the *YANG* process.

Three: Subjective idea of machine mutuality rather than objective idea of machine autonomy. Functioning of the ideal heat machine in one sense supports but in a holistic perspective opposes the mechanistic claim that any system may be totally understood in terms of the interactions among it parts. On the one hand, the ideal machine converts the energy flux of a potential into a sequence of autonomous cycles where each cycle consists of a Yang phase that produces an elemental task work and a Yin phase that produces a potential in an energy coupler to produce the Yang phase. In this view, the machine appears to have the intrinsic power to convert the energy related to a potential, for example, potential energy, into a sequence of autonomous work tasks that taken together equals the work goal of the machine. However, this view breaks down in two ways. Just as the idea of child implies parent and likewise parent implies child, so also machine work and energy flux imply one another. The machine is not an autonomous entity that converts energy into machine work. Likewise an energy flux may produce some work, but it does not produce machine work. Rather, an autonomous energy flux and an autonomous machine must enter a collaborative mutuality in order to produce machine work. A paraphrase of what a Taoist would say is that once this collaboration is established, the work task will occur on its own accord. Humans using machines cannot make things happen. All they can do is create the collaboration between any energy flux and an appropriate energy coupler; then the task will occur without further human effort. Likewise, just as *response* exists only in relation to a *stimulus* and vice versa, in each work cycle, the Yang phase exists only in relation to a Yin phase which legitimately is called *Yin* only if it leads to a Yang phase. Thus, in relation to the operation of a machine, the idea of autonomous parts must be supplemented with the idea of mutuality and with the subjective ideas of irreversible time and events.

The ideal heat machine – the prototype for all other machines – in itself cannot accomplish anything. Rather *this machine collaborates with an aspect of its environment.* The energy coupler, in this case, the ideal gas, is oriented so that it can convert the flow of heat energy from a "heat source" to a "heat sink" to accomplish a specific task. Neither the heat energy flow nor the autonomous machine can accomplish the desired task work, but the collaboration of these two factors can. This insight is a special case of the postmodern perspective that the meaning of a thing or event is partially determined by context. The meaning of any machine partially depends on the environmental work and/or heat potentials it exploits. Moreover, the flow of energy down an isolated potential always is from higher to lower energy levels; never the reverse. Of course, this energy flow occurs over time. Thus, time associated with any machine is irreversible just as humans' subjective experience of time is irreversible. Mechanistic time, on the other hand, is reversible just as distance is. Thus, machine dynamics implies irreversibility of time and correspondingly also implies some degree of subjectivity in contrast to the dogmatic objectivity of the mechanistic perspective, (Pribor, 1999)[7].

Four: Mutuality of Order and Chaos. The ideal heat machine introduces another subjective insight which is the idea of *mutuality of Order and Chaos.* In so many words in 1865 Clausius saw this mutuality in relation to phases 2 and 4 of one cycle of an ideal heat machine. The decrease of internal energy in phase 2 means that the ideal gas now has less ability to do work. Clausius called this change an increase of entropy where *entropy* is a measure of a system's *inability to do work.* Correspondingly, the increase of internal energy in phase 4 means that the ideal gas has regained a greater ability to do work, and this change is represented by a decrease in entropy. The ability to do work may be subjectively understood as degree of organization represented by the word, *Order.* The decrease of the ability to do work is a decrease of Order which we may represent as *Chaos.* Thus, entropy changes in the universe involve an increase of entropy where Order goes to some degree of Chaos or a decrease of entropy where Chaos goes to some degree of Order. Thus, Order and Chaos always imply one another for all changes in the universe; there is a mutuality between Order and Chaos.

Energy represents a potential for expression of an event that is work or heat or work and heat. Energy as a potential is order and *flux* is the transformation of energy into work and/or heat. Flux is analogous to force interaction resulting in the loss of one state of motion wherein velocity changes and thus represents chaos. Perhaps an easier way of seeing this is in relation to a pencil falling off a table to the floor. The pencil on the table has potential energy due to the force of gravity. As the pencil falls toward the floor, it progressively converts potential energy into kinetic energy. This conversion is the flux of the potential energy which is ambiguous. From one point of view potential energy is lost and of course simultaneously the potential disappears which means order becomes loss of order, that is, chaos. From another point of view, the chaos of the loss of potential equals to the creation of new order, in which during the free fall this new order is the emergence of kinetic energy. Thus, the simultaneity of these two points of view produces two mutualities: (1) the mutuality of order and chaos and (2) the mutuality of chaos and creativity.

---

7 Pribor, D. 1999. "Transcending subjectivity double binds of science." Proceedings of the Institute for Liberal Studies, vol. 10, pp. 27-37.

The functioning of any machine points to a more radical mutuality of order and chaos. An ideal machine converts a large potential in nature into the sum of autonomous work cycles. This is analogous to breaking up a ten foot-long stick into ten one foot-long pieces, but from the point of potential to do task work, this analogy breaks down. Without a machine the large potential in nature can be converted totally into work (when the conversion is done infinitely slowly). But with an ideal machine the large potential is converted into a specific task work, which is less than the total amount of work that the potential could express. This is because each of the work cycles produced by the ideal machine consists of a Yang phase that produces a small unit of task work plus a Yin phase that produces work to prepare the energy coupler for another Yang phase. Thus, when a large potential in nature is converted into machine work, there is an overall loss of order and therefore an increase in chaos. The order of the large potential is greater than the order of the machine work. Of course a non-ideal machine generates even greater chaos in that that some of the order of the large potential is converted into heat.

Five: Hierarchal mutuality of chaos and machine creativity. The mutuality of order and chaos associated with an ideal machine points to another subjective, profound insight about nature. The order of a potential in nature and the order of machine work may be represented by the same mathematical-physical symbols, that is, units of energy, but the orders are *qualitatively* different. The autonomous potential in nature is not a potential to accomplish a particular task; it has no machine purpose. It only becomes a "task potential" when it collaborates with an appropriate energy coupler. This collaboration is not determined by the physical laws that specify the potential in nature nor by the laws that specify the operation of an energy coupler. There are an indefinite number of ways the collaboration can occur resulting in the efficiency of the machine varying from low to high. Csanyi makes an equivalent evaluation of this situation in stating that there is no algorithm for building the most efficient machine[8]. Machine work results from machine structure, which in turn contains within its definition the subjective idea of collaboration between a flux and an energy coupler. An autonomous potential in nature does not contain or in any way refer to this idea of collaboration. Therefore, order of a potential in nature and order of machine work are related but fundamentally different. Another way of stating this difference is that a machine has an organization pattern that can convert flux into machine work. Energy flux drives the *translation* of machine organization into machine work.

The orders associated with energy flux and machine work respectively are different but are fundamentally related by a hierarchal transformation. A machine transforms the order of a potential into machine work by dividing a potential into a sum of machine work cycles. An "old order" of a potential goes into chaos which is the flux mirroring this order. The flux collaborating with the machine organization creates a "new order" which is the machine work. During the process the old order is incorporated into the new order, which is thereby said to be at a higher level of order. That is, a potential in nature is incorporated into a machine work event, but as stated earlier, the potential in nature does not "contain" the machine work. This is what a two-level hierarchal system, AB, is. A incorporates B but B does not contain A. Even though the phases of an ideal heat machine are reversible, each work cycle is irreversible. This is because heat flow equal to flux is unidirectional; it only flows from high to low. It would take another machine exploiting a different potential in nature to make the flux occur in the opposite

---

8 Csanyi, V. 1989. *Evolutionary systems and society.* Durham, N.C.: Duke University Press, pp. 8-10.

direction. This unidirectional flow in collaboration with a machine generates a two-level hierarchal arrangement in nature.

The above insight about hierarchal orders may be generalized to a statement about hierarchal mutuality of chaos and machine creativity. Machine creativity is the process of a machine work emerging from the chaos of a flux associated with an autonomous potential in nature. The overall process is Order, Chaos, New Order that includes a modified old order. That is, Order represented by a potential goes to Chaos represented by flux. The flux is converted into work cycles each generating an element of machine work. The sum of these elements equals total machine task, which sometimes may be represented by a new autonomous potential in nature that is less than the old potential from which it came. There is an overall increase in entropy in nature as a result of this machine creativity. The increase in entropy represents the transformation of some order into chaos in nature that is coupled to the "machine creation" of a new order.

Six: Machine creativity involves a narrative interpretation of an historical process. If we focus only on the mathematical description of machine creativity, then machine work is not fundamentally different from order of a potential, and therefore, this creative process does not generate a hierarchy of orders. However, if we focus on an *interpretation* of what is happening during machine creativity, then we see that this is an historical process that produces a new two-level hierarchy of orders. During this historical process, new order emerges from chaos. Thus, there is a mutuality of chaos and creativity.

A description of an historical process is a story interpretation of events. The story interpretation or "narrative knowing" uses metaphorical concepts that relate to some objective pattern in nature to one's subjective experiences of nature. The subjective aspect makes metaphorical concepts ambiguous sometimes leading to contradictory interpretations. This, I believe, is because the experience of reality is at times paradoxical. Of course if one assumes that modern science is and must always be only utilitarian, then any metaphorical conceptual interpretation of nature must be replaced by a logical, conceptual model that has been tested for validity. However, *modern science is not only useful; it also provides valid metaphorical interpretations of nature.*

Seven: Machine creativity points to a new, higher narrative way of understanding science. Machine creativity may be described from two different, hierarchal perspectives. The metaphorical conceptual narrative perspective not only interprets machine creativity but provides a way to have empathetic, participatory understanding of nature. This partially subjective knowing cannot guide objective action, but it can be reduced to logical, conceptual knowing, such as mathematics and measurement, which is utilitarian. The metaphorical narrative knowing is ambiguous not only in content but also in level of knowing. On the one hand, it is a primordial kind of knowing that can guide the formation of a rational hypothesis. On the other hand, when a hypothesis becomes a valid theory, then a metaphorical, narrative understanding of it is a higher level of knowing. It is higher for several reasons. The metaphorical conceptual understanding of a rational theory gives one the openness to see possible connections with other areas in science and with non-science areas of knowledge. This type of knowing provides ordinary language as a way to communicate with thinkers in diverse disciplines; that is, it partially overcomes the fragmentation of knowledge in postmodernism. This type of knowing also actuates the spiritual-mystical aspect of being human. The mutuality of the two levels of knowing is what Heidegger refers to as the mutuality of calculative thinking and meditative thinking. In the introduction to Heidegger's book, *Discourse on Thinking*, John Anderson summarizes Heidegger's view of this mutuality as follows. Meditative thinking

> Requires two attributes… two strands which man can take, and which he calls *releasement toward things* and *openness to the mystery* …
>
> He also drives home the importance of such thinking to man's very being, claiming indeed, that even the ultimate meaning of the calculative thinking of modern science and its humanly significant applications are discerned in and through meditative thinking. But fundamentally Heidegger is urging his hearers and readers toward a kind of transmutation of themselves, toward a commitment which will enable them to pass out of their bondage to what is clear and evident but shallow, on to what is ultimate, however, obscure and difficult that may be.[9]

Chapter 2
PHYSICAL MUTUALITIES

Law 1: No Event Is Autonomous

Rather every event is mutually related to some other event. For example, Newton's laws of motion which implies the conservation of momentum make up one aspect of the foundation of modern science. Momentum is the quantity of directional motion of an object at some point in space; that is, momentum = mass x velocity. Then, according to the conservation of momentum, if some object increases its momentum at a point in space, some other object must have decreased its momentum by an equal amount. The event of increasing momentum is never autonomous but always simultaneously involves another event during which momentum decreases. Likewise, the conservation of energy, another aspect of the foundation of modern science, involves mutuality of events. Energy represents the potential quantity of motion plus the actual quantity of motion an object has at some point in space. Then, according to the conservation principle, if some object increases its energy at a point in space, some other object must have decreased its energy by an equal amount. Again, as is the case with momentum events, there are no autonomous energy events. As our ideas for describing the universe become more complex, the mutuality of events becomes even more obvious and undeniable.

The scientific revolution (1500 – 1700) was the first movement of Western thinking toward the Buddha's insight of dependent co-origination. Physics then and now abandons the Aristotelian ideas of being, substance, and causation. Force, which equals mass times acceleration ($F = ma$) of some body at a point in space and moment in time, is not a cause of motion as many textbooks erroneously proclaim. Force, always and everywhere, is an interaction between two things. This interaction is quantitatively defined in terms of the quantity of mass of one object involved in the force interaction times the rate of change of the velocity, which is something like speed, of the object. This definition of force, in turn, depends upon the

9 Heidegger, M. 1966. Anderson, J. & Freund, F. H. ed. *Discourse on thinking*. New York: Harper Torchbook, Harper & Row Pub., pp. 12-13.

mutualities of :(1) conservation of mass, (2) conservation of momentum, and (3) forces of action and reaction and on what may be called the *suchness* of inertia.

Newton proposed that every material body in the universe has a quantity of inertia called *mass* that expresses itself in a way that I call *dependent co-origination with respect to motion*. If there were only one mass-object in the universe or if it were possible to totally isolate one mass-object from all others, then two statements eternally describe its state of motion. If the mass-object is not moving, that is, velocity = 0, it will never move. If the mass-object is moving at some constant velocity, then it always will move at that constant velocity. The mass-object is *such-that-it-is* which is a state of motion independent of any notion of causality or it being a type of substance, that is, the mass-object is not an entity with an essence that determines (causes) its existence to manifest as a particular state of motion. The independent mass-object simply is a particular state of motion (velocity = 0 or velocity not equal to 0). There is no "why" and it has no purpose. But Newton proposed – and all physicists to this day agree – that all mass-objects are interdependent. This means that mass-objects in physical contact with one another influence one another's motion *not* by causation but by the mutualities described in laws of conservation such as the conservation of momentum. Newton's so-called three laws of motion specify this influence during physical contact. When mass-objects are separated, they influence one another's motion in a way as specified by Newton's law of universal gravitational force. All the interdependent mass-objects are interdependent in their motion. Yet, each mass-object is equally in itself, of itself, without one being prior to the other. Also there is nothing whatsoever more substantial or more real which grounds the interdependence of all mass-objects. As a group, scientists did not come to a full realization of this last statement until the late 1700s. Pierre Laplace (1749-1827) gave expression to this realization with his pronouncement: "We no longer need the God hypothesis."

Law 2: Mutuality of Order and Chaos

The order of the universe may be represented by potential events. A potential event may be expressed by some mutuality as described in law 1. For example, Newton's law of universal gravitation quantitatively specifies the potential for any two objects in the universe to move toward one another; the potential is called a force of gravity. Likewise, the most fundamental understanding of energy is that it is a potential for some energy event to occur. In physics there only are two types of energy events. One is *work* which quantitatively specifies motion of an object in an interval, and the other is *heat* which quantitatively specifies the transfer of thermal energy that results in the rise of temperature at some one or two points in space. Correspondingly, the universe is described in terms of work potentials (which include potentials for changing momentum) and heat potentials. Though the modern theory of quantum mechanics and a modern interpretation of Einstein's general theory of relatively are based on assumptions that contradict one another, both theories acknowledge the validity of the second law of thermodynamics. This law may be understood in terms of potential events wherein every potential event at a point in space and/or moment in time will tend to expresses itself as some mutuality as described by law 1. When the potential event does express itself, the original potential event disappears, for example, when a rock falls to the ground, it loses its potential to fall to the ground. Likewise, when a bottle of milk looses thermal energy so as to come to the same temperature as that of a refrigerator, it looses its potential to decrease its temperature in these circumstances. Thus, Order represented by a potential is also a potential for a loss of Order which we refer to as Chaos, and as Order decreases, Chaos simultaneously always increases to the same extent.

One of the original formulations of the second law of thermodynamics spells out this mutuality in terms of *entropy*. Entropy is a concept created to represent the quantitative increase or decrease of any potential event. Entropy is defined in such a way that when the quantity of the potential event increases, such as, greater force of gravity or greater temperature difference between two objects, entropy decreases and Order increases. Conversely, when the quantity of the potential event decreases, that is, an increase in Chaos, the entropy increases. According to the second law of thermodynamics, the entropy of any object or system in the universe tends to increase. This means that the isolated potential event associated with that object tends to decrease and correspondingly, a decrease in Order is simultaneously replaced by an increase in Chaos. Thus, the second law of thermodynamics really is a statement about the mutuality of Order and Chaos anywhere in the universe. All isolated systems in the universe tend to change:

1. From Order to Chaos
2. From lower Entropy to higher Entropy
3. From networks of potentials to one or more potentials decreasing (sometimes to zero)
4. From non-equilibrium to equilibrium
5. From structure to less structure
6. From maintained separation to union = gross harmony

Machine Creativity, Law 3: Mutuality of Chaos and Creativity

At the most elementary level as described by physics, *creativity* is the emergence of a new potential event = new Order. The mutuality of Chaos and Creativity means that New Order always is coupled with some Chaos resulting from the decrease or disappearance of some potential event representing Old Order.

For example, any machine has at least one energy coupler. If the machine has more than one, the interconnection among two or more couplers may be thought of as a single complex energy coupler. An energy coupler takes in energy and focuses it on accomplishing a particular task defined as the creation of a new potential event = a New Order in the universe. For example, a hammer takes in energy from the human arm and hand muscles and focuses it on pounding a nail into a block of wood – perhaps connecting one piece of wood with another as in a cabinet. This outcome represents a New Order that like any order in the universe will tend to break down into chaos. The energy coupler goes through one or more (usually many) cycles where each cycle consists of a YANG phase and a YIN phase. In the YANG phase the energy coupler starts out as a particular potential event represented by a particular value of entropy. When the potential event is expressed, entropy increases as the quantity of the potential event decreases. The mutuality of energy events – Law 1 – shows up as a decrease of the energy of the energy coupler which equals to the increase of energy of one or more other objects. When this Order of the energy coupler is lost and replaced by Chaos, some New Order = the task of the machine is *created* and so we have Chaos-Creativity mutuality. In the YIN phase the energy coupler receives energy from some other objects in the universe. As a result, the energy coupler decreases its entropy as it increases its quantity of potential event = increase in Order. The mutuality of energy events – Law 1 – shows up as an increase in energy of the energy coupler which equals the decrease of energy of some other objects in the universe. As the quantity of the potential event of other objects in the universe decreases (entropy increases) and thus go to Chaos, the quantity of the potential event of the energy coupler increases, i.e., it is re-Created. Thus, the YIN phase also expresses Chaos-Creativity mutuality

Every machine expresses *machine creativity* as a result of a mutuality between the machine and its environment. In the YIN phase, Order in the environment goes to Chaos in such a way that the quantity of the potential event of the energy coupler, that is, energy coupler Order, is re-Created. Then, in the YANG phase, the energy coupler Order goes to Chaos in such a way that a task is accomplished. The task is the New Order that is Created by the YANG phase. Machine creativity also may be represented by a mutuality between Logos = Order = potential event and Eros = drive to express a potential event. The word, *Eros*, comes from the word, *erotic* which relates to the human sexual drive in its various manifestations. The sex drive is a metaphor for the tendency of any potential event in the universe to be expressed. Correspondingly, it seems fitting to use the word, *Eros*, to represent this tendency for any potential event to be expressed in the universe. Beginning with this terminology I use machine creativity to introduce new fundamental terms that will be useful in other situations.. During the YIN phase, environmental Order (Logos) goes to Chaos and the drive to do this is called *Eros-chaos*. Eros-chaos is coupled with, that is, is mutual with, re-Creating Order (Logos) of the energy coupler and the drive to do this is called *Eros-order*. In the YANG phase the energy coupler Order (Logos) goes to Chaos and again the drive to do this is Eros-chaos. The Eros-chaos is coupled with creating task Order (task Logos) and the drive to do this Eros-order, see Figure 2.1. The mutuality of Chaos and Creativity may be represented as the mutuality of Eros-chaos and Eros-order. Generalization 4 & 5 about the ideal heat machine provides another way of stating this law. An energy flux is Eros-chaos where Order goes to Chaos as energy is transferred from one place to another. The collaboration between an energy coupler and flux produces potentials that represent Eros-order. Eros-order in each cycle of a machine creates machine order = continuous cycles + work task for each cycle. Thus, the mutuality of Eros-chaos and Eros-order generates the mutuality of flux and machine order.

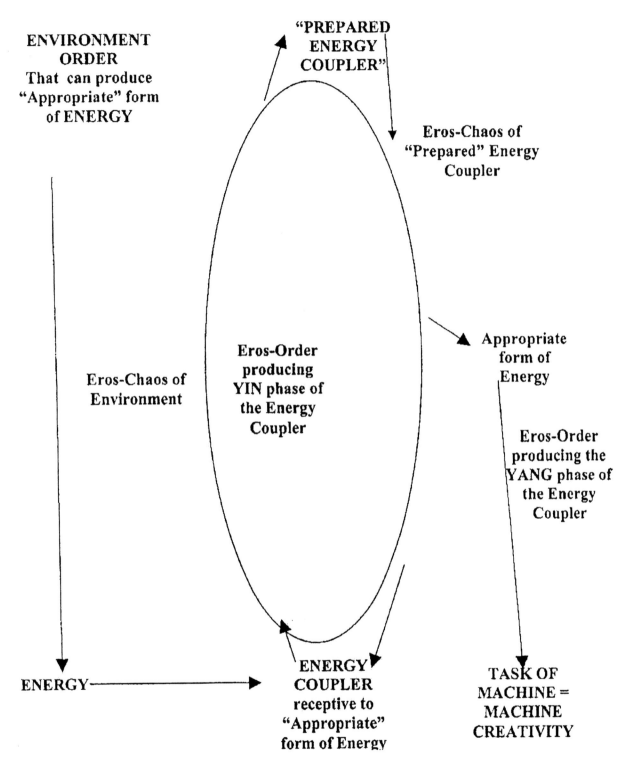

FIGURE 2.1 Mutuality of Chaos and Creativity of an Energy Coupler

Law 4: Hierarchal Mutuality of Chaos and Creativity

Let us call "order in a particular system" *Logos*, then the loss of Order may be called *Chaotic Logos*. Eros-chaos drives some Order (Logos) in the universe to Chaotic Logos where the chaos makes potential events available to be expressed, which before the chaos could not be expressed. Eros-order coordinates the expression of one or more of these potential events with remaining Order of the chaotic Logos to produce a New Order that includes aspects of the old order. Thus, Eros-chaos is mutual with Eros-order in such a way that a modified Old Order is incorporated into a New Order. The New Order expresses *emergent properties* which have the following four aspects: (1) these properties only can be described in terms of the whole higher order pattern. (2) These properties cannot be understood in terms of (reduced to) interactions among components; that is, the properties cannot be understood in a gross, mechanistic way. (3) These properties cannot (or could not) be predicted to occur from knowing the components of the new pattern. (4) The properties are the expression of immanent potentials of a set of apparently autonomous entities that interact in a context that actualizes the immanent potentials. One can know the existence of these immanent potentials only as a result of knowing their manifestation that shows up as emergent properties. One can know the context for the expression of these immanent potentials (and thus count on their expression when the appropriate context is set up), but one never can know the *how-why* the context leads to their expression. Knowing the *how-why* implies a mechanistic description, and the expression of emergent properties is a non-mechanistic and indeed, a non-rational process. The overall process of *hierarchal mutuality* is: Old Order, Chaos, New Order expressing emergent properties and incorporating aspects of the old order.

For example, under certain conditions, such as, very high temperatures, the Order in a population of hydrogen gas molecules and oxygen gas molecules goes to Chaos = Chaotic Logos. The hydrogen molecules each consisting of two hydrogen atoms chemically bonded to one another, break up into hydrogen atoms. Likewise, oxygen molecules, each consisting of two oxygen atoms chemically bonded to one another, break up into oxygen atoms. Then, the one negative electron orbiting the positive nucleus (a proton) of each hydrogen atom breaks away from the hydrogen atom producing chaotic hydrogen Logos = positive hydrogen atom (= positive hydrogen ion). Each of the free oxygen atoms has a potential to receive into its structure, that is, its orbital system, two electrons. The now available electrons that broke away from the hydrogen atoms go to the oxygen atoms thus producing chaotic oxygen Logos thus producing negative oxygen atoms each with two negative sites (= negative oxygen ions). The resulting chaotic hydrogen & oxygen Logos makes available new potential events. Eros-order leads to the expression of one of these new potential events. Namely, two positive hydrogen atoms (ions) are received by the two negative sites of each negative oxygen atom (ion). The resulting chemical bonding between each negative oxygen atom and two positive hydrogen atoms produces a water molecule, H_2O, see figure 2.2. The population of water molecules has emergent properties that could not be predicted from knowing the properties of hydrogen gas or oxygen gas or a mixture of hydrogen and oxygen gases.

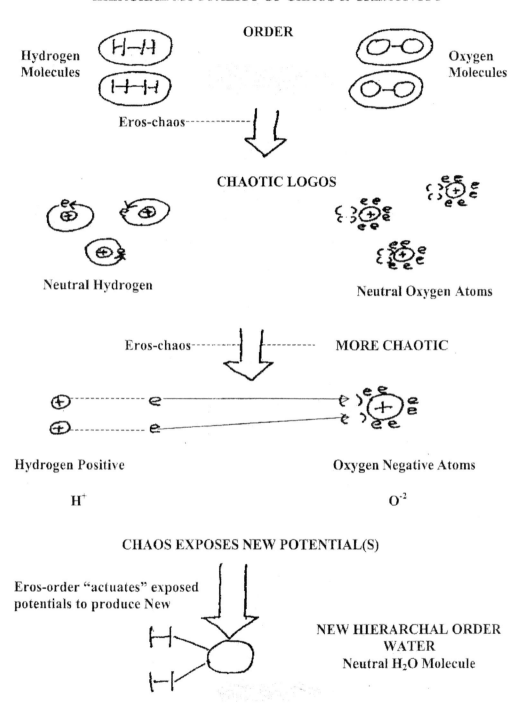

FIGURE 2.2 Chemical Reaction Producing Water Representing the Hierarchal Mutuality of Chaos and Creativity

Law 5: Physical Individuation Is an Order, Chaos, New Order Process.

Individualism is an acquired degree of uniqueness of an entity that enables it to enter into new types of interactions. For example, each electron (fundamental unit of negative charge) in a cloud of electrons and each proton (positive charge) in a cloud of protons has no uniqueness. Under the right conditions a particular electron may break away from its population and a particular proton may break away from its population. Now the more unique electron and the more unique proton have the possibility of interacting with one another. If and when they do so, the electron will begin to orbit the proton thus producing a hydrogen atom. That is, the electron and the proton each acquired a degree of uniqueness expressed as their mutual interaction in forming the hydrogen atom. The overall process involves breaking away from the low level order of a population to acquire autonomy and then entering into a mutual interaction to form a new structure, a new order. This process of individuation involving going to a higher level of individualism and autonomy exemplifies the pattern of transformations that gave (and gives) rise to all levels of organization in the universe. Things achieve some degree of separateness and autonomy by breaking from one level of order and thus going into one corresponding level of chaos. The separated, autonomous things in chaos then may mutually interact to produce a new level of order. The overall transformation process may be represented by the sequence: Old Order, fragmentation producing Chaos, from which emerges a New Order, which includes the modified fragmentations of the Old Order. More briefly the pattern is: Order, Chaos, New Order. The characteristic of the universe that allows for transformations of this sort are patterns of mutuality, laws of mutuality. These laws of mutuality interrelate all "parts" of the universe so that parts of the universe only are relatively autonomous.

UNIT II PRE-CIVILIZED PHASE OF HUMAN EVOLUTION
Chapter 3
ANIMAL CONSCIOUSNESS

FEELINGS AND ANIMAL CONSCIOUS COMMUNICATION

Analogical Understanding of Animal Consciousness

I take the view that animal consciousness is identical to conscious feelings which involve subjective orientations which, of course, cannot be observed. Those who believe in these subjective behaviors attribute feelings to other humans and to some non-human mammals via an analogy. The person may reason as follows: "I know what it is like to see a tree and upon realizing that you have eyes and a brain similar to mine, I confidently presume that you see this same tree in approximately the same way as I do. In like manner, I know what it is like to feel anger or fear, and I realize that I show these feelings via certain facial expressions, body gestures, and physiological arousal. Therefore, when I see another human or a non-human animal show analogous types of facial and body expressions and physiological arousal, I presume these other animals, human and non-human alike, also experience feelings analogous to mine." Joseph Le Doux, takes a similar view:

> [C]onscious feelings … are in one sense no different from other
> states of consciousness, such as [perception of roundness] …
> States of consciousness occur when the system responsible for
> awareness becomes privy to the activity occurring in unconscious
> [I believe this term should be *non-conscious*; *unconscious* is a type
> of consciousness not ordinarily available to ego consciousness.]
> processing systems…There is but one mechanism of consciousness
> and it can be occupied by mundane facts or highly charged
> emotions.[1]

The above quote implies that feelings are conscious emotions, where emotions may be defined in terms of behavior mechanisms for survival. Again according to LeDoux:

> [T]he basic building blocks of emotions are neural systems that
> mediate behavior interactions with the environment, particularly
> behaviors that take care of fundamental problems of survival. And
> while all animals have some version of these survival systems in
> their brains [instincts], I believe that feelings can only occur when

---

1. Joseph LeDoux, *The Emotional Brain, the Mysterious Underpinnings of Emotional Life* (New York: Touchstone, 1996), p. 19.

a survival system is present in a brain that also has the capacity for consciousness.[2]

Besides instinctive survival behaviors instinctive adaptations also may include physiological arousal, but these only occur in vertebrates with a homeostatic body plan.

ANIMAL HOMEOSTATIC BODY PLAN

Homeostatic refers to the ability of any system to make adjustments so that a particular quantitative characteristic of the system, for example, temperature, will stay near (fluctuate around) some steady state value, which is the human's steady state body temperature that is $98.^0F$. The animal homeostatic body plan includes the ideas of internal milieu, blood-internal milieu interaction, and coordinated organ systems.

Internal Milieu

Single-celled organisms were the first forms of life to appear on earth. Random association of single cells into colonies had the adaptive advantage of size. Another competing, single-celled organism that can "eat" (engulf) cells could not subdue a colony. Cell colonies led to the emergence of bona fide individual organisms consisting of many cells. In addition to increased size, such organisms were able to cope with the challenges of the environment more efficiently by means of cell specialization. Most of the cells, called vegetative cells, deal with getting chemicals and energy from the environment in order to stay alive. A few cells became specialized for reproduction of the multi-cellular organism. The vegetative cells take care of the chemical and energy needs of the reproductive cells.

The multi-cellular adaptation resulted in organisms having an internal environment, also called *internal milieu*, as well as an external environment. All the surface cells of the multi-cellular organism provided the boundary between the individual and the external environment. The spaces between the cells and the fluid filling these spaces, called *tissue fluid*, are the internal environment, the internal milieu, of an organism. Very often this fluid filled space, a microfilm around each cell, can only be seen with the aid of an electron microscope. In simple organisms, the internal milieu is only one cell removed from the external environment from which nutrients can move in and to which wastes can move out. However, with the increase in multi-cellularity, the internal milieu becomes more distinct and isolated from the external environment. This means that for many cells of the organism, the only environment they deal with in their whole lives is the fluid that surrounds them. If these cells are to continue to live, the whole organism must regulate the interconnected fluid internal milieu to make it suitable for cellular life.

Blood-Internal Milieu Interaction

Individual animal cells are like car engines; they take in oxygen and glucose (analogous to gasoline) and chemically combine them to produce energy to carry out life processes. As this is occurring wastes, thermal energy, carbon dioxide and water are excreted into the internal

---

2. Joseph LeDoux, *The Emotional Brain, the Mysterious Underpinnings of Emotional Life* (New York: Touchstone, 1996), p. 125.

milieu. Animal cells also take up many other nutrients. After a while the internal milieu becomes depleted of nutrients and polluted by wastes so that it no longer is suitable for animal cells to continue to survive. The only hope for the continued survival of multi-cellular animals is some mechanism for the renewal of the internal milieu making it suitable again for these cells to survive. Continually circulating blood provides this mechanism. Namely, as blood from small arteries (arterioles) flows through capillaries, which are bathed in the internal milieu just as all cells are, the flowing blood takes up excess heat energy, water, carbon dioxide and other wastes from the internal milieu and simultaneously re-supplies the internal milieu with oxygen and glucose. Thus, the internal milieu is *renewed* by the blood in capillaries flowing past it, but now the blood is polluted by the wastes it picks up and depleted of oxygen and glucose.

Coordinated Organ Systems

In smaller organisms such as the earth worm, the circulating blood is continually depleted of nutrients and polluted by wastes as a result of interactions with the internal milieu. Nevertheless, the blood is continually renewed sufficiently rapidly because of its proximity to the digestive tube and to the external environment. In larger organisms, especially those that are more active, the blood must be renewed in a more efficient manner. Animals with backbones, that is, vertebrates, have developed organ systems that keep properties of blood close to preset values (by means of negative feedback regulation). For example, the respiratory system regulates oxygen and carbon dioxide concentrations; the digestive system regulates the concentration of glucose along with many other nutrients; the kidney removes wastes and poisons and regulates concentrations of water, various ions, and other substances. These homeostatic mechanisms for renewing the blood lead to the efficient continuous renewal of the tissue fluid. Cells are able to survive in vertebrates because the circulatory system has become elaborate enough to help coordinate the activities of all organs that indirectly modify the internal milieu by modifying the blood, or that directly modify the internal milieu; i.e., the actions of the lymphatic and immune systems. In less complex animals (invertebrates), some of these organs, for example, excretory organs, operate independent of the circulating blood. In vertebrates, blood moving in a closed circulatory system serves as an intermediary between the external environment of the animal and the internal milieu of its cells.

Various organ systems, such as the respiratory system, digestive system, and kidney, continually interact with the external environment to *renew* the blood that is polluted by wastes and depleted of nutrients, such as oxygen and glucose. The *renewed* blood as a result of being kept flowing past the tissue fluid (internal milieu) renews it, and thus, this tissue fluid is continually maintained suitable for the cells surrounded by it to stay alive, see figs. 3.1 and 3.2. A portion of the central nervous system (the hypothalamus) regulates the flow of blood to all tissues in the body, and it also regulates all other organ systems so that the properties of the blood, for example, temperature, concentration and volume of water, concentration of oxygen, carbon dioxide, and glucose, are kept near constant values. As a result, the tissue fluid is maintained suitable for cells to stay alive. This overall pattern of interactions among tissue fluid, circulating blood, other organ systems, and a regulatory portion of the central nervous system is called *homeostatic body plan.*

In summary, the human body consists of trillions of cells each surrounded by tissue fluid which supports cellular life. Each cell continually pollutes the tissue fluid and depletes it of nutrients. Circulating blood continually reestablishes the life supporting properties of tissue fluid, but in doing this, the blood is polluted and depleted of nutrients. Four organ systems, (1) the integumentary system (skin), (2) the respiratory system, (3) the digestive system, and (4) the urinary system, carry out exchanges between the external environment and the blood resulting in the reestablishment of the blood's life-supporting properties. The lymphatic system removes excess water and debris from the tissue fluid, and the immune system in conjunction with the lymphatic and circulatory systems, removes foreign substances, such as bacteria, viruses, and poisons see figure 3.1.

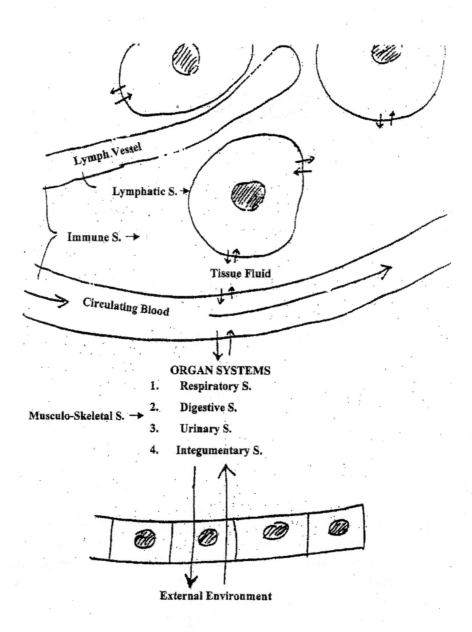

FIGURE 3.1 Schematic Representation of Homeostasis in Relation to Tissue Fluid and Blood

The neuro-endocrine system interacts with all other organ systems to establish homeostasis, in which various properties of the body are kept at near constant values. A portion of the brain called the limbic system is associated with the experience and expression of feelings. Feeling is the connecting link between human consciousness and animal consciousness, which is manifested through animal feelings. Animal feeling, in turn, is the connecting link between the animal psyche and the non-conscious body. Thus, human feeling is the connecting link between the human psyche and the non-conscious human body, see figure 3.2.

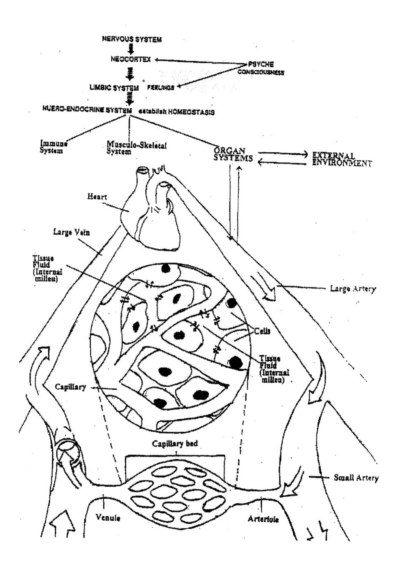

FIGURE 3.2 Homeostatic Body Plan of Mammals

FEELINGS ARE CONSCIOUS EMOTIONS

Emotions

Vertebrates, like the invertebrates, have nervous system circuits that coordinate the expression of instincts. Vertebrates also have circuits contained in a portion of the nervous system called the ANS (the autonomic nervous system) that coordinate homeostatic adjustments in the animal's body. Instincts are inherited behaviors that may be modified by learning or are themselves involved with programmed learning. Vertebrates also have neuron circuits that coordinate increased activity of the circulatory and homeostatic organ systems with some instinctual behaviors, especially drives, for example the sex drive. That is, situations that stimulate attack, escape or courtship and mating behaviors also stimulate the animal to prepare to carry out these behaviors. This uniquely vertebrate, instinctual behavior pattern is called *emotion*, which may be defined as the association of increased alertness and increased activity of the circulatory and homeostatic organ systems with a particular instinctual behavior.

Some neuron circuits, for example, the primitive limbic system, in conjunction with the ANS and portions of the brain stem, bring about emotion via three steps: (1) A stimulus which ordinarily activates a particular instinctive behavior also stimulates the limbic system complex to inhibit the motor programs that carry out that behavior; (2) the limbic system then stimulates the ANS to bring on physiological arousal, such as increased sensitivity to sensations, increased muscle tone, increased blood sugar, oxygen, and various hormones, increased blood pressure and blood flow to muscles, and many other changes; (3) after animal arousal, the motor programs are dis-inhibited, thus leading to a particular instinctual behavior. For example, when a fish is stimulated to attack, it also becomes physiologically aroused so that it can vigorously carry out this attack behavior. The association of physiological arousal with instinctive attack behavior is "attack emotion." In mammals, this behavior is called rage. When rage overwhelms a human, causing him/her to carry out acts of violence, the person is said to be temporarily insane. Fish, amphibians, and reptiles exhibit raw emotions from time to time whereas mammals and birds usually incorporate emotions into a behavioral innovation called feeling. However, in humans feeling may degenerate into a raw emotion.

Feelings Equal Animal Consciousness

As indicated above, emotions may be defined in terms of overt behaviors and physiological changes that can be observed and studied via an experimental approach. Feelings that incorporate emotions also have objective aspects. However, feelings also involve subjective orientations, which, of course, cannot be observed. As one might expect, this subjective aspect of feelings makes this topic controversial, especially to those many scientists who desire to define all ideas totally objectively. Many thinkers use the terms emotions and feelings to refer to the same phenomenon. Some writers attempt to define emotion equivalent to feeling in totally objective terms; others, such as LeDoux, introduce subjective aspects. There is not the same kind of consensus about the meaning of emotion and feeling as there is about other behavioral concepts such as reflex, innate behavior, or programmed learning. The view presented here is that feelings are not equivalent to emotions, but rather are emotions associated with subjective orientations or dispositions toward objects. All vertebrates have emotions, but only birds and mammals have emotions incorporated into feelings. Feelings have (1) perceptive, (2) experiential, and (3) expressive aspects. The perceptive aspect is an animal's imagination of his/her orientation (intention) to some aspect of the environment. One's understanding of the concepts of perception, imagination, and orientation must come from reflection on one's subjective experiences of various feelings that express different intentions. For example, anger expresses one type of orientation, fear another, curiosity still another.

The experiential aspect of feelings is that feeling orientations always are associated with direct or remembered experiences of pleasure or pain. Of course, the terms pleasure and pain also are subjective and not describable by mechanistic analysis. A scientist must project his own subjective experiences onto observed situations that mimic his experiences. This, however, can be done in a controlled experimental manner. In 1953, Dr. J. Olds designed an ingenious experiment that demonstrated areas in the brain related to the experience of pleasure. After anesthetizing rats, he inserted electrodes into various parts of their brains and cemented them into their skulls. After the operation, the rats moved around as if nothing had happened. Olds put these rats in cages where, if each rat pushed a lever, the implanted electrodes would stimulate the surrounding nervous tissue. When the electrodes were implanted in certain well-defined areas, the rats seemed to gain "pleasure" from pressing the lever. They ignored hunger and thirst in their pursuit of "self-excitement." Often the rats would continue pressing the lever until overcome with exhaustion. Other neuro-physiologists also discovered other areas that when stimulated led to apparent rage, fright, and "pain." These areas were designated as punishment centers. Most of these so-called "pleasure" (reward) and "pain" (punishment) centers are located in the portion of the brain near the primitive limbic system (that is, in the midbrain and hypothalamus).

Feelings are expressed through: (1) a change in the pattern of muscle tone, (2) physiological arousal, and (3) a chosen behavior. A change in pattern of muscle tone shows up in primates as a change in facial expression, body posture, and the way the animal moves. The physiological arousal is the same as found in emotions, except that the arousal ranges in intensity from low to very high. The chosen behavior may consist of gestures along with some overt activity or suppression of any overt activity.

TRIUNE BRAIN IN HIGHER MAMMALS

Though invertebrates are capable of some degree of learning, they are dominated by instincts that may be modified to become learned but somewhat unalterable habits. Mammals, on the other hand, are dominated by complex learned behavior patterns that incorporate some instincts but repress others. For example, a dog may be taught not to attack any stranger that approaches his territory. These learned behaviors are quite flexible in the young though they tend to fossilize into unalterable habits as the animal grows older. The increased capacity of mammals to learn is correlated with an increase in size of the brain, especially of the cerebrum. In contrast to other vertebrates, mammals have a network of interconnecting nerve cells, called *cerebral cortex*, at the surface of the cerebrum. The cerebral cortex is the major landmark for locating the different animal behaviors of emotions, feelings, and complex learning.

The primitive limbic, system which coordinates the expression of emotions, is located in the interior of the cerebrum in all vertebrates except mammals. In mammals, this system is located in that part of the surface of the cerebrum known as the primitive cerebral cortex. This primitive cortex is close to or on the side of each cerebral hemisphere, which faces the brain stem upon which the hemispheres sit. The newly evolved nervous network involved with complex

27

learning is called neocortex (neo means new). Neocortex is the lateral portion of each cerebral hemisphere. In primates, the neocortex makes up most of the cerebral cortex. A part of the cerebral cortex between the primitive cortex and the neocortex is called *mesocortex* (meso means in the middle). The mesocortex has been shown to be involved with the expression of feelings. Thus, all but the most primitive mammals may be thought of as having "three brains in one" referred to as the triune brain: (1) the emotional brain consisting of the primitive cerebral cortex in association with the brain stem and the spinal cord, (2) the feeling brain consisting of the mesocortex, which has many connections with the emotional brain and the learning brain, and (3) the learning brain consisting of the neocortex see Figure 3.3. For the most part in all developmentally normal, healthy, non-human mammals, these three brains are integrated in the functioning of a unified triune brain. Feelings that incorporate emotions integrate with complex learning to make up unified behavior patterns.

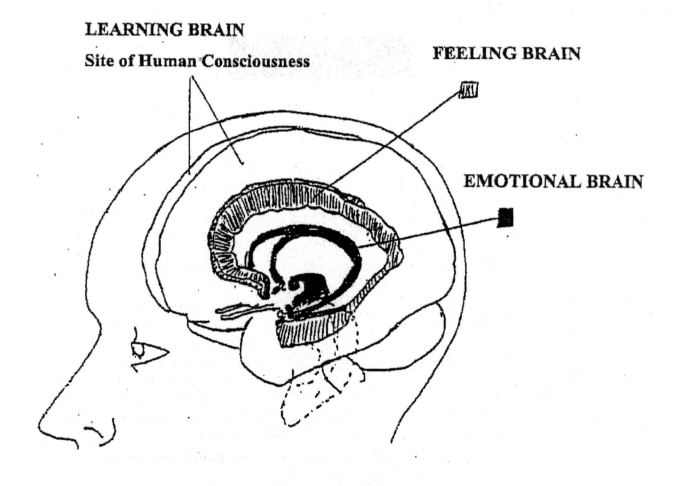

FIGURE 3.3 The Triune Brain of Humans

The above set of statements is a modified version of Paul MacLean's theory. Le Doux argues that the anatomical aspect of this theory no longer is tenable:

> Emotion is only a label, a convenient way of talking about aspects of the brain and its mind…. In a similar vein, the various classes of emotions are mediated by separate neural systems that have evolved for different reasons. The system we use to defend against danger is different from the one we use in procreation, and the feelings that result from activating these systems – fear and sexual pleasure – do not have a common origin. There is no such thing as the "emotion" faculty and there is no single brain system dedicated to this phantom function.[3]

I accept LeDoux's critique, but we still may use the idea of a triune brain even though there are no specific, recognizable brain structures corresponding to each brain. Rather there are separate neural systems scattered throughout the central nervous system that regulate the expression of emotions. We may refer to this set of separate neural systems as the *emotion brain.* The set of neural systems that associate awareness with emotions is the *feeling brain.* The set of neural systems that process information independent of emotions or feelings is the *learning brain.*

Chapter 4
EVOLUTION TO THE BODY EGO

ANIMAL SUBJECTIVE CONSCIOUSNESS

Mammals and birds have evolved a triune brain that orchestrates conscious feelings that promotes a more individuated way of communication among individuals in a society. This feeling awareness is channeled into social communicative actions to structures of dominance hierarchy, territoriality, individuated mate selection, and bonding within family or in small animal groups. Thus, these networks of communications represent what may be called unreflective animal social consciousness.

However, studies of modern monkeys and chimpanzees show that they form larger societies that involve complex interactions within the society that are not structured by the collective consciousness of the group. For example, East African vervet monkeys show complex social interactions. They classify social relationships into types and thus are able to generate appropriate behavior in complex social situations. For example, a young monkey chooses different appropriate behaviors when interacting with its sibling, or with its mother or with a dominant male monkey who may be its biological father. They also create a classification of sounds to denote different objects associated with events. For example, they use different sounds for denoting leopard alarm, snake alarm and eagle alarm. Each type of alarm sound is associated

---

3. Joseph LeDoux, *The Emotional Brain, the Mysterious Underpinnings of Emotional Life* (New York: Touchstone, 1996), p. 16

with specific behaviors. However, they apparently do not see similarities among their classified sounds. For example, they do not see the similarity between a leopard posing a danger to them and the appearance of some new different kind of danger. As a result, they do not see "leopard sound" as a basis for inventing a new category of sound to represent a new type of danger. Seeing similarities among diverse experiences is a mental process that, in turn, depends on an individual being aware of himself having some denotation knowledge. Monkeys do not have knowledge of their knowledge, which is to say that they do not have self-consciousness.[4] The complex social behaviors of these monkeys suggest that they have the beginnings of an animal subjective consciousness.

The chimpanzees show even more complex social behaviors than various species of monkey. This is significant because genetic analysis indicates that humans' closest ancestor is the immediate evolutionary ancestor of modern chimpanzee. This genetic closeness leads one to assume that chimpanzees are precursors to behaviors of hominids from which modern humans emerged.[5]

Bownds summarizes the evolution from chimpanzees to humans as follows.

> The earliest *anthropoids*, the monkeys and apes, appeared in Africa during the Oligocene epoch about 30 million years ago. They usually have single births, the newborn are largely helpless, and the young depend on parental care during a long period of growth and maturation. There is now general consensus that the hominid and chimpanzee line diverged from a common ancestor about 5 million years ago, whereas the gorilla and the orangutan split off at least 9 million and 12 million years ago, respectively.[6]

Bownds describes the social behaviors of chimpanzees, which are socially the most advanced of the non-human primates, as follows. They form societies of 30-40 individuals that occupy a home range for several years. Offspring are strongly bonded to their mothers and are socialized by her, and other members of the society, which often is a particular male. These societies show human-like behaviors such as males cooperating during hunting and sharing of meat. Two different societies sometimes share a feeding site. Chimps have created a classification of 35 different sounds, and they communicate by means of these classified sounds and by complex visual displays. Chimps are not only territorial, but there is documentation of one troop of chimps eliminating another troop in a slow and systematic way. In other more positive ways chimps show human-like behaviors. Socialization in the social unit leads to reciprocity involving giving, trading, and revenge. They seem to anticipate punishment for not obeying the rules of their society and are aggressive toward those who disregard the rules of reciprocity. Societies of chimps have been observed to give special treatment to disabled and injured members of their society. Within the context of their social hierarchy chimps initiate

---

4 M. D. Bownds, *The Origins and Structures of Mind, Brain, and Consciousness* (Bethesda, Maryland: Fitzgerald Science Press, 1999), p. 79.
5 M. D. Bownds, *The Origins and Structures of Mind, Brain, and Consciousness* (Bethesda, Maryland: Fitzgerald Science Press, 1999), p.78.
6. M. D. Bownds, *The Origins and Structures of Mind, Brain, and Consciousness* (Bethesda, Maryland: Fitzgerald Science Press, 1999), p. 71.

complex alliances and peacemaking, which appears to involve negotiation. Individual chimps exhibit feelings such as being happy, sad, angry, tired, embarrassed, and interest in or repulsion from some other individual.[7]

These complex social behaviors in monkeys but more especially in chimpanzees, suggest that these primates have evolved a higher level of primate intelligence associated with animal subjective consciousness. Merlin Donald's theory of the main stages of cognitive evolution in primates, as described by Bownds, calls the stage represented by monkeys and chimpanzees, *episodic intelligence*. Episodic intelligence is existential knowing equal pure subjectivity in humans. However, in non-human primates who probably do not have *self-consciousness*, this kind of knowing is not a basis for deep reflection that brings one to a personal engagement with an aesthetic experience or a spiritual insight as can occur in humans. At the same time, episodic intelligence is the basis for the continuity of human existential knowing with non-human primate knowing, as described by Polanyi:

> The ineffable domain of skillful knowing [existential knowing] is continuous in its inarticulateness with the knowledge possessed by animals and infants, who, as we have seen, also possess the capacity for reorganizing their inarticulate knowledge and using it as an interpretative framework…. We may say in general that by acquiring a skill, whether muscular or intellectual, we achieve an understanding which we cannot put into words and which is continuous with the inarticulate faculties of animals.[8]

Bownds describes this kind of animal intelligence as *episodic intelligence* that lives in the present. A chimp or a human baby exhibit this type of intelligence because there is no awareness of a self that can interfere with the awareness of the external world manifesting as autonomous episodes. There is no awareness of an underlying self that gives continuity of awareness of one event after another. Rather the chimp or baby is aware of one episodic event "seen" as disconnected from many other possible awareness events. The individual goes out to the event rather than internalizing the event by creating a representation of it and then becoming aware of this representation. Thus, episodic intelligence is a kind of fusion of the knower and some external, seemingly autonomous aspect of nature. I call this *participatory consciousness*, which causes the animal to fuse with and thereby become identified with one autonomous episode. Because there is no self-awareness to give continuity to these autonomous experiences, the individual appears to have no unique identity but rather at any one moment takes on whatever episodic identity the external environment thrusts upon the animal.

However, as indicated earlier in chapter 3, expression of feeling always is associated with memory. A large memory bank interacting with episodic awareness gives the animal a degree of subjective feeling consciousness. The memory bank stores single events, a sequence of events, and complex sets of social relations. All these stored memories are associated with concrete subjective feeling evaluations ranging from very painful to very pleasurable. The animal's memory bank connects any here and now awareness with remembered past events and social

7 M. D. Bownds, *The Origins and Structures of Mind, Brain, and Consciousness* (Bethesda, Maryland: Fitzgerald Science Press, 1999), pp. 80-81.

8. M. Polanyi, *Personal Knowledge* (New York: Harper Torch book, 1964), p. 90.

relationships. This connection provides the basis for choosing behaviors to be carried out in the immediate future. Thus, while the chimp or human baby is continually only aware of the concrete autonomous event of the here and now, by means of the memory bank, the individual is always connected to its past and to its immediate future[9].

When self-consciousness emerges in humans, episodic awareness not only is modified and incorporated into a new kind of knowing, it also is progressively overshadowed by this new self-conscious knowing. At the higher stages of self-consciousness humans may evolve to a momentary or permanent higher state of consciousness where the awareness of self is subordinated to achieving some goal that transcends pleasure/pain or good/bad evaluations. The athlete or performer exhibiting "peak performances" exemplifies this higher-level momentary episodic awareness. The Zen master/mystic exemplifies the semi-permanent or permanent higher-level episodic intelligence.

TRANSITION FROM APES TO HOMINIDS

Primitive Hominids

There is extensive evidence that humans and currently existing anthropoid apes, such as chimpanzees, evolved from a common ancestor. Some descendants of our ancestor apes evolved into hominids that further evolved into *Homo sapiens* no sooner than about 200,000 years ago. Extant anthropoid apes and *Homo sapiens* have many common features not only at the anatomical and physiological level but even at the level of molecular structure. For example, there is a 98- percent identity in non-repeated DNA in humans and chimpanzees, which would allow the possibility for these two species to mate and produce a viable hybrid. At the same time *Homo sapiens* exhibit radically new features: (1) human consciousness associated with free will, and (2) a social organization structured by culture that is transmitted from one generation to the next by teaching rather than by genetics. Consequently, the evolution of culture is relatively autonomous though dependent on the evolution of the biosphere.

In spite of these two radical differences between *Homo sapiens* and apes, humans are brothers and sisters to apes in the time table of evolution. What characteristics in our ancestor apes and their anthropoid descendants could be drastically modified to produce the distinguishing characteristics of humans? The most common answer is that all higher primates have a well-developed neocortex, which enables them to make extensive associations among diverse sensations and to learn new patterns of behavior. The use of primitive tools by extant anthropoid apes suggests that tools requiring further expansion of the neocortex associated with development of "intelligence," were pivotal to the divergence of hominids from ancestor apes. However, "... the earliest recognizable tools [used by hominids] are only about two million years old."[10] But, the hominid and chimpanzee line diverged from a common ancestor about five million years ago. Study of reconstructed pelvises of fossil skeletons show that the marked expansion of the hominid neocortex took place only during the last two to three million years. Thus, some other factor besides tools or an expanded neocortex was the connecting link between ancestor apes and primitive hominids.

---

9 M. D. Bownds, *The Origins and Structures of Mind, Brain, and Consciousness* (Bethesda, Maryland: Fitzgerald Science Press, 1999), pp. 78-79.
10 C. O. Lovejoy (1981). *The Origin of Man* <u>Science</u>, 211

In contrast to other vertebrates, mammals (and birds) have animal consciousness which is the basis for a feeling communication among them. To some extent most mammals express via changes in muscle tone, physiological arousal, and behavior how they feel about a particular situation, such as fear, anger, or curiosity. Each individual of a species is programmed to interpret and learn more about these body cues. In exhibiting these body cues, animals communicate to one another how they feel. This communicated feeling awareness which is partly instinctive and partly learned is the expression of animal consciousness. The learning aspect becomes most predominant in the reciprocal, mutually reinforced displays that produce bonding between the communicating animals. Because of the high degree of learning and individuality involved, bonding is one of the highest expressions of animal consciousness. In higher primates the mother and offspring form a matrifocal unit held together by strong feeling bonds. "The matrifocal unit of chimpanzees continues throughout the life of the mother, as do sibling ties."[11]

Evolution of Hominids

The earliest fossil remains of hominids, 2.9 to 3.0 million years old, indicate that the species called *Australopithecus afarensis* had the character pattern of bipedal locomotion, small brains, no use of tools, and nuclear family social structure.[12] *Homo habilis*, who emerged about 1.6 million years ago, mark the beginning of proto-human social consciousness differentiating into *control consciousness*, involving using tools, hunting, and making fires and *participatory consciousness*, which is akin to and continuous with feeling awareness. This hominid species evolved into *Homo erectus*, which emerged 1.6 million to 100,000 years ago. *H. erectus* had a somewhat increased cranial capacity, used tools, and left behind cave burials which implies the concept of death. That they could form any concept at all indicates the first stage of developing a language. However, there are no fossils that indicate they had religion or art.[13] I surmise that the beginnings of a conceptual language indicates the beginning of awareness of "I versus not-I," which implies the emergence of a primordial ego consciousness. Between 100,000 to 35,000 years ago "Neanderthal man," now considered to be a subspecies of *Homo sapiens,* left fossils indicating an expanded neocortex and artifacts indicating a consolidated culture.

    Hominid subjectivity. Bownds points to the emergence of new traits in hominids as represented by *H. erectus*. Skeletal fossils indicate that they had more complex facial musculature, which suggests they had a wider range of feeling expressions with more subtle variations than chimps with only episodic intelligence. These more complex feelings probably were supported by a greater number of types of sounds and calls in which the sounds could vary in volume, pitch, tone, and emphasis. The fossil records indicate that this species had a larger brain, and they made more elaborate tools, used fire, and had seasonal camps. *H. erectus* immigrated from Africa to Eurasia.[14]

---

[11] C. O. Lovejoy (1981). *The Origin of Man*. <u>Science</u>, 211

[12] Marilyn French, <u>Beyond Power</u> (New York: Belles-Letters, Inc., 1985), pp. 141-159.

[13] Marilyn French, <u>Beyond Power</u> (New York: Belles-Letters, Inc., 1985), pp. 163-175

[14] M. D. Bownds, *The Origins and Structures of Mind, Brain, and Consciousness* (Bethesda, Maryland: Fitzgerald Science Press, 1999), pp. 89-90.

Mimetic intelligence: an expansion of existential, episodic intelligence. Merlin Donald's theory, as described by Bownds, proposes that ancestor chimps evolved to hominids that express mimetic intelligence. My interpretation of Donald's theory is that this intelligence is associated with greater subjective learning/knowing that is a precursor to or is the first stage of self-consciousness. Mimetic intelligence involves learning by imitation that in some way is represented in the nervous system and then stored in the animal's memory bank. Because evolution is hierarchal, modified hominid mimetic intelligence is embedded in the modern human mind. According to M. Donald, it is seen in pre-linguistic children, illiterate deaf-mutes as well as in those who lost language but retained social skills and communicate by mime. Thus, this hominid cognition resembles the extra-linguistic features of the modern mind.[15]

Mimetic intelligence is an expansion of "existential episodic intelligence" to a "proto-conceptual knowing" that is the basis for the emergence of what I call metaphorical-conceptual knowing. Metaphorical conceptual knowing implies conceptual language, but represents the intermediary between subjective, non-logical, non-linguistic, existential knowing and objective, logical, language knowing, that is, what I designate as essential knowing. This indicates that mimetic knowing is the necessary foundation for the emergence of metaphorical-conceptual knowing, which in turn is the basis for the emergence of logical conceptual knowing. This analysis fits very well with Merlin Donald's theory that represents the epigenesis of primate cognition as occurring in the four stages: episodic, mimetic, mythic – what I call metaphorical-conceptual – and theoretic – what I call essential – knowing. In this epigenetic series, as is true of all epigenesis, each higher level of cognition modifies and incorporates the lower-level type of cognition, and episodic cognition modifies and incorporates feeling awareness found in most mammals and birds.

Bownds' analysis further supports these epigenetic relationships. According to him, in agreement with M. Donald's theory, a kind of non-linguistic conceptual knowing produces mythic intelligence in humans. A view consistent with that of Bownds' view is that word learning involves attaching words to preexisting concepts of objects, actions, collections, and social situations. These non-linguistic concepts emerge from mimetic intelligence. This is indicated by the fact that isolated deaf people who have not learned sign language show an understanding of spatial relations and associated skills, can carryout monetary interactions and can pantomime narratives. Likewise patients with brain lesions that disrupt language usually can communicate by sign or body language thus using mimetic skills. On the other hand, destruction of parts of the brain controlling mimetic skills also destroys conceptual communication.[16]

Origin of Language

Each new addition to material hominid culture -- the use and refinement of tools, invention of containers for carrying food, the beginning of organized hunting, and the harnessing of fire -- separated hominids more and more from nature as a result of their having increasing control over it. The final major break from nature was the creating of concepts to represent experiences of nature. A burial indicates a ceremony and ritual that imply the creation of the concept of death.

---

15 M. D. Bownds, *The Origins and Structures of Mind, Brain, and Consciousness* (Bethesda, Maryland: Fitzgerald Science Press, 1999), p. 90.
16 M. D. Bownds, *The Origins and Structures of Mind, Brain, and Consciousness* (Bethesda, Maryland: Fitzgerald Science Press, 1999), p. 91.

"There are remains of cave burials from the third ice age, 150,000 to 100,000 years ago."[17] The invention of language gives a formal structure to contain diverse concepts. Since language is learned during a critical period in ones childhood or not at all and since even today mothers carry on extensive non-verbal communication with their infants, it is possible that "... mothers cooing to, warning, and teaching their youngsters developed a symbolic system that was then built on and transmitted by the children."[18]

The above statement is consistent with Bownds' description of the epigenetic emergence of language. Although language emerged from hominids around 50,000 years ago, the brain mechanism underlying speech may have evolved over a mush longer period of time. Mimetic intelligence, which emerged from episodic intelligence, adds the more extended communication of social procedure and rituals. Building on this, conceptual language may have originated from oral reinforcement of symbolic gestures used in feeling communications. Thus, brain pathways underlying mimetic and conceptual language communication could have developed in sequence but for a time also in parallel to one another. Though it is possible to dissociate mimetic and conceptual language intelligence, the linguistic communication enriches and reinforces the mimetic communication.[19]

This book suggests that individual, control consciousness emerges from social, participatory consciousness. This is consistent with the idea that language emerges as a social adaptation. Natural selection operating on hominids may have driven the emergence of conceptual language that more effectively distinguishes different types of complex social organizations. Hominid alliances based on grooming began to interfere with other social activities. Conceptual language may have been selected as an adaptation to replace grooming as the basis of alliances. Conceptual language can distinguish kinship, marriage, and lineage in hominid cultures. Conceptual language allows the naming of these social institutions. Thus, a culture defined by mimetic intelligence may have undergone a social adaptation producing language communication.[20]

The invention of language is the culmination of hominids evolving into "proto-humans" with "group ego" (see next section), and then into humans with the distinguishing characteristics of individual ego self-consciousness and free will. This occurred anywhere from 200,000 to 35,000 years ago. In any case primitive humans had a control consciousness epitomized by language which separated them from nature. They also had a participatory consciousness, which allowed them to feel at one with nature and with one another. It was the basis for a nuclear family involving the cooperation of mother and father in the parenting of offspring. Later, when families became extended families that merged to form clans, the social organization was egalitarian. Participatory consciousness predominated, which meant that the sense of connectedness with nature and with one another was greater than the sense of separateness. Anthropologists refer to the various species of hominids as *archaic humans*. Modern humans develop ego-consciousness in association with language using consensually chosen vocal sounds

---

17 M. French, *Beyond power* (New York: Belles-Letters, Inc., 1985), p. 38

18 M. French, *Beyond power* (New York: Belles-Letters, Inc., 1985), p. 38

19 M. D. Bownds, *The Origins and Structures of Mind, Brain, and Consciousness* (Bethesda, Maryland: Fitzgerald Science Press, 1999), p. 94.

20 M. D. Bownds, *The Origins and Structures of Mind, Brain, and Consciousness* (Bethesda, Maryland: Fitzgerald Science Press, 1999), pp. 95-96.

to represent categories of experience, that is, for example, the vocal sound to represent the experience of fire or danger or mother or father.

FEELING ACTION SELF

Transition to Group Ego

Where, when, and the context in which the beginnings of human self-consciousness first emerged are not known. We may surmise that Eros-chaos (see end of chapter 2), drove some hominids bonded to one another to collectively experience a participatory, feeling, vague self-awareness. This group self-awareness brought these individuals out from being embedded in nature. Of course we cannot know what this experience of first awakening is like, but we may surmise that it is something like when a person is embedded in a life situation and then is able to mentally step back and look at oneself in that situation (a process described in chapter 27 as *mindfulness*). In doing this one not only sees new possibilities of what can be done, but he/she also sees oneself as able to actualize some of these possibilities. This breaking away from being embedded in some structure and then actualizing newly seen possibilities is a pattern in all instances of individuation. Individuation of human self-consciousness is associated with the differentiated, hominid, control consciousness progressively displacing and eventually overshadowing participatory consciousness. The emergence and differentiation of symbolic language is the mechanism by which this displacement occurs. However, the first emergence of self-consciousness may be prior to the emergence of symbolic language. That is, a transformation of hominids occurs analogous to the emergence of life from non-life, described in chapter 2. Separation from being embedded in nature meant that the first self-conscious humans became aware of their animal conscious episodic experiences and their mimetic intelligence. Prior to this transforming event, hominid family members and male hominids forming hunting parties of the clan – consisting of several genetically related families – were in participatory union with one another and with nature. Over many centuries these family and hunting party groups in a non-self-conscious way created tools and other innovations that contributed to their survival and evolutionary advantage over other animals. Human self-consciousness emerged when these groups began to become aware of themselves creating innovations.

The dawn of human self-consciousness, then, presents a kind of group-ego, self-awareness that is structured by action and innovations related to action. This group self-awareness may be something like the interactions within a group of jazz musicians that participate with one another and improvise playing new themes in the context of this mutual engagement with one another. The resulting music is magical in that it is action that proceeds, so to speak, from the soul of the group rather than from any language communication before or during the session. Likewise, the group self-consciousness of the first humans producing action innovations for survival may be said to have a magic structure. Gebser describes this magic structure as the means by which the first self-conscious humans break from their embeddedness in nature. That is to say, the first emergence of self-consciousness corresponds to humans spontaneously creating a magic structure initially focused on making tools. The self-conscious making and using tools enables humans to have a "sleep-like" awareness of having some control over nature and therefore some break from being embedded in her. The evolution of humans' magic control over nature leads to greater human independence.

Gebser also describes this magic structure as a group-ego self-consciousness that is aware of shared episodic experiences of actions such as making tools and hunting. Nature is vaguely seen as made up of autonomous, episodically experienced events. Group-ego episodic experiences drive individuals to become engaged and identified with one another in a clan and with events in nature. This beginning of self-consciousness is not yet centered in any individual human but rather is spread out over the world of phenomena. Anthropologists call this *participation mystique*. Nevertheless, this spread out group-ego consciousness is associated with individuals as members of a clan separating from and independent of nature by exerting some control over her. This leads to humans becoming aware of a "will to independence" associated with self-consciousness. This will to independence is expressed by witchcraft and sorcery, totem and taboo, which lead to increasing self-consciousness. Impulses and instinct associated with group-ego consciousness and the will to independence enables humans to participate with one another in a clan that copes with nature now vaguely seen as "Other." Nature as "Other" continually threatens humans' material existence and tends to cast a spell on humans so as to entice them to be again embedded in her. Humans create their own spells as part of their magic structure, which helps them to make things using the materials of nature and thus maintain some independence of her. Thus, each human in participating in this magic structure becomes the maker.[21]

In becoming "the maker" humans are beginning to separate themselves from Nature by developing control conscious. This control is further developed when magic humans are confronted with animals who threaten their survival. The magical human objectifies the threatening animal by disguising himself as that animal or by making a drawing of it. Thus, by partially disengaging from the animal now seen as a threatening other, the magical human gains some power over it.[22] Also, magic humans objectify various objects by casting a spell on them. This centers the human will on an object thereby centering human psychic energies.[23]

Thus, there are at least five essential characteristics of magic humans. He/she:

1. is egoless.
2. has a sleep-like awareness of isolated, episodic events; that is, sleep-like awareness of existential engagement with and thus merger with objects or isolated events.
3. has non-mindful, Zen experiences; that is, the episodic events are *not* understood to be aspects of some whole system that exists over time; so each experience is spaceless and timeless.
4. merges, by means of participatory consciousness, with Nature during each experienced event.
5. focuses on breaking from this merging with Nature by entering group, non-mindful, creative participatory dialogue with Nature that starts what may be called the magic, learning cycle.

---

21 J. Gebser, *The Ever Present Origin* (Athens, OH: Ohio Uni. Press, 1953, English trans., 1985), p. 46.

22 J. Gebser, *The Ever Present Origin* (Athens, OH: Ohio Uni. Press, 1953, English trans., 1985), pp. 46-47.

23 J. Gebser, *The Ever Present Origin* (Athens, OH: Ohio Uni. Press, 1953, English trans., 1985), p. 48.

Magic learning cycle. There is a fundamental discrepancy in magic human's episodic mergers with Nature. On the one hand, Nature seems to "cast a spell" on human consciousness in which each human, separated from Nature by a sleep-like, egoless, self-awareness, is drawn to participatory, conscious union with Her. The spell cast by Nature may be thought of as an innate Will-to-openness to engagement with Nature associated with Eros-chaos that drives each human to lose his/her newly emerged, partial separateness from Nature; that is, one tends to lose individuality stemming from "magic self-consciousness." On the other hand, there is an innate drive within each magic human to resist total merger and subsequent loss of "magic self-consciousness." This innate drive is Eros-order that is associated with Will-to-power which expresses magic humans' Will to control Nature rather than be totally overwhelmed by Her.

The opposition of Eros-chaos, Will-to-openness to Eros-order, Will-to-power is transcended by a mutuality between these two drives. As is true in the non-human world (see chapter 2), this mutuality leads to creativity – in this case magic creativity that achieves several goals. 1) It gives magic humans some control over Nature; (2) it not only preserves but progressively leads to greater "magic self-consciousness" as a result of control consciousness progressively displacing participatory consciousness; (3) it leads magic humans to partially transcend the fundamental ambiguity of human self-consciousness wherein this ambiguity is manifest at all levels of pre-personal and personal human individuation; the ambiguity is represented by participatory consciousness versus control consciousness and Eros-chaos vs. Eros-order; (4) it specifies magic humans as action self with group-Ego; that is, action self expresses magic which is "doing without Ego-consciousness" or "doing without Ego-knowledge"; (5) it puts magic humans on the path to the emergence of a consolidated center of self-consciousness in each individual. This new center is the *mind self* that expresses itself through Ego consciousness exerting some degree of control over Nature.

The magic learning cycle may be divided into six steps as follows:

1. Eros-chaos draws members of a group to partially merge with Nature in a sleep-like conscious episodic engagement with some object or event.
2. Spontaneous, innate Will-to-openness associated with Eros-chaos allows the magic group to "see" possibilities for creative, magic control of some aspect of nature. This is analogous to members of a jazz group seeing new possibilities of variation on a theme of music.
3. Eros-order drives members of the group to resist total merger with nature so as to maintain some separate individuality by objectifying aspects of the episodic experience.
4. Will-to-power associated with Eros-order leads the magic group to actualize one or more of the possibilities seen during the episodic experience.
5. The magic group via Eros-chaos episodically experiences the outcome of actualizing one of the possibilities to which Will-to-power led.
6. Group Will-to-openness "sees" whether the Will-to-power leading to "magic doing" increases the group's control of nature. If it does, the group may see new possibilities for exerting greater control. If the Will-to-power action does not increase control, the group will start new learning cycles until one of them produces an increase in control.

Initially the magic learning cycles do not require language. On the one hand, humans ' deep merger with Nature explains their possessing powers of telepathy, heightened, natural, sensory apperception, divination and "powers of second sight." Gebser led me to realize that the first emergence of self-consciousness occurred before the emergence of artificial symbolic language. "Communication between members of the group-ego, the We, does not as yet require language, but occurs to a certain extent subconsciously or telepathically."[24]

Neanderthals. Neanderthals were the "... first people known to bury the dead."[25] There also is evidence that they developed a religious cult centered on the bear. They invented the spear and axe used for hunting, which became a more important food source to supplement gathering. These cultural factors, in conjunction with an expanded cranial capacity and a larynx, indicate they had a language. Not much can be said about their social structure except that there probably was division of labor based on sex; the males hunted and domesticated dogs, and females took care of the children, prepared food, and performed other tasks associated with the home base. Thus, I surmise that Neanderthal social consciousness associated with spoken language differentiated into: (1) participatory consciousness expressed in their religious cult centered on the bear, and (2) control consciousness expressed in inventions of tools and division of labor based on sex. The further differentiation of a primordial ego consciousness associated with the development of spoken language probably only manifested a primitive differentiation of control versus participatory consciousness.

Homo sapiens sapiens. Primitive human ecology. The other human subspecies, H. sapiens sapiens, produced the much more advanced Paleolithic culture beginning around 40,000 years ago. They invented spear-throwers, harpoons, and possibly the bow and arrow. There is some evidence that they recorded the changes of the phases of the moon. Some of the most spectacular signs of their culture are the cave paintings discovered in Spain and southern France. At the very least these paintings indicate a very keen sense of aesthetics and skill. Some scholars interpret these paintings to have religious and magical significance. Leroi-Gourhan, director of the Sorbonne's Center for Prehistoric and Protohistoric Studies states:

> ...Paleolithic art expressed some form of early religion in which feminine representations and symbols played a central part.... Characteristically, the female figures and the symbols he interpreted as feminine were located in a central position in the excavated chambers. In contrast, the masculine symbols typically either occupied peripheral positions or were arranged around the female figures and symbols.[26]

Other cave findings are in line with Riane Eisler's proposal

> that the vagina-shaped cowrie shells, the red ocher in burials, the so-called Venus figurines, and the hybrid woman-animal figurines

---

24 J. Gebser, *The Ever Present Origin* (Athens, OH: Ohio Uni. Press, 1953, English trans., 1985), p. 58.

25 M. D. Bownds, *The Origins and Structures of Mind, Brain, and Consciousness* (Bethesda, Maryland: Fitzgerald Science Press, 1999), p. 180

26 Riane Eisler, *The Chalice & the Blade* (San Francisco: Harper & Row, 1988), p. 6.

earlier writers dismissed as `monstrosities' all relate to an early
form of worship in which the life-giving powers of woman played
a major part.[27]

Gerda Lerner also suggests that the cave paintings and sculptures indicate the pervasive veneration of the Mother-Goddess.[28] In line with a biological trend found in all higher primates, the human mother had great power over infant and child. Because of a relatively short life span, it was essential for group survival for early humans as well as hominids to have as many offspring as possible, resulting in women's devoting most of their adulthood to nursing and caring for the young. In both males and females accepting this necessity, it is reasonable that both would tend to construct beliefs, mores, and values around the female being the source of life and of creativity. Furthermore, "It would follow that women would choose or prefer those economic activities which would be combined easily with their mothering duties."[29]

At the same time there was a further differentiation of control consciousness and participatory consciousness. Spoken language guided technologies that increased human survival in some way. The language-technology behavior complex converted nature into an object divided into aspects that could be manipulated to accomplish some human goal. A stone, for example, could be hammered into a tool useful for hunting. In objectifying nature in this way, humans separated themselves from her in order to control her. Participatory consciousness in this period of history led spoken language to guiding the expression of feelings, which enabled humans to live in harmony with nature. First and foremost, harmonious living required bonding and nurturing to occur in the nuclear family. Harmony beyond the family required an ecological oral tradition about Nature, such as the interrelations among diverse plants and animals in a particular region. Moreover, the drive for harmony included art forms, rituals, ceremonies, magical, mythologies and religious practices that venerated nature and worshiped the Mother Goddess in all her manifestations. Thus, *we may define ecology of primitive humans as the symbolic expression of that aspect of participatory consciousness that is directed to networks of interrelationships.*

Masculine versus. feminine group-ego consciousness. Though both sexes participated in creating objectification of nature and participatory ecology, males specialized in the technologies associated with hunting, warfare, and animal husbandry. Females specialized in producing and raising children and in the technologies associated with maintaining the household. Females had menstrual cycles, which came to be associated with the phases of the moon and the blood that flowed had mystical power because blood symbolized life. As a result of this and the fact that women give birth to new life, females were more closely associated with nature than males. This difference was reinforced by the sexual division of labor. Males dominated nature through technology; females subordinated technology to nurturing within the nuclear family.

Both males and females helped to create and participated in the artistic and religious dimensions of ecology. However, because women were more closely associated with cyclic processes in nature and with reproductive creativity, they were held in higher regard than men.

---

[27] Riane Eisler, *The Chalice & the Blade* (San Francisco: Harper & Row, 1988), p. 6

[28] Gerda Lerner, *The Creation of Patriarchy* (New York: Oxford University Press, 1986), p. 39.

[29] Gerda Lerner, *The Creation of Patriarchy* (New York: Oxford University Press, 1986), p. 41

The sexes had separate complementary roles in Neolithic culture, but women were closer to, and more directly embodied, the creative power of the Mother-Goddess. This feminine power, however, is not equivalent to domination. In fact, current archeological evidence strongly indicates that at no time did women dominate men. Because of the division of labor by sex, *males focused on developing control consciousness, Will-to-power, and objectification of nature. Females focused on developing participatory consciousness, Will-to-openness, and participatory ecology.*

Magic structure of individual, self-consciousness. The growing human fetus is a differentiation process determined by genetic factors and interactions within the mother's womb, and, as such, is an aspect of Nature like all other non-human systems. At some point in time, perhaps in the seventh month of pregnancy, the fetus develops primordial animal consciousness. I propose that further developmental interactions involve the mother's feelings and sounds from the outside world. These feeling interactions prepare the conscious, animal fetus to take on a vague sense of itself separate from but in union with its womb environment. This "spiritual mutation" that may occur before birth is the emergence of the primordial magic structure of self-consciousness. The feeling interactions leading up to it constitute the hominid-like phase of human evolution. The emergence of this magic structure indicates that the developing human individual is radically different from all other types of living organisms. Biological evolution involves populations of individuals changing over many generations. In contrast, each human individual is a developmental-evolutionary process extending from conception (the zygote) to the death of the organism.

The magic structure emerged in hominid societies before the emergence of human symbolic language. In the pre-human phase the individual fetus differentiates structures necessary for the baby to speak, but it also differentiates networks of nerve cell connections in the brain for it to manifest the vague, proto-magic structure of self-awareness. Just as in telepathic communication among individuals in primitive magic tribes, there may be telepathic communication between mother and fetus and between the fetus and other aspects of the mother's environment, such as music, sounds of violence, feeling dimension of communication among many humans. The mother's womb is to the fetus what the natural environment is to "magic man." Thus, the fetus "experiences" events in the uterus such as movement of the uterus and the fetus especially "experiences" the cataclysmic event of birth. All of these feeling experiences are stored in a non-linguistic way in the body of the fetus, probably especially in patterns of muscle tension.

EMERGENCE OF BODY EGO

Baby's Impossible Project

Before birth the fetus differentiates a disposition to participate in "group-ego" consciousness. After birth, the fetus, now a baby, begins episodic engagement with and thus merger with objects or isolated events. These engagements lead the baby to participate in group-ego consciousness with its caretakers. Eventually one caretaker group-ego, usually involving the mother, will be to the baby what the nuclear family and later, the tribe, was to primitive, magic humans. At first mother not only participates in the baby's group-ego consciousness, she also is Nature that casts a spell on the baby. The baby at times separated from being totally, psychically embedded in mother is drawn to participatory, conscious union with her, thus loosing its separateness. But there is an innate drive – Eros-order associated with Will-to-power – that resists total psychic merger with mother. From the time of birth the baby is driven to attempt to control mother who represents Nature rather than be totally overwhelmed by her. Eros-chaos, Will-to-openness drives the baby toward participatory oneness with mother. Simultaneously, Eros-order, Will-to-power drives the baby toward separation from mother and toward control of her. At first, mother is overwhelming so that the baby temporarily merges with her or with Nature during sleep and loses self-consciousness. The impossible project for the baby is to establish a base magic consciousness that can progressively increase control over mother representing Nature.

The Baby's Magic Learning Cycle

If the mother consents to bonding with her baby – Nature drives her to do this though her free will enables her to go against this drive – the baby-mother group-ego engage in magic learning cycles that produce a mutuality that transcends the opposition of Eros-chaos, Will-to-openness and Eros-order, Will-to-power. This leads to "magic creativity" that achieves several goals: (1) The baby gains progressively more control over its body episodically perceived at this time as part of Nature; (2) the baby not only preserves but increases "magic self-consciousness" as a result of control consciousness progressively displacing participatory consciousness; (3) even as the intensity of the bonding between mother and child increases, the child separates more and more from mother thereby increasing the tension of the child wanting to control the mother and the mother wanting to draw the child to her; (4) especially after the baby develops a sense of its physical separateness from Nature (between 5 – 8 months after birth), the child becomes an action self in that it continually learns greater body control and new activities; (5) the child in participation with others, especially mother, begins to create for itself "proto-language."

The action self becomes the body ego. The magic structure of the child generates what is called an *action self*, in that its consciousness is spread out, so to speak, in creating and controlling various activities, such as crawling, standing, walking, eating. At the same time as the child is co-creating with its family a language for itself, the child is objectifying its separateness from Nature and correspondingly attempting to exert greater control over Nature represented by mother and other members of the family. Between 1½ and 3 years of age, the child "verbalizes" its awareness of a willfulness expressed in feelings that "belong" to the action self and are different from the willfulness expressed in feelings of mother and of other family members. Often this verbalization takes the form of "No!" Thus, besides being drawn to episodically participate in external Nature represented by mother, the child has a participatory awareness of its physical self that not only carries out actions but "has" willfulness expressed in feelings. The child identifies via participatory awareness with being a physical self. Thus, the action self child becomes a *body ego*.

Summary of the emergence of the body ego. Eros-chaos in collaboration with a human society drives the feeling action self with no language to be disposed to see that its non-conceptual episodic experiences can be represented by spoken words, for example, mama, dad, table, car. Eros-order, again in collaboration with a human society drives the child to learn these spoken words and thus begin to develop a proto-language. This proto-language, in turn, enables the child to represent to itself its magic self-consciousness. On the one hand, the child already has some control over its body and on the other hand the child episodically participates in its body doing things by itself independent of others, such as mother. This episodic individual self seeing itself doing things and simultaneously identifying with the body that is doing various actions, produces an episodic self identity: "I am this body that is doing things." Moreover, the child is aware of itself having the willfulness associated with feelings that are different from any one else's feelings. This episodic, subjective awareness represented by a word, that is, the child's first name, becomes a permanent self identity: "I am this body self with this name who has willful feelings and performs actions associated with these feelings."

Chapter 5
POLAR MIND SELF

SOCIAL PSYCHOLOGICAL ASPECTS

Ambiguity and Eros Stress of Primitive Tribes

As males individuated to greater control objectivity as a result of creating weapons for hunting, they became more separate from females, who focused on differentiation of participatory subjectivity. As a result, the males and through them, the primitive society began to experience the stress of Eros ambiguity. On the one hand, Eros-order drove males to form and maintain the nuclear family unit by participating in it. On the other hand, Eros-chaos of primitive human individuation drove males to separate from the nuclear family in order to develop technology, skills, virtues, such as courage, patience, tolerance of pain, and male bonding within the hunting group. From the perspective of social structures, the male value sphere associated with hunting was becoming disconnected from the female value sphere associated with the nuclear family. Survival of primitive societies depended on these two social institutions – hunting band and nuclear family – being coordinated with rather than fragmented from one another. The emergence of familialization of the male became the evolutionary way of coordinating these two institutions.

Solution to Eros Stress: Emergence of Patriarchal, Polar Mind Self

Familialization of the male converts the egalitarian nuclear family into a hierarchal structure. Males took on the role of father defined in a social-psychological sense which, of course, included the biological aspect of fatherhood. The social sense of this produced a patrilineal kinship system in which a human traces his/her descent through the father and his ancestry. The extended family is organized on the basis of male descent and inheritance. Matrilocality was replaced by patrilocality, in which a new couple's residence and/or activity centered around the residence of the husband's family or tribe. These social traditions differentiated into exogamy, the custom or law requiring marriage only outside one's own tribe, clan, and so forth. The psycho-social aspect of the father role made the father the head of a hierarchal structure that replaced the egalitarian structure of the hominid family.

Ken Wilber, quoting Habermas, summarizes how this solves the problem produced by Eros ambiguity. The individuation of some archaic humans to primitive modern humans produced the social differentiation of male society of the hunting party versus food-gathering females and the young they nurtured. This social differentiation was the further separation of control consciousness represented by males from participatory consciousness represented by females who took care of the young and together with the older children gathered food. This differentiation of male functions versus female functions was a necessary adaptation for survival, but it produced social conflicting tensions, what I call *social Eros stress* of the nuclear family. Males transcended this family stress by transforming to a new self-identity. Males transcended being the outward-directed hunter that contrasted with the inward-directed, participatory females. Males retained the status of being an individuated warrior-hunter who identified with other males "born again" as a warrior-hunters. The family stress was overcome by the male further individuating to modifying his warrior status to the father role in the family. With this more complex warrior and father status he could participate in the egalitarian relations within the hunting band and in the participatory relations of the nuclear family, which now became a hierarchal, egalitarian family. As a result, social labor functions of hunting and gathering food were integrated with raising and nurturing the young.[30]

At first there was an egalitarian relationship between males and females who participated in this hierarchal family structure. Wilber notes that the male as hunter and father was not seen as a higher status than the female as mother and nurturer of children. This new social innovation of a more complex, hierarchal family structure resulted from the evolutionary drive of humans to survive. Back then as is true now, "a house divided against itself cannot last." Continued human survival depended on family harmony. The male father status was a further elaboration of the males' greater physical strength and mobility complementary to female procreation and greater emphasis on participatory nurturing of the young.[31]

Another way of describing this social transformation is in terms of the emergence of Mind that generates ego consciousness that defines itself by polar relations. Individuals in a non-patriarchal tribe had a body ego that enabled them to express individual feelings in the context of a group ego. Males strongly bonded with all participants in the family group ego and strongly bonded with all participants in the hunting party group ego. The adult males experiencing ambiguity of body ego definition produced social stress for the tribe. Male bonding producing participatory subjectivity and nurturing in the family was at odds with bonding in the hunting party that led to independence and control objectivity. That is, participatory subjectivity leading to fusion was at odds with control objectivity leading to independence. Eros-chaos produced this ambiguity in individual males. This broke open a new possibility of a new kind of self-consciousness, which we designate as Mind with individual ego consciousness; that is, group ego was transformed into individual ego. The individual ego could *imagine* itself in different polar relations, the most fundamental of which is ego in mutual relation to "inner self." This "inner self" was vaguely perceived as the source of one's magical, feeling imaginations, but the source itself was mysterious. From our vantage point we say that these primitive tribe people discovered *soul*.

Eros-chaos opened up the possibility of a new kind of consciousness, which is individual ego imagination that involves non-conceptual, metaphorical thinking. Non-conceptual, metaphorical thinking put into words is the beginning of poetry and myths. Eros-order brought people in primitive tribes to tell stories about themselves and about their perceptions of their world. In terms of the Eros stress of the tribe, males now could and did imagine themselves in different polar relations. The individual, via having an ego, now could see these polar relations as different and separate. The recognition of their distinct separateness meant that the male could define self in terms of the ability to participate in divers polar relations rather than in terms of a single bonding context. In particular, the male could see himself as two mutually compatible

30 K. Wilber, *Sex, Ecology, Spirituality* (Boston: Shambhala, 1995), p. 156.
31 K. Wilber, *Sex, Ecology, Spirituality* (Boston: Shambhala, 1995), p. 157.

polarities: head of family vs. inner self and hunter/warrior vs. inner self. Thus, Eros-order created familialization of the male. The egalitarian family became hierarchal, egalitarian patriarchy.

Egalitarian Patriarchy of Neolithic Cultures

Neolithic culture, which extends from 8,000 to 3,500 B.C., extended the differentiation of human awareness into control and participatory consciousness, and extended the division of labor according to sex. On the one hand, Eros-order associated with control consciousness was expressed by the invention of horticulture of barley and wild forms of wheat, domestication of sheep, goats, cattle, and pigs, and the invention of ground stone tools, weaving, and pottery. These people developed skills necessary for architecture, the ability to plan outlay of towns, and an economy based on trade of raw materials.

On the other hand, Eros-chaos associated with participatory consciousness was expressed by the creation of an advanced religion with symbolism and mythology. As was true of the peoples that produced the cave paintings, goddesses were worshiped. According to some interpretations of extant fossils, this religion was both monotheistic and polytheistic. The Great Mother Goddess was the source of all things, and she is depicted in many forms. Sometimes she is depicted as pregnant or in the process of giving birth. At other times she is shown as part animal and part human, which may be interpreted as symbolizing humans' participatory unity with nature. She is accompanied by powerful animals such as the leopard and the bull. Though in much later cultures the bull symbolizes masculine power, in this context the bull is a manifestation of the ultimate power of the Goddess. Other manifestations of the Goddess are snakes, eggs, butterflies, crescent of the moon, and the swift-growing bull's horns, all of which symbolize transformation. Thus, all things in nature participate in the ongoing processes of life, death, rebirth.

Excavations of Neolithic urban settlements such as Catal Huyuk indicate a peaceful, egalitarian social structure that is consistent with a predominance of a participatory consciousness. Catal Huyuk, located in present-day Turkey and consisting of 6000 to 8000 persons, was founded in 6,250 B.C.

> The absence of streets, a large plaza, or a palace and the uniform
> size and furnishing of the houses led Mellaart [who excavated this
> city] to speculate that there was neither hierarchy nor a central
> political authority at Catal Huyuk,...[32]

Mellaart also argues that there was no military caste and no evidence of warfare over a period of 1000 years. However, Gerda Lerner cautions:

> Mellaart's observations about the absence of warfare at Catal
> Huyuk must be measured against the abundant evidence of the
> existence of warfaring and militant communities in neighboring
> regions. And, finally, we cannot omit from consideration the
> sudden and unexplained abandonment of the settlement by its

---

[32] Gerda Lerner, The Creation of Patriarchy (New York: Oxford University Press, 1986), p.33

45

inhabitants ca. 5700 B.C., which seems to indicate either military defeat or the inability of the community to adapt to changing ecological conditions.[33]

In Neolithic culture men and women were separate but equal. Men, who probably first domesticated animals, took care of the animals, hunted, and worked the fields especially after the invention of agriculture. Women prepared food, wove cloth and made pottery all ancillary to their primary biological role of producing and nurturing children. This function was necessary for group survival and was venerated. Expressions of control consciousness differentiated more in association with males, and expressions of participatory consciousness differentiated more with females. While neither sex nor type of consciousness dominated the other, participatory consciousness as manifested in Goddess worship gave unity and meaning to Neolithic culture. It also was a bridge to Nature for evolving human societies that were more rapidly separating themselves from Her.

Pre-Conventional Morality of Neolithic Cultures. In primitive societies males and females were united in their participation in religions based on the worship of the Great Mother Goddess. Men and women followed rules of behavior such as those governing patriarchal dominance relationships as a result of a shared participatory subjective commitment to them. The rules were aspects of a shared magical-mythical perspective of the world. Wilber describes this state of affairs as *preoperational thinking* that leads to a *pre-conventional morality.* Preoperational thinking works with images, symbols, and concepts but not yet with complex rules and formal operations. In some manner humans at this stage of individuation converted their perceptions into internalized representations of sensory objects. The action self that generates this type of knowing goes out to and identifies with episodic events in nature and with the body that by sensations provides objects of this episodic knowing. Thus, the action self sees itself as a "Body self (ego)." Correspondingly, the moral rules for behavior concern physical pragmatic activities as well as having experiences of pleasure and avoiding experiences of pain. Concrete self-awareness of successful pragmatic activities and experiences of pleasure are associated with the words, good and right, and the self-awareness of unsuccessful pragmatic activities and experiences of pain are associated with the words, bad and wrong. Those people who proclaim and enforce these "physical, moral rules" bring to themselves physical power.

This type of knowing, which Habermas, Gebser, and Piaget called "magical," involves self-conscious episodic awareness in which the knower and the known become fused to some extent. As a result of this fusion, the internalized representation of sensory objects often are confused with or even identified with events in the external world. When this happens, humans believe that they can magically alter the physical world by means of voodoo, exoteric mantra, the fetish, magical ritual, "sympathetic magic," or magic in general. By looking from the side of nature at this fusion of the knower and the known, one can appreciate why these primitive humans perceived physical objects as being alive and possessing explicit personal intentions. This is called *animism.*[34]

---

[33] Gerda Lerner, The Creation of Patriarchy (New York: Oxford University Press, 1986), p. 35.

34 K. Wilber, *Sex, Ecology, Spirituality* (Boston: Shambhala, 1995), p. 165.

The earlier description of egalitarian Neolithic societies suggests the way the dominance patriarchal relationships may have arisen. So long as the creative power of the Goddess embodied in women was revered by both sexes, Eros-chaos primarily associated with women kept in check Eros-order primarily associated with men. If the reproductive functions of women were to become objectified and manipulated by humans, this delicate balance would be broken, especially if technology also became more sophisticated. The coming together of several events probably led to this disintegration of male-female mutuality; that is, Eros-order, Eros-chaos mutuality.

In some places increasing populations led to increasing competition for resources rather than cooperation to obtain them. In both horticultural and agricultural societies, children were valued for being able to gather food or work in the fields. Women then were valued to some extent as objects analogous to domesticated animals that accomplish tasks for society, in this case, produce children. At the same time the evolution of hunting techniques produced two profound changes in society. First, the more advanced tools for hunting, such as the bow and arrow and spear, also could be used in organized warfare. Second, as hunting and warfare became exclusive male occupations, males began to bond to one another while helping one another to develop great skill in using weapons and discipline involving control of fear and repressing pain. The male bonding and long training necessary to become a good hunter-warrior made adult males feel very separate from nature and from women and children. A band of hunter-warriors could steal women from another tribe and thus gain more child producers. Women, in turn, would begin to depend on men for protection. These practices reinforced objectification of the reproductive powers of women and dependency of women on men's learned abilities and virtues needed for fighting.

As neighboring tribes began fighting one another, it occurred to some of them to form alliances by exchanging women and giving daughters to marriage to seal political bonds. Women rather than men were exchanged because women by their biological natures had the abilities to: give sexual pleasure, produce children, nurture children, adapt to a foreign culture, and remain with that society out of loyalty to their children rather than run away as men would do. The power in women once revered by both sexes now was objectified and under control of men. However, not all men had this power; only those who after long training were admitted into the group of males that became a hunting band. Marilyn French points out that even societies where women also hunt, there emerges a male cult and ritual that excludes women. These cults created myths and pre-conventional moral rules that gave males physical power over women and singled out sacred objects, such as totems or weapons or tools, which women are not supposed to touch or, sometimes, even see.[35]

These gradual changes of males exerting control over other animals, "mother earth" via agriculture, other tribes, and females, would tend to produce new kinds of myths and religious perceptions to represent these new experiences. Power innate in nature and in women became somewhat subordinated to male's learned power. This evolution toward dominance patriarchy was associated with the gradual transformation of episodic knowing to the more abstract metaphorical, conceptual knowing that co-emerged with the first civilizations. As a result, according to French, male control over women and nature became more abstract, separated from

---

35 Marilyn French, *Beyond Power* (New York: Belles-Letters, Inc., 1985), p. 77.

context and thereby a good in its own right. Correspondingly, local immanent gods described by magical myths became transcendent gods of power described by a civilization's metaphorical, conceptual myths. For those males that were admitted to the hunting cult, male identity was redefined as he who has the ability to exert control. The societies evolving toward civilization began to describe in myths and celebrate by initiation rites the second birth of boys into manhood. These myths and rituals defined reborn males as fundamentally different from and superior to females, who are born only once in the biological process of birth.[36]

At first after a battle, the victorious tribe would kill all the surviving members of the defeated tribe. Later a new patriarchal innovation, slavery, gradually emerged:

> Historical evidence suggests that this process of enslavement was at first developed and perfected upon female war captives; that it was reinforced by already known practices of marital exchange and concubinage. During long periods, perhaps centuries, while enemy males were being killed by their captors..., females and children were made captives and incorporated into the households and society of the captors.... Most likely their [women and children] greater physical vulnerability and weakness made them appear less of a threat in captivity than did male enemy warriors.[37]

Lerner notes further:

> As subordination of women by men provided the conceptual model for the creation of slavery as an institution, so the patriarchal family provided the structural model.[38]

In the patriarchal family the father has absolute power over the wife and children. Thus, the first gradual emergence of patriarchy follows a sequence of "survival innovations"; first, the objectification of females' sexual and reproductive services; second, the enslavement of female captives which later became the basis of the institution of slavery; and finally, the institutionalization first in the family and later in the city-state of women being inferior to men.

INDIVIDUAL, PSYCHOLOGICAL ASPECT

Eros Stress of Body Ego-Awareness (the Body Ego's Impossible Project). The social awareness of the mother-child society makes possible conscious conflict between its members. Each vies for control of the other, but, of course, since mother is so much more powerful, she usually wins and thus becomes "the Terrible Mother." These conflicts between mother and child result from an opposition of wills; the mother wills one thing and the child wills another. Will is another unique characteristic of humans. Once a human has acquired some level of self-awareness, he/she can experience will, but will is not the same as nor an aspect of self-

---

36 Marilyn French, *Beyond Power* (New York: Belles-Letters, Inc., 1985), p. 77.

[37] Gerda Lerner, The Creation of Patriarchy (New York: Oxford University Press, 1986), p.78.

[38] Gerda Lerner, The Creation of Patriarchy (New York: Oxford University Press, 1986), p. 89.

awareness. It is difficult to describe what will is because it is more fundamental than self-awareness. That is, it is possible to have will without having self-awareness, but once one has self-awareness, he/she also possesses will. We in Western cultures tend to think that self-awareness or ability to know aspects of the world is the most fundamental aspect of being human – note, language merely is a way by which we express what we in some way know. Therefore, it is difficult to pinpoint the experience of will. In fact, we only experience the result of will which is action. Having experienced action, we reason back to the idea that there must have been or is will that led to the action. In other words, will is something like a potential, such as the potential for a set of keys in one's hand to fall to the ground. We do not "see" the potential in the set of keys, but we know from past experiences that when we stop holding the set of keys up, it will fall to the ground. As indicated in chapter 1, the expression of any potential in nature may be thought of as the expression of a drive, which I referred to as Eros. When focusing on a system losing order and thus going to chaos, the drive producing this is Eros-chaos. When focusing on a system accomplishing a task or creating order in another system, the drive producing this order is Eros-order.

Subjective and social control communications result in the child developing, to some degree, an *autonomous* body ego. The child establishes a preliminary gender identity; that is, "I am like dad, a male;" or "I am like mom, a female." As a result of having a somewhat autonomous body ego, the child also is able to have a social consciousness. On the one hand, within its feeling participation with mother it is aware of its separate self, bonded to and loved by mother but at the same time totally dependent and overawed by her. She is not only "the Good Mother" but also "the Great Mother." On the other hand, the newly emerged Body Self induces the child to affirm its separateness and say "no!" to demands made by mother. The self-awareness of the body ego indicates that it has the potential to "bend back on itself," what may be called reflection, and thereby produce the phenomenon of a "self" that has awareness of itself. This phenomenon is unique to humans; no other living or non-living system manifests reflection to produce self-awareness. We may say there is a Feeling Center that is the "generator" of all levels of self-consciousness but is not specified by or limited to any particular level. It is, so to speak, a center without a self that generates self-consciousness.

However, self-conscious humans can uncouple these Eros drives and prevent creativity. This choosing to couple the Eros drives and be creative and individuate, or to uncouple the Eros drives and be non-creative is an expression of human free will. Thus, the will associated with the Eros drives of potential or actual self-conscious humans is fundamentally different from the will associated with Eros drives manifested in the non-human aspects of reality. Humans express free will; all other systems or things in nature do not express free will. Will is a center in humans associated with Eros drives that lead to the emergence of self-consciousness. After that event, will is expressed as some degree of free choice and as further individuation to higher levels of self-consciousness. The two-year old child can choose to say no to its mother or yes to her; the child has some degree of free will. At this point the child is driven to individuate further so that a mind self emerges (see next section). The child cannot resist the drive to further individuation, but later at a higher level of individuation, adolescence, the human has a correspondingly higher level of free choice that enables him/her to resist further individuation. As indicated earlier, the Feeling Center in humans is that from which self-consciousness emerges. It, then, is clear that Will is identical to the Feeling center. The will aspect of the Feeling center is the Eros drive to human individuation, which manifests as progressively higher levels of free choice.

After its first emergence, the body ego, in normal situations, becomes stabilized by the time the child is three years old. The child learns new behaviors, such as walking, eating, "going potty," and he/she takes delight in the things he can do, all of which contribute to his sense of autonomy. The child begins what Piaget calls preoperative cognitive development, in which mental pictures of experienced objects are replaced by paleosymbols, and he/she creates for himself/herself a magical perspective of his/her cosmos where he/she becomes the center of this magical universe. This egocentricity coupled with being acknowledged by family members usually produces a lull from the chaos of the first emergence of the body ego. The *challenge of the impossible project* disturbs this lull and prepares the child for the emergence of mind.

Psychic energy of the Eros-chaos from the Feeling Center surges through the body ego, actualizing the participatory consciousness to produce in the child polymorphous sensations, super-aliveness, and numinosity. It is futile to try to explain what these terms mean to most adults, who a long time ago have repressed the body ego that produces these kind of experiences. One way to reenact them is to regress to early childhood by means of psychedelic drugs or other psychic techniques, but the danger of mental destabilization can be too high a price for such experiential knowledge. There are other safer ways of obtaining it and if one does so, he/she knows what the child is experiencing much of the time. Psychic energy of the Eros-order from the Feeling Center actualizes the subjective control consciousness of the body ego to further develop the child's language skills in association with preoperative cognitive skills expressed in control behaviors. All this enhances the child's autonomy and egocentricity. In the context of his social control consciousness, the child thinks of itself as the center of the cosmos and correspondingly he/she wants to be served -- waited upon -- by all other feeling selves and to control what they do, especially in relation to this "child-god."

In the context of social, participatory consciousness early childhood is characterized by a *participation mystique*, a kind of collective consciousness which modern anthropological research (Levy-Bruhl, Cassierer, Neumann) attributes to primitive societies. The *participation mystique* is a pre-personal phase of human consciousness, which is "...a vague indissociation of self and group, self and nature, self and animals (hence totemism, the clan-self, kinship, etc.)."[39] Neumann suggests the following characterizations of it: (1) sense of union with nature; (2) at home with the body and in touch with feelings (via participatory consciousness); (3) bisexual, no problems with sexual drives; (4) sense of innocence; (5) all is Good; (6) no feeling of loneliness; (7) sense of one's wholeness and integration (the triune brain functions as an integrated system); (8) humility; (9) pleasure principle; (10) avid imagination and creativity (via the autosymbolic process); (11) playfulness; and (12) enthusiasm.[40] This "indissociation of self and nature" is according to Jean Houston the basis of a young child's sense of wonder and creativity.

> The engaging, growing child is in a state of continuous creation of
> mutual relations with the environment. He is in a state that I have
> termed *psychoecology*, by which I mean that he has leaky margins

---

[39] K. Wilber, *Eye to Eye: the Quest for the New Paradigm* (Garden City, New York: Anchor press/Doubleday,1983), p. 204

[40] Erich Neumann, *The Origins and History of Consciousness* (Princeton, New Jersey, Princeton Uni. Press, 1973)

with the world at large. The nervous system of a child flows into and is systemic with the systems of nature, so nature is experienced sensually as self and cosmos, the one continuous with the other.[41]

However, psychic energy of Eros-chaos wells up to disturb this fragile peace. In actualizing the participatory consciousness it overwhelms the subjective control consciousness, resulting in destruction of the body ego's stability and sense of autonomy. Thus, these invasions by psychic energy continually confront the newly emerged body ego with a subjective, impossible project. The child's ego identifies with the old physical self (the body) via participatory consciousness and at the same time as a result of Eros-order, it seeks to be autonomous and to control the psychic energies coming up through the physical self. That is, its subjective participatory consciousness, which identifies with the physical self, is continually in direct conflict with the Eros-order of the body ego, which tries in vain to maintain an autonomous subjective control consciousness.

Likewise, psychic energy actualizing the social, participatory consciousness causes the child to bond with family members, especially with mother, who is experienced as the Great Good Mother and who is all loving and is loved by the child. At the same time, the child driven by the Eros-order associated with its social control consciousness continually asserts itself by disobeying mother and demanding to be taken care of on the basis of the child's wishes rather than on those of the mother. In rebuking the child's disobedience and resisting its attempt to manipulate her, the mother elicits hate from the child, who now also sees her as the Terrible Mother. Thus, the child has a *social impossible project* resulting from the opposition between Eros-chaos, of social participatory consciousness and Eros-order of social control consciousness. The child seeks to be intimate with mother and at the same time seeks to express its autonomy by disobeying her and attempting to control her. The child is consumed with intense love and dependent awe of its mother perceived as the Great Good Mother, and is also obsessed with intense hate and fear of its mother perceived as the Terrible Mother.

Emergence of Mind as the Polar Self

As indicated in the previous section, the baby in participatory union with mother forms a "group ego" with her. This "group ego" enters magic learning cycles that enable the baby to gain greater control of its body and to learn various behaviors, such as feeding itself. Later, with the continued participation of mother in the group ego, the child begins to learn proto-conceptual language. The beginning of differentiation of symbolic language – spoken words that symbolize things – leads to the emergence of the body ego. The emergence of the body ego is an evolution toward the child having an internalized self-awareness. However, this "proto-inner self" still is embedded in the feeling awareness interactions of the physical self with the external world. That is, by means of language the body ego is beginning to objectify to itself its feeling interactions with the external world. The "itself" of the body ego's awareness is a beginning of a subjective self being aware of the external world, but this subjectivity is not yet clearly demarcated from that of which it is aware. After attaining a certain level of differentiation of proto-conceptual

---

[41] J. Houston, *Life Force: the Psycho-Historical Recovery of the Self* (New York: Dell Pub. Co., 1980), p. 83.

51

language, the 4½ to 5-year old child spontaneously creates for itself metaphorical non-conceptual insights that separate the child's subjectivity and the objects of its awareness into a polarity. As a result, the child wakes up to an awareness of an inner subjective self separate from but in some relation with an outer, objective non-self, that is, the objects of its body ego's feeling perceptions. The inner self is aware of external objects either by participatory union with them or by an awareness of its separateness from but control over – or lack of control over – them.

Before this awakening the child could objectify its own feelings as separate from those of its mother, and it also could use words to represent and thus objectify objects of its feeling perceptions. With the emergence of metaphorical non-conceptual thinking and a "polar self," the child can objectify an inner subjective self separate from but in mutual relation with objects of its subjective, feeling perceptions. For the first time, the child has an inner subjective self all to itself, so to speak, rather than a self spread out in a group ego or embedded in feeling perceptions. That is to say, the child now has an individual self-consciousness. What enables this great awakening to an individual self-consciousness is the emergence of a new psychic center called *mind*. The mind "sees" the polar relation and in doing so, objectifies the inner subjective self in mutual relation with an objective non-self. The inner subjective feeling self is the body ego in communion with the Feeling Center that generated it. The mind "sees" the body ego in communion with the Feeling Center as well as "seeing" objects of feeling perceptions, and the mind objectifies by metaphorical, non-conceptual language this polar relation. In this context the body ego is the inner subjective self and the mind is the "objective self", meaning that it objectifies by means of language all objects of perception including one's perception of an inner self.

The mind expresses a center of self-consciousness that at this stage identifies with its great awakening vision of polar mutuality. The ego consciousness of the mind identifies itself as a polar self that metaphorically "comprehends" an inner subjective self interacting with an outer, objective non-self. It is useful for understanding further individuation of the polar self to superimpose a secondary polarity onto this fundamental existential polarity of inner self versus non-self. The secondary polarity is the body representing the inner subjective self versus the mind that metaphorically represents non-self. (Note: the language representation of the subjective inner self by the mind is not at all the same as the inner subjective self in itself.)

Solution of the Impossible Project

With the emergence of a mind identified as a polar self, the child "solves" the impossible project by transcending polar oppositions. First of all, the mind displaces the body ego as the center of human self-awareness. In the new polar order the body ego is seen to be a psychic center through which the inner self – what some thinkers call *soul* – has participatory dialogue with the external world of non-self. As a result of this new polar order, the mind "objectifies" the body ego and thereby can begin to regulate the expression of its willful demands. The mind can begin to see that it is not identified with any particular feeling expression of the body ego but rather "has at this time" a particular feeling, which the mind can choose to express or not express. In other words, Eros-chaos leads the body ego to manifest some intense feeling, and Eros-order leads the Mind Self to express or repress the expression of the feeling. Thus, the new polar order begins to differentiate into the mind, dominating, regulating, and repressing the body ego. That is, the self of the mind regulates the inner self expressing feeling interactions through the body ego with the external world of non-self.

Further differentiation of the mind via development of language skills consolidates and stabilizes self consciousness so that it gains greater control over the body ego and its expression of feelings. At first the child still views mother either as participant in the polarity: Child—Great Mother or as participant in the polarity: Child—Terrible Mother. Eventually the mind self gains sufficient strength to acknowledge and tolerate having contradictory orientations toward mother, involving love of the Great Good Mother and hate of the Terrible Mother. Thus, the child polar self becomes capable of a mutual polar relationship with mother, rather than the previous adversarial relationship in which the child was resisting union with or trying to control mother. This potential for mutuality only is expressed if the child first breaks away from group-ego, participatory union with mother and with other family members. Such a break also implies that the child moves out of being the center of his/her magic world and into polar relations with Nature.

Emergence of the Triune Mind

The young child after the emergence of a polar self still is an embodied mental self-consciousness. The mind is driven by Eros drives from the Feeling Center to develop cognitive skills that it uses to repress and control the body ego. However, the child still has "leaky margins with the world at large." Consequently, psychic energy from the Feeling Center through the body ego constantly reestablishes participatory consciousness that disrupts control consciousness of the mind. This draws the child into his/her internal fantasy world and entices him/her into having somewhat intimate relationships with others, sometimes even with strangers. The war between the body ego's Eros drives and the mind Eros drives has been only partially won by the controlling mind. The body ego continues to reassert itself, thus bringing on new conflicts, but from here on, unless the child or later the adult becomes psychotic, the battles will be between a differentiating mind and a rebellious body ego.

Though the infantile mind has partially repressed the body ego with its identification with the physical self, a remnant of participatory consciousness of the body ego remains associated with the mind self-consciousness. On the one hand, this participatory consciousness provides subjective intuitions and paleosymbols that are used by the mental self to begin differentiating experiential learning. On the other hand, the child=s participatory consciousness is receptive to the psychic energy from the Feeling center coming to it through the body ego. This psychic energy maintains the integrated functioning of the triune brain (see section on emotions and feelings in chapter 3). By means of this upwardly projected energy, the mind becomes differentiated into three mind centers: (1) Thinking Mind Center corresponding to the perceptual learning brain, (2) Expressive Mind Center corresponding to the feeling brain, and (3) Action Mind Center corresponding to the emotional brain.

At first these three mind centers focus energy from the body ego to produce primitive, mental learning cycles. The Thinking Mind Center takes in concrete sensations and converts them into subjective feeling perceptions that represent the sensations. The Expressive Mind Center translates the subjective feeling perceptions into a unified magical-mythical non-information interpretation of them. The interpretations are instances of tacit (existential) knowing; they are non-conceptual. The Thinking Mind Center translates these magical-mythical interpretations into concrete language representations, that is, the thinking mind translates non-information knowing into vague information knowing. The language representations consist of metaphorical, paleoconcepts. The paleoconceptual aspect of the language representation is only vaguely objective in that the words are not precise and they are not logically organized. The child may decide to act in accord with his/her language-represented interpretation. When this happens, the Action Mind Center translates the concrete language representation into a particular behavior.

UNIT III EMERGENCE OF CIVILIZATION AND THE PERSONA SELF

Chapter 6
TRANSFORMATION FROM PARTICIPATORY, EMBEDDED KNOWING TO METAPHORICAL, CONCEPTUAL KNOWING

PARTICIPATORY, EMBEDDED KNOWING

The first phase of the emergence of the persona self involves the transformation *from* participatory embedded knowing to metaphorical, conceptual, disengaged knowing. Prior to this transformation humans are to various degrees embedded in Nature. Anthropologists refer to this as *participation mystique.* Individual ancestor apes have an episodic awareness of a feeling interaction between themselves and some other ape or some other aspect of nature. They have an awareness of their total embeddedness in nature; there is no sense of a "self" that is separate but closely tied to nature. With the emergence of the feeling action self, humans are aware of their participatory embeddedness in a family or in a tribe, and at the same time they have a vague awareness of nature or another tribe as "other." This "other" is ambiguous in that humans have a participatory affinity with it and yet they are vaguely aware of actions they must take to keep this other from destroying them. Eros-chaos drives them to participatory affinity with the other and Eros-order drives them to separate from the other and exert some "magical" control over it. Individuation to the body ego is a differentiation toward a more individualistic ego – rather than a total group-ego – awareness of affinity with and separation from the "other." With the emergence of the polar self, humans transform the ambiguity of the "other" into an ambiguity within themselves, that is, there emerges polar self ambiguities.

The basis for all these internal ambiguities is that each human has an awareness of a ego self that is "driven" by feelings that come from a source within the individual. The polar self is somewhat disengaged from its feelings. The human no longer *is* a particular feeling; rather he/she *has* a particular feeling in a particular circumstance. The newly emerged mental self sees itself as having an inner source that drives it to external actions. The inner source is the Feeling center aspect of one's individual soul as described in Chapter 5, p. 49. The mental self is able to represent its awareness to itself. The representations can be expressed by various artistic expressions, such as drawings, and by proto-conceptual language. The self "sees" itself in different polar relations with the "other;" so it "knows" the "other" as an aspect of its relation to it. The self does not "see" any particular "other" as a "part" of some grand, larger "Other." Thus, the polar self's knowledge of the "Other" is not an object "out there" with a separate existence and characteristics that make it different from the knowing self. That is to say, the self's anthropomorphic knowing is non-conceptual. It does not manifest the most fundamental categories represented by the concepts of whole versus parts of a whole.

However, the polar self knowing does recognize different polar relations versus similar polar relations. The ideas of different versus similar implies the conceptual categories of *same* versus *different.* This polar self's recognition of different versus similar is a more differentiated, self-conscious recognition of the feeling action self's recognition of diverse episodic feeling events. This suggests that the polar self is evolving toward a "proto-conceptual knowing" of whole versus part and same versus different. Moreover, polar selves have an anthropomorphic understanding of time and events. These humans tell stories in which there is an interconnected sequence of events. There is no conceptual understanding of individual events occurring *in* time. The vague sense of duration and sequential interconnection among episodic happenings implies that this story knowing, also called *mythical knowing*, is an individuation toward conceptual understanding of time, events, and causation that connects one event with another. Thus, the polar self's anthropomorphic, proto-conceptual, polar knowing of "others" and the "Other" via myths may be designated as *participatory, embedded knowing*. The transition to the next higher level of self-consciousness involves creating concepts to represent categories.

ELEMENTARY, OBJECTIVE, CONCEPTUAL KNOWING

Individuation from proto-conceptual, participatory embedded knowing to non-embedded, conceptual knowing is a mystery to each of us. In order to conceptually represent to ourselves what is going on while this individuation occurred in each of us, we needed to already have conceptual knowing. Not yet having this type of knowing, we could not realize what was happening while we were individuating toward it. But now as a result of having logical, conceptual knowing, the self of each of us can separate itself from any anthropomorphic knowing and analyze itself as an object analogous to any object in nature. From this higher vantage point, it is possible to imagine the emergence of conceptual knowing corresponding to the first emergence of conceptual language.

Non-civilized humans experience aspects of the world episodically. These experiences are in some way represented and stored in the "knower." They are existential perceptions that can be associated with an arbitrary symbol, for example, a word. Individuation to Mind with a subjective self corresponds to the "knower" "seeing" similarities among several perceptions, which then become a domain of similarities in the mind. Individuation to the persona self in a civilized society enables the "knower" to conceptualize these similarities, which is to say that the knower creates and stores a pattern called a *concept*. Thereafter, each existential perception that relates to the same domain of experiences will be "seen" as exhibiting this pattern. Correspondingly, all these existential perceptions will be seen as identical. When this occurs, the knower sees each experience as an instance of a concept. The individual, existential uniqueness of each experience is sacrificed for the sake of seeing patterns; that is, conceptual knowing. At this point conceptual knowing still is a subjective process occurring within the individual knower.

However, growing up in a civilized society always leads to some of these subjective concepts being associated with a word in an artificial symbolic language.

Many humans may have similar subjective concepts. The consensually agreed upon word associated with these similar, subjective concepts represents all of them as if they were identical, thus producing *objective, conceptual, language knowing* of the world. This objective knowing within the context of civilized society leads each knower to see the world as a "whole" made up of relatively autonomous "parts." The parts are things or events that are conceptually known. This type of knowing conceptualizes concrete, physical experiences and produces sentences such as "I ate the apple," and "the book is on the table." Ideas in these types of sentences are called literal concepts because they are understood to be objective representations of the world as a whole made up of relatively autonomous parts that exist and interact with one another independent of any human knowing them. However, any concept emerges from self-conscious, episodic experiences, each of which includes a subject-object polar interaction. These polar interactions are more real than the conceptual knowing derived from them. Put in another way, conceptual knowing produces the illusion that there is a world "out there" with a structure that is defined by interactions among the parts making up the world. This structure supposedly exists independent of humans and can be known by humans. This knowing process produces valid conceptual, language knowledge. This illusion is *valid* because it is necessary for one's day-to-day functioning in civilized society, but the conceptual knowing that supports this illusion is derived from human's more fundamental polar knowing of world. This polar knowing indicates that we do not know the "objective world in itself," but rather we conceptualize our polar knowing interactions with it. (This insight about the way humans know is a core insight of one interpretation of quantum mechanics that includes Heisenberg's uncertainty principle and a resolution of the particle-wave paradox.)

This first phase of the emergence of an *objective ego* – so named because it now understands the world by means of objective, conceptual, language knowing – is what Piaget calls *concrete operational thinking*. The drawings of my son, Andy, illustrate the transition from *participatory embedded knowing* – also called *non-conceptual mythical knowing* – to *conceptual mythical knowing*, which Piaget calls *concrete operational thinking*. Figures. 6.1a and 6.1b represent Andy's non-conceptual mythical thinking about his subjective world of experiences.

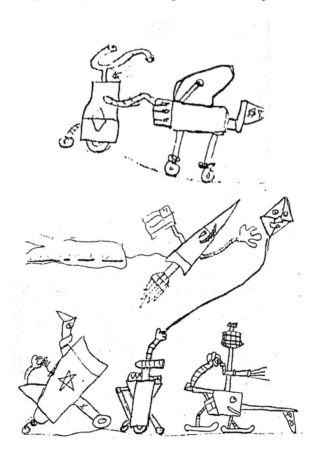

FIGURE 6.1a Andy's non-conceptual mythical knowing (participatory embedded knowing) of the world that includes a portrayal of his oldest brother, Don, who use to practice on the piano 5 to 6 hours each day

FIGURE 6.1b Andy's figures representing non-conceptual mythical knowing (participatory embedded knowing), which includes a drawing of a "birdman" who represents Matt, my third son, with the silly grin he use to have before he totally rejected this clown aspect of his early childhood.

Figure 6.1c is a drawing that Andy created several months later. It represents Andy's newly emerged conceptual, mythical understanding of the world corresponding to his beginning to become civilized at the age of 7 years old. Now he could begin to understand *persona* expectations and the meaning of rules of behavior. He, therefore, was scolded or punished if he did not fulfill these expectations or if he did not obey these rules.

FIGURE 6.1c Figure representing Andy's transformation of thinking from participatory embedded knowing, also called <u>non-conceptual</u> mythical knowing, to <u>conceptual</u> mythical knowing

MODERN THEORY OF METAPHORICAL, CONCEPTUAL KNOWING

Ever since Socrates, concepts were thought of as "real" in themselves or as ideas created by or already present in the mind that exactly corresponded to autonomous entities, that is, parts that make up the whole of Nature. For example, Aristotle's theory of knowing that was incorporated into Thomistic psychology of medieval philosophy states that every autonomous thing is a duality. Every finite being is an *essence* that *exists*. The individual essence has a "structure" or order that is its innate nature. This innate nature consists of one or several forms. Human knowing involves converting a form that determines a concrete, individual, material being into a conceptual modification of the mental self. For ex ample, a particular concrete slab may have a circle form. This circle form is abstracted from the concrete thing and begins to exist as a conceptual modification of the mental self of a human knower. This mental concept now exists as an aspect of a human's self-consciousness and has the same structure as the form that determines the material being. This idea, stemming from medieval philosophy of the reality of concepts, was the basis of natural law and of the idea of a liberal education. These perspectives guaranteed that humans could have absolutely true knowledge of the world and that knowledge was a Good in itself.

Catholic colleges and universities still teach this perspective as a basis for their commitment to natural law and a liberal education. Many non-Catholic academicians whom I have talked to also are committed to natural law and/or to some version of a liberal education, but they are unaware and/or are unwilling to become aware of the assumptions about knowing that underlie these commitments. But for physicists beginning in the 16th century, for biologists after Darwin's theory of evolution (in the mid 19th century), and for modern thinkers influenced by science, conceptual, language knowing is a social phenomenon manifested in individual humans. This social phenomenon is called *nominalism*. Each member of a group of humans point to the same aspect of the world each is perceiving. No two perceptions are the same, but they all agree that a particular arbitrary symbol will thereafter represent this aspect of nature and represent the overall pattern by which each individual perceives it. Thus, the diverse subjective perceptions of a particular aspect of the world are focused into a single, conceptual, objective representation of that aspect. In this manner what we now call ordinary language knowing emerges and via language communication represents the structure of a society. A human born into such a society is influenced by ordinary language communication to individuate to a mental self with conceptual language knowing.

According to the modern theory of metaphor as summarized by George Lakoff,[1] metaphor emerges after the emergence of conceptual knowing and then profoundly enhances the power of ordinary language. Contrary to classical theories of language, metaphor is not a type of linguistic expression; rather, it is a higher level of knowing than elementary conceptual knowing. Lakoff defines metaphor as a general mapping across conceptual domains. That is, metaphor is a higher level of conceptual knowing in which one conceptualizes one mental domain of experience in terms of a different mental domain of experience that is "seen" to have in some ways the same internal structure.

Thus, metaphorical knowing is a higher order of elementary objective conceptual knowing. All conceptual knowing, in turn, is a mutuality between self-conscious subjective,

---

1. George Lakoff, "The Contemporary Theory of Metaphor" in: *Metaphor and Thought*, 2nd ed., A. Ortaony ed. (Cambridge: Cambridge Univ. Press, 1992)

episodic knowing and objective knowing involving social consensus. As summarized in the section on the first stages of becoming civilized, each human experiences aspects of the world episodically. These experiences are in some way represented and stored in the "knower." They are existential perceptions that can be associated with an arbitrary symbol, which is a word. Individuation to the polar self corresponds to the "knower" "seeing" similarities among several perceptions which then become a domain of similarities in the mind. Individuation to the persona self in a civilized society enables the "knower" to conceptualize these similarities which is to say that the knower creates and stores a pattern called a *concept*. Thereafter, each existential perception that relates to the same domain of experiences will be "seen" as exhibiting this pattern. Correspondingly, all these existential perceptions will be seen as identical. When this occurs, the knower sees each experience as an instance of a concept. The individual, existential uniqueness of each experience is sacrificed for the sake of seeing patterns; that is, conceptual knowing. At this point conceptual knowing still is a subjective process occurring within the individual knower. However, growing up in a civilized society always leads to some – perhaps most if not all – of these subjective concepts to be associated with a word in an artificial symbolic language. Many humans may have similar subjective concepts. The consensually agreed upon word associated with these similar, subjective concepts represents all of them as if they were identical, thus producing objective, conceptual, language knowing of the world. This objective knowing within the context of civilized society leads each knower to see the world as a "whole" made up of relatively autonomous "parts." The parts are things or events that are conceptually known.

Metaphorical knowing is the process of "seeing" similarities between two domains of experience, each represented by a concept. The knower starts out seeing that the two concepts, that is, the two patterns, are quite different. Metaphorical knowing emerges when one sees similarities between the two patterns and then formulates a new higher-level concept, the *metaphorical concept*, that represents the similarities between the "lower level concepts." The metaphorical concept then leads to new insights in the following way. While one sees similarities between concepts A and B, one may know a lot about A and not much about B. The metaphorical concept is the pattern: *Some structure relationship in A may be thought of as the same as some structure relationship in B*. As a result of the metaphorical concept, one understands aspects of B in terms of what one knows about A. Lakoff formulates this idea as: The structure of A is mapped onto B, where A is the source domain and B is the target domain. However, this mapping is not an arbitrary, mathematical or prepositional mapping. Rather the mapping is the core of communication between some person, George, who already sees the similarities between A and B and another person, Don, who knows A but does not know much about B. By means of the metaphorical concept, George gets Don to understand B in a new way, that is, he now understands B in terms of what he already knows about A.

For example, a core idea of modern biology is molecular communication, such as hormone action on a cell that responds to the hormone in a particular way. One instance is the hormone adrenalin, stimulating some muscle cells to contract. How can anyone visualize molecular communication when no one has ever seen a molecule, let alone molecules communicating with each other? One can do so by means of the metaphorical concept, *key-lock interaction*. A, representing *key-lock interaction*, is mapped onto B, representing *hormone-cell interaction*. A key is a source pattern that fits into an appropriate lock having a *complementary* receptor pattern; the key pattern is like one piece of a puzzle fitting into a complementary piece of a puzzle, which is like the lock receptor pattern. When the key interacts with the lock, such as when the key fitting into the lock and then being turned to the right, something happens, for example, the key-lock interaction in a car leads to the engine starting. In like manner, a hormone is like a key that fits into a lock, which is like a receptor site in the membrane of a cell. Hormone-receptor site interactions in some muscle cells lead to muscle contraction. Lakoff would represent this metaphorical concept as: HORMONE-CELL INTERACTION is KEY-LOCK INTERACTION. It turns out that this type of metaphorical concept is valid for many areas of modern biology, for example, nerve cell interactions (neural transmissions), virus host interactions, and immunological interactions.

This metaphor involving key-lock interactions also applies to human communication, and in particular to teaching. Using Lakoff's notation I propose that: TEACHING-LEARNING is KEY-LOCK INTERACTION. KEY representing conceptual information sent out by a teacher fits into and further interacts with a LOCK representing the conceptual domain of a learner that can recognize incoming information. When the learner "processes" the interaction between incoming information and the receptive mental domain, there emerges a new insight. This creative process is fundamentally mysterious, but nevertheless, the metaphorical concept does tell us something about what is going on. If the learner does not have the appropriate receptive mental domain, no new insight will emerge. Conceptual knowing becomes dynamic and potentially creative when one understands it as a mutuality between subjective, experiential knowing and objective, conceptual, language representation of experiential knowing. A dynamic concept involving this mutuality is malleable. It can "deform" its pattern so that it can interact with other seemingly contrary ideas to form a new, more powerful concept. However, if the concept is thought of as a rigid, totally objectively true or valid representation of an unchanging structure of the world, then it cannot "adapt" to incoming contrary concepts. The conceptual receptor of the mind will be like a lock that is not complementary to and not receptive to a key representing incoming new ideas. The learner with totally objective true or valid conceptual knowing only can accept compatible incoming ideas. In this situation the mind is like an incomplete puzzle. The only ideas that are received and processed are those that fit into the empty spaces. The present mental pattern is enhanced but not fundamentally changed. Thus, such a learner cannot create new insights.

Human metaphorical knowing becoming literal knowing is a major turning point in human evolution. This emergence of civilization is a shift in emphasis from subjective, participatory consciousness associated with Eros-chaos to objective, control consciousness associated with Eros-order. As will be described in Chapter 7, the emergence of persona self associated with civilization overcame a threat of human extinction. It also produced a platform for humans to individuate to a still higher level of self-consciousness. However, at the same time, this shift partially alienated humans from participatory awareness of their individual *soul* in union with absolute Reality, which I call SOURCE. This shift also partially cut off humans from spontaneous creativity and from further individuation.

LITERAL, CONCEPTUAL KNOWING LEADS TO THE IDEA OF PERSONA

Eros-stress of Patriarchal, Primitive Tribes

After the emergence of the polar self, which is the first level of differentiation of the mind that emerges in the mythical stage, primitive people began to differentiate control objectivity via metaphors organized into magical-mythical story representations, such as poetry, of the world.

Each tribe in which the polar mind self emerged became more consolidated in its beliefs. This led to a more precise understanding of its pre-conventional morality that correspondingly produced more individuated control over social behavior. In particular, the shadow pole of consciousness in any one mythical tribe contained many aspects of the cultures of the neighboring tribes. As a result, the magical-mythical story and its associated pre-conventional morality of each primitive tribe became more incompatible with a different story of any other tribe. Several tribes living side-by-side, with incompatible magical-mythical stories competing for scarce resources, began waging war with one another. This could have progressively weakened all of the tribes to the point where all of them would become extinct.

The stress to human survival resulting from neighboring tribes fighting one another is analogous to the Eros-stress of the ambiguity of the male group-ego of hunting parties versus the male group-ego of the nuclear family; Eros-stress in this case led to the emergence of egalitarian patriarchy. Stress also arouse within tribes due to their successful expansion in a given area. This stress came from problems of land scarcity and population density or problems having to do with unequal distribution of wealth – a natural tendency that co-emerged with the polar self of the mythical structure of society. There was no morality independent of tribal tradition to resolve these conflicts; nor was there any authority within the tribes to judge who had the greater claim to some social wealth, such as, the land or the cattle that was is in dispute.

Solution of Eros-stress: Emergence of Metaphorical, Conceptual Myths

As mythical humans developed metaphorical, conceptual thinking, they became capable of seeing that many different magical-mythical tribal myths refer to the same system of categories, for example, sky-earth, god-goddess, good-evil, persona mask-shadow. Over time these categories, now represented by metaphorical, conceptual words, could be organized into a single, overarching story that is believed to be literally true. Moreover, based on this overarching story, many tribes could adopt a new set of polar oppositions represented by persona mask vs. shadow. The members of all the tribes would accept the definitions of many different types of persona such as masculine, feminine, wife, husband, mother, father, son, daughter, king, priest, general, soldier, hierarchy of professions, slave, foreigner. Each type of persona defined as *Good* would be associated with a shadow aspect defined as *Evil*. The definitions of diverse types of persona-shadow opposition polarities would come from an elaborate set of laws concretely and operationally defined in terms of metaphorical concepts. Such a system of laws is the beginning of *conventional morality*. Later, with the transformation of the metaphorical, conceptual mind to a rational, autonomous self associated with the formulation of the patriarchal perspective, concrete, conventional morality will be transformed into an abstract, ethnocentric, conventional, moral code. The rise of Islam from 622 to 632 A.D. in Saudi Arabia illustrates the "spiritual mutation" to the first stage of civilization involving metaphorical, conceptual thinking.

Social Function of Myths Believed to Be Literally True

Metaphorical concepts that are organized into stories believed to be literally true accomplish two goals: 1) people hearing the story can relate it to their particular subjective experiences; 2) people who choose to believe the story are united by a shared commitment to an unambiguous, objective, "literal truth." The shared story, in turn, can generate unambiguous, shared rules of behavior. Wilber, incorporating Habermas's research, provides another way of describing how literal myths can unite diverse tribes into a unified nation-state. These literally true myths united people beyond mere blood lineage, and as a result, clans that use to or could fight one another while competing for resources, now cooperate with one another because of committing to the myth and the moral code derived from it.[2]

Wilber then quotes Habermas who further elaborates this idea. Because the central myths of the newly emerged civilized societies were understood by metaphorical, conceptual knowing, members of diverse clans with conflicting traditions could see patterns within the myth being similar to patterns in their particular tradition. This metaphorical, conceptual connection of diverse traditions produced several consequences. One, a political order emerged that organized a society in which its members could belong to different lineages. Kinship relations were replaced by political relations. Second, citizens participated in a collective identity based on a common ruler, which replaced the figure of a common ancestor. Third, pre-conventional morality was incorporated into conventional morality specified by metaphorical, conceptual rules of behavior that apply to diverse traditions and thus are more universal than the clan-dependent pre-conventional rules. Correspondingly, judges who stated and interpreted the moral code had the legitimized power in conjunction with a police force (army) to administer rule by law. Fourth, citizens could develop specialized functions – judge, soldier, farmer, artists, priests associated with the myth, and so on – which could be integrated to maintain and enhance the survival of this larger, more complex society.[3]

Wilber/Habermas further described the first and second consequence. My interpretation of Wilber's views is that there was a gradual mental transformation from embedded polar self experiences to partial disengaged, metaphorical, conceptual thinking. This led to "societies organized through a state" in which tribal identities were not lost but became secondary. Collective social identity switched from membership of individuals on common descent to belonging in common to a territorial organization. The key factor in this transition was that all "citizens" identified with the figure of a ruler, who could claim close association with the gods and goddesses of the new myth, which were incorporated into it and integrated to produce an expansion of the world of gods.[4] Wilber goes on to propose that within this – what I am calling metaphorical, conceptual – "mythological structure," personal identity switched to a "a role identity" in a society of common political, not genetically related ruler.[5]

Chapter 7
THE PERSONA SELF

OVERVIEW OF EMERGENCE OF PERSONA SELF

Eros-chaos producing survival stress to primitive, patriarchal tribes brought at least a few individuals to a more abstract type of thinking called metaphorical, conceptual thinking. Eros-order brought these individuals to create a metaphorical, conceptual myth about the creation of humans and their role in the unfolding story of the universe. This myth, in turn, provided the

---

2 K. Wilber. *Sex, Ecology, Spirituality* (Boston: Shambhala, 1995), p. 169.

3 K. Wilber. *Sex, Ecology, Spirituality* (Boston: Shambhala, 1995), p. 169.

4 K. Wilber. *Sex, Ecology, Spirituality* (Boston: Shambhala, 1995), p. 171.

5 K. Wilber. *Sex, Ecology, Spirituality* (Boston: Shambhala, 1995), p. 171.

basis for defining personas, that is, roles that various humans play in society and for specifying moral codes and rules of behavior. Over time many individuals in the same and in different tribes came to this higher level of thinking and committed themselves to this story and to the personas and rules of behavior derived from it. Thus, Eros-order consolidated many tribes into a single civilized society. However, Eros-chaos leading to participatory subjectivity opened one to the expression of the centers of the action self, feeling body self, and the polar self. These expressions of the lower levels of consciousness sometimes opposed the personas and rules of behavior dictated by the socially accepted, literally true myth. In attempting to control such expressions, Eros-order leading to control objectivity created the *Individual, Collective Unconscious*. That is, one aspect of self consciousness, the *objective self*, blocked out from self consciousness any expression of the lower consciousness centers that opposed socially prescribed identity and behavior. These lower centers continued to function but the results of their functioning was not acknowledged by the self. In effect, human consciousness was divided into two compartments: persona self consciousness and the compartment containing the other three lower level centers of consciousness. Since the self consciousness dominated the other compartment, this other compartment is referred to as the *unconscious*; it is unconscious with respect to the maintained exclusivity of self consciousness. Because this compartmentalization of human consciousness is a universal phenomena for all humans that individuate to the persona self, the second compartment is called *Individual, Collective Unconscious*.

PERSONA, THE CORE IDEA OF CIVILIZATION

Persona is a social innovation that subordinates the polar self, embedded in concrete experiences and the rational, autonomous self to the rule of law in civilized society. When primitive humans developed metaphorical, conceptual thinking, they began to see the world and human societies in a mechanistic way. That is, the world is seen to be made up of relatively autonomous, interacting things, and every concrete thing is in some category, such as, flower, tree, cat, dog, human, member of my tribe, enemy. There are laws of Nature that determine how types of things interact. Likewise, a human society is made-up of relatively, autonomous individuals who tend to satisfy their desires for survival, pleasure, status, and power over others. The literally true religious, political myth that brought together several tribes into one civilized society specifies how humans should interact with one another and what roles each human should play in various circumstances in society. In acquiring metaphorical, conceptual knowing, primitive humans developed a higher level of abstract knowing and a greater opportunity for choice, that is, greater free will. This greater freedom of choice promotes the possibility of individuals opposing the mythical-religious laws that specify human behaviors in myth-regulated societies. Personas, as defined earlier, overcome this problem.

The metaphorical concept of a machine indicates how conflict among relatively autonomous individuals can be overcome. The unity of a civilized society – unlike the participatory, holistic unity of a tribe maintained by family allegiance – is due to its being like a machine. All humans in a civilized society are types of "parts" of the society machine that fit together to produce a specified social structure. This maintained structure enables the social machine to accomplish tasks such as producing agriculture, building cities, making war, and doing other tasks specified by the ruling class of the society. Each human from the time of birth onwards is socialized to create for himself/herself a masculine or feminine persona that incorporates other "self-defining" social functions in various specified circumstances. For example, the feminine girl persona later becomes a wife defined by various functions, and later a mother defined by a new set of functions. Likewise the masculine boy persona becomes a husband defined by various functions, and later a father defined by a new set of functions. The net effect is that, by means of socially prescribed personas, each human defines himself/herself as a functioning part of a civilized society.

Individuation of humans from a polar self to what I call a persona self transforms humans from being embedded in polar experiences of Nature to become separated, autonomous parts of a civilized social machine. Individuals still express participatory subjectivity by participating in the religious-cultural rituals and by creating art, for example, music, dance, sculpture, and so on, that represent the ideals of their religious culture. Some personal interactions not covered by law also give expression to participatory subjectivity. However, the emergence of *persona* as the core of civilization is the great turning from some mutuality between Eros-chaos, participatory consciousness and Eros-order, control consciousness to hierarchal dominance of control consciousness over participatory consciousness. This turning led to the major social-cultural double bind of all civilized societies. On the one hand, it eliminated the Eros-stress within individuals and within societies of the diverse conflicting ambiguities. The diverse personas are the pillars that support and maintain stability of individual egos and social structure. On the other hand, creativity requires mutuality between Eros-chaos and Eros-order, see chapter 2. The linear dominance of Eros-order over Eros-chaos required by personas prevents creativity. When a society enters into severe stress, the society must individuate to a new perspective. This is a creative process that is blocked by commitment to personas. On the other hand, loosening commitment to personas – or worse yet, what postmodernism calls deconstruction of persona, for example, the feminist movement – upsets individual and social stability.

Through the ages, civilized societies managed to individuate to some extent by tolerating some "social mutants." These are individuals who reject personas in order to be creative. Usually they are ostracized from society, imprisoned or executed, but not before their creativity contributes to social individuation. In the last few centuries in Western cultures – and especially today – these "social mutants" are tolerated. But, often they also become so unstable that they destroy themselves one way or another, by addictions, suicide, or psychosis. I maintain that the current world cultural crisis requires large numbers of people to become "social mutants" who, through creative dialogue, transform patriarchal civilizations to trans-patriarchal civilizations.

EMERGTENCE OF TWO PSYCHIC WORLDS

As described earlier, the persona self that co-emerged with metaphorical, conceptual thinking stabilized civilized societies by creating in each human an Individual Collective Unconscious. This divided the persona self into an *external self* that expressed personas appropriate to particular situations as dictated by society, and an *internal self* that interacted with the newly created psychic world of the Inner Self. The Inner Self consists of SOURCE manifesting soul, which continues to manifest feeling action self, body ego, and polar self, and all the contents of the Individual Collective Unconscious. The other psychic world is the consensual interpretation of reality as contained in the language of a society and prescribed by the hierarchy of power in a society. The Inner Self of each human contains infinitely many possible insights and dispositions that oppose the structure of the consensual interpretations of what may be called the "outer social world." However, this potential for conflict between the two worlds is partially overcome by the acceptance as literally true of a grand overarching myth and its associated religious rituals and ceremonies, and by the acceptance of other related social institutions such as patriarchy extended from the nuclear family to all institutions of a civilization.

The external self "saw" society as like a machine in which each autonomous part interacts with all the other parts in such a way that the machine accomplishes the tasks of society as dictated by its rulers. Because of this mechanistic view, the external self is associated with objectivity expressing control consciousness and with Eros-order that establishes and maintains order. The internal self "looks" to the Inner Self and thereby provides new possibilities for self expression. Because of this looking inwards, the internal ego is associated with subjectivity that takes in existential knowing while expressing participatory consciousness of events. Because this is a break from the domination of the external self, the internal self in doing this is associated with Eros-chaos and therefore with the possibility of initiating creativity.

INDIVIDUAL, PSYCHOLOGICAL ASPECTS OF THE PERSONA SELF

The Great Repression

Michael Washburn, who elaborates a modified version of Carl Jung's psychoanalytic theory, associates the first emergence of the Mind with repressing the body ego into what he calls the unconscious dynamic ground equivalent to what Jung calls the Collective Unconscious. Washburn calls this "the great repression." This designation presupposes the prior existence of the unconscious, dynamic ground into which the body ego may be repressed. Washburn, in agreement with Jung, claims that archetypes that guide human individuation are in the dynamic ground and after the emergence of the mind, each human is the polarity of the unconscious, dynamic ground versus the self consciousness of the mind. Human individuation for both Washburn and Jung is the emergence of ego consciousness that progressively expands and deepens by bringing into ego consciousness psychic factors and archetypes in the dynamic ground. For these thinkers the Self archetype, which guides this process, is the psychological equivalent to the religious idea of God or to what I call *SOURCE*. In radical contrast to Washburn and Jung, I propose that each human is a developmental-evolutionary process generated by SOURCE, which is the manifestation of God or Brahman or Emptiness or Ground of Being. Each human during the course of its individuation creates the polarity of: conceptual self versus Individual, Collective Unconscious. The great repression is the formation of this polarity. That is, metaphorical, conceptual knowing becomes the center of self consciousness that is opposed to and represses mythical and magical types of awareness. When metaphorical, conceptual knowing differentiates into pure conceptual knowing, the created polarity becomes the dichotomy: rational self consciousness versus Individual, Collective Unconscious.

The formation of the unconscious – conscious polarity into "the great repression" is the creative event that civilizes the magical-mythical child. As described earlier in relation to personas, this event is the unwinding of the Uroboros (snake eating its tail, see cover of this book) to become a linear hierarchal structure of the persona ego dominating the lower centers of self-consciousness. At the same time, this event leads to an individual double bind. On the one hand, the great repression produces a stabilized self consciousness that exerts some control over the individual's expression of disrupting feelings and over some aspects of the individual's physical and social environment. At the same time the great repression prevents any further individuation or other kind of creativity so long as one remains so repressed. On the other hand, if one breaks from the great repression, that is, deconstructs persona, he/she becomes open to further individuation and other types of creativity. At the same time, one destabilizes self consciousness, which results in anxiety and suffering, and self destabilization that could lead to mental illness, mental breakdown, and self-destructive behavior.

Collective Non-Conscious versus Individual Collective Unconscious

Within each human a psychic center in union with the SOURCE generates the emergence and further individuation of self-consciousness. Because all levels of self-consciousness include a feeling aspect, this psychic center is called the *Feeling Center*. Because all levels of self-consciousness include a Will aspect, which involves Eros-chaos associated with Will-to-openness and Eros-order associated with Will-to-power, this psychic center is called *Will*. One may recognize archetypes such as described by Carl Jung, for example, persona, animus, anima, shadow, Mother, Father, and Hero, which guide human individuation to progressively higher levels of self-conscious individuation. I propose that the Feeling Center that also equals Will also should be called the *Collective Non-conscious*, which contains these "seeds" of self-conscious individuation. The various oppositional ambiguities of the patriarchal, polar self emerged as a result of the unfolding of some of these "seeds," that is, archetypes, of self-individuation. These diverse ambiguities are stored – remembered -- by the body of the body ego, either with no association with any type of human symbolic language or in association with proto-conceptual (magical) language or metaphorical (mythical) language. In any case these ambiguities will not be "seen" by the mind primarily structured by autonomous, mechanistic thinking, that is, the self consciousness of the conceptual, autonomous mind. Therefore, I propose that these unseen ambiguities are stored in what may be called the Individual, Collective Unconscious. Psychic activities such as dreams or conscious poetry or magical practices or special types of body message may bring one to remember some of these ambiguities as they are brought to conceptual self consciousness. Some of these "remembered" ambiguities may be seeds for further human individuation, which is what psycho-analytical psychology and depth psychology facilitate, or are seeds for conceptual, self -conscious creativity. I caution the reader to note that my idea of the *Individual Collective Unconscious* is not a deep center in each human before the emergence of the magic structure of the body ego. Rather, it comes into being in the course of the differentiation of the magical and mythical structure of self-consciousness. Also, it does not contain archetypes of self-conscious individuation but becomes associated with some archetypes that are contained in the *Collective Non-conscious*. Thus, in dramatic contrast to Carl Jung, I claim that each human is not the fundamental polarity of the Collective Unconscious versus Ego Consciousness. Rather, each human is a creative process generated by SOURCE through a hierarchal unfolding of archetypes in the Collective Non-conscious, where this center is the individual soul in union with SOURCE.

Transition to Persona Self as a Life, Death, Rebirth Process

Life. The child identifies with a stabilized polar self that is able to regulate feeling expressions of the body ego; also, the child has a partial separateness from parents. Early childhood is characterized by a participation mystique. After developing proto-conceptual thinking, the child via the emergence of a Mind, represses the body ego and partially represses his participatory mystique, but the magical-mythical perspective of the mind is continually influenced by the body ego so as to produce a "mental participation mystique."

Death. As the young child acquires greater language skills corresponding to metaphorical, conceptual thinking, the mind begins to further differentiate and separate more completely from the polar self and the body ego. As the child begins to further identify with this new mental, that is, spiritual, aspect of itself, he/she breaks away more radically from the *participation mystique* associated with the body ego. At this point the mind starts to create a "self-image" and the polar self and the body ego become more completely repressed into what later will become the Individual, Collective Unconscious or what Freud called the id with its "primary processes." For the rational adult, invasion into ego consciousness of elements of the Individual, Collective Unconscious or id will seem mysterious and terrifying. However, the child still has enough connections with it to experience wonder at the outer world from which he/she is beginning to separate. Images from the *participation mystique*, wonder, and metaphorical, conceptual thinking enable the child to build story representations of his reality. The child has moved from polar mythical to metaphorical, conceptual thinking.

Metaphorical, conceptual representations of reality produce a further separation from *participation mystique*. This further separation produces in the child the experience of alienation and abandonment by the once all supporting unitary feeling consciousness of the body ego represented as the Good Mother.

> The myths and fairy tales of childhood almost always tell of
> separation and abandonment. These motifs clearly bear the charge
> of the stress of separation from the nursery culture and the warm
> symbiosis with the mother. And so the child wanders, lost in the
> forest, abandoned in the river, and attempts a task far beyond his
> strength.... [to stabilize] a heroic new consciousness [which] is
> emerging... [and which has] prodigious energies and precocious
> cunning.[6]

Thus, late childhood is the beginning of a more self-conscious awareness of an inner war between the mind and the inner self, that is, the war between ego consciousness and the psychic energies emerging from what will become the Individual, Collective Unconscious. The protagonists of this conflict in early childhood are symbolized by baby boy versus the Great Mother. After the transformation to late childhood expressing literal, metaphorical, conceptual thinking, the conflict is represented as son-lover against the Great Mother, now thought of as Fate. This separation and consequent war may be likened to the separation of negative and positive charges. The positive mind, that is, the self consciousness, is drawn into the negative unconscious body ego, but the Eros-chaos, Eros-order mutuality that differentiated the polar mind self from the body self not only maintains but increases the separation. As a result, the mind is energized as it moves away from the unconscious body ego, thus increasing the polarity between the two, that is, between the mind and the body ego mostly repressed into the Individual, Collective Unconscious.

The child is bewildered by where all these energies are coming from. Often he/she is overwhelmed by the excesses of feelings that degenerate into rage and fear and then guilt about his/her own willful self-centeredness that maintains this separateness. The child defies any

---

[6]. Jean Houston, *Life Force: The Psycho-Historical Recovery of Self* (New York: Del Pub. Co.,1980), p.80

authority and satisfies his/her desires even when they are at the expense of the legitimate rights of others. At the same time the child feels alienated from and betrayed by the family and friends whose rights he/she has violated. In order to maintain his/her separateness, the child may undermine any attempt of others, such as family, to draw close to him/her.

> Then, on conquering [rejecting, defying the other], he is seized with remorse. Overwhelmed and shocked by the excesses of his own assertiveness, he falls into grief, for still he is tied to the common bonding of the symbiotic communion that he has only recently left. He blames the gods for his irrational behavior, but strongly suspects that there is something working in him that, although out of bounds, is still his own.
> Thus he feels shame and, not knowing what or who is causing his behavior, moves between paranoid rage and a shame that takes full responsibility. With such anarchy without as well as within, he is frequently driven to suicide. He has lost face, that face which once had been the common face, mirroring all others in the co-conscious society, and now is the mask of a hard-won ego [in the persona mind self]. With no tradition to support and sustain this new "face," with no rites to wash away its violations of taboo, there is no recourse but to self-destruct....
> To bear the melancholy of divided consciousness [persona mind self opposed to the body ego], the hero [child] often seeks relief in the transient unity of mind to be found in drugs or intoxicants.[7]

Rebirth. Traditionally, rebirth comes when the child in association with a stabilized, literal, metaphorical, conceptual perspective, such as traditional world religions, is taught and forced to obey rules and prescribed roles in its community. According to Wilber:

> ... a fairly coherent mental-ego [mind self] eventually emerges (usually between ages four and seven), differentiates itself from the body [body ego that identifies with the body] (after the Oedipal stage), transcends the simple biological world, and therefore can to a certain degree operate on the biological world... using the tools of representational thinking [via proto-conceptual, mythical thinking]. This whole trend is consolidated with the emergence (around age seven) of what Piaget calls "concrete operational thinking" [elementary, conceptual thinking that differentiates into metaphorical, conceptual thinking] [8]

---

[7]. Jean Houston, *Life Force: The Psycho-Historical Recovery of Self* (New York: Del Pub. Co.,1980), p.76

[8]. K. Wilber, *Eye to Eye The Quest for the New Paradigm* (Garden City, N.Y.: Anchor Press/Doubleday, 1983), p. 89.

This literal mythical thinking allows the child to delay the body's immediate discharges and postpone instinctual gratifications. The child uses verbal insertions to sublimate the body ego's emotional-sexual energies into more subtle, complex, and evolved activities. As the persona mind self incorporates verbal laws and identifies with prescribed roles, it becomes attuned to the rights, opinions, and roles of others, though as yet it has no capacity to distinguish true from false roles. The mental self becomes internally stabilized by the incorporated codes of conduct and externally stabilized via roles as a result of entering the community of other viewpoints.

In particular, rebirth includes the transformation from infantile gender identity that is primarily biological, sexual identity to *child gender identity*. On the basis of body image, body sex organs, and attitudes/behavior of others, the child establishes gender identity as a fundamental characteristic of his or her personality. During this process, the child also associates cultural expectations deemed appropriate for one gender and not for the other. In other words, gender sex becomes associated with gender role, so that male identity is culturally transformed into masculinity, and female identity, into femininity. Before 1900, virtually all cultures had clearly defined prescriptions for masculine versus feminine roles. The same prescriptions were taught and enforced in the family, neighborhood, work situation, town/city, and society as a whole. There were few if any ambiguities, and most individuals went on to develop stereotypical masculine or feminine gender/role identities. Of course, there always were and are exceptions to this rule.

This traditional way of stabilizing the persona mind self has its price. The child's sense of wonder and his creativity is diminished, but his mythical thinking is sparked by psychic energy from the Feeling center and the self-consciousness centers repressed into the Individual, Collective Unconscious. This psychic energy still can find expression in society's secular and religious myths, art, and ceremonies. Schools could enhance the child's genius for dramatizing speculation that could form the basis for rational world building that he/she will be taught in adolescence and adulthood. Unfortunately, science-technology has greatly helped to undermine all cultural institutions, especially the family and educational systems. The destabilized child is left with an undisciplined creativity that often is self-destructive.

INTERNAL CONFLICTS OF HUMAN INDIVIDUATION

The Individual, Collective Unconscious Versus the Mind Self

The first fundamental internal conflict in the developing human results from the opposition between the mind self and the body ego. As indicated earlier, Eros drives of the body ego guided the developmental-evolution of proto-conceptual knowing associated with language that represents feeling insights of the body ego. This process culminated in the emergence of a new center of self-consciousness, the mind self. Initially Eros drives of the Mind Self which direct experiential learning cycles produce progressively more complex language knowing in the thinking mind center and action directed by the action mind center. As a result, the mind ego "sees itself" as progressively more separate and different from the body ego, which only has non-conceptual feeling self-awareness. The Eros drives of the body ego continue to maintain the stability of the body ego, which includes the coordination of the activities of the physical self and the integration of the feeling self-awareness with the triune brain. Because of the physical self, the body ego is tied to a developmental-evolutionary process that expresses aging and eventual death of the physical self.

The Eros drives of the mind self direct the further differentiation of language knowing and activities, and coordinates knowing and behavior to maintain psychic stability, what may be called mind self homeostasis. This initially occurs with the emergence of the persona self that produces the great repression: the ego-mind self dominates and represses: (1) the magic consciousness of the action self, (2) the feelings of the body ego, and (3) the subjective imagination of the polar self into the Individual, Collective Unconscious. As a result, the ego-Mind Self dominated by the thinking and/or action mind center "sees itself" as more and more separate from the Individual, Collective Unconscious and correspondingly, more and more autonomous. In contrast, the Mind Self dominated by the expressive mind center "sees itself" as less separate from the Unconscious, less autonomous, and correspondingly, more spontaneous and creative. However, such individuals are less able to maintain mind self homeostasis. In this context the centers in the Unconscious of Eros-chaos are expressed as will-to-Egoness and the complementary drive, Eros-order, is expressed as will-to-control. In effect the Unconscious seeks to be the center of self-consciousness of the child, which, of course, disrupts the Mind Self's homeostasis (psychic stability). Will-to-Egoness drives the child away from conceptual knowing and the will-to-control drives the child to focus on subjective feelings and fantasies rather than maintain the ego-mind self psychic stability. The mind self focused on maintaining its psychic stability continually attempts to repress these feelings and fantasies into the Individual Collective Unconscious. However, an upsurge of psychic energy of Eros drives of the Unconscious continues to challenge the psychic stability of the mind self, which responds by repressing them all the more. In the early stages of its development, the mind self's stability often is overwhelmed by up-surges of the Eros drives of the Unconscious. This is why social structures, such as laws and social roles involving gender identity, are needed for each human to survive and begin to stave off these invasions. It only is when the mind self, by means of experiential learning cycles, evolves to a rational autonomous self that it can function effectively without being forcibly subordinated to these social structures. What happens is that the mind self internalizes some of these external structures so that it commits itself to follow some social laws and creates for itself its own vision and rational guidelines for behavior. However, the rational autonomous self still succumbs to invasions of the Individual, Collective Unconscious from time to time.

The Individual, Collective Unconscious versus the mind self produces a double bind involving creativity and human individuation. On the one hand, the ego-Mind Self must continually repress the Individual, Collective Unconscious in order to maintain psychic stability. But the Eros drives of the Unconscious flow into and energize the Eros drives of the Mind Self to direct experiential learning cycles. The creativity of these cycles is what drives human individuation toward higher levels of consciousness. Thus, repression of the Unconscious prevents creativity and human individuation. On the other hand, not repressing the Unconscious leads to a breakdown in Mind Self psychic stability – the extreme form of this is insanity or depression leading to suicide. Less extreme forms of Mind Self instability may prevent creativity or allow creativity but disengage it from driving human individuation.

The Individual versus Collective Social Consciousness

The individual versus collective social consciousness leads to a double bind that also involves creativity and human individuation. As indicated earlier, the 1½ - 3 year-old child begins to learn a language expressing paleoconcepts as a result of its participatory feeling identification with its caretakers, for example, its mother. The emergence of language in this context of a collective social consciousness, that is, group-ego equal the mother-child consciousness, leads to the emergence of a body ego consciousness separate from the mother-child collective consciousness. After language skills in the context of this collective consciousness have further differentiated in the 4½ - 5 year-old child, a (language) mind self consciousness emerges separate from the collective, social consciousness of the family and of the larger society in which the child is a member. In order for the body ego child to survive, he/she must subordinate his/her individual consciousness to the structures of the caretakers' social consciousness. Likewise, in order for the language mind self child to maintain psychic stability, he/she must subordinate his/her individual consciousness to the collective consciousness of the society of which he/she is a member. However, via the creativity of the experiential learning cycles, human individuation directs the child to evolve to greater individual autonomy, which causes the child to rebel against and further separate from any collective social consciousness. Thus, in each stage of human individuation – until one enters the trans-personal phase – all humans confront the individual versus collective social consciousness double bind. On the one hand, in each stage of human individuation, the evolving pre-personal or personal human must subordinate himself/herself to structures of some collective, social consciousness. This subordination is necessary for survival, for psychic stability, and for a grounding for further human individuation. But these social structures close off the evolving human from the openness to chaos that allows for the creative emergence of new structures that produce new levels of individuation. On the other hand, the human who rebels against the structures of collective social consciousness in order to be creative and thus facilitate individuation to a higher level of consciousness, also jeopardizes his/her survival, psychic stability, and necessary grounding for further individuation.

The Internal Conflict of the Mind Centers

This internal conflict leads to a double bind that involves the opposition between the two objective mind centers (the thinking and action centers) and the subjective expressive mind center within the Mind Self. As indicated earlier, the action mind center directs actions that objectively represent non-conceptual, subjective insights. For example, the young child has a non-conceptual, subjective understanding of how to walk. However, the action of walking is an objective representation of this non-conceptual, subjective knowing. It is objective because any two or more people can reach a consensus that the child is indeed walking rather than not walking or doing something else. Likewise, the thinking mind center produces a conceptual, objective representation of a non-conceptual, subjective perception. For example, each person's subjective sensations may lead to the non-conceptual, subjective perception of what humans agree to call a tree. The concept of treeness represented by the word *tree* is objective because any two or more people speaking the same language can reach consensus that the symbol *tree* does indeed represent what each person subjectively perceives as a tree. Thus, the action mind center and the thinking mind center deal with creating conceptual, objective representations of non-conceptual, subjective representations of perceptions.

The expressive mind center, as described earlier, translates subjective feeling perceptions into non-conceptual, subjective interpretations. In the young child just after the emergence of a Mind Self, these interpretations are unified magical-mythical ways of understanding. In later stages of individuation these interpretations become progressively more abstract, non-conceptual ways of understanding. In all stages of pre-personal and personal individuation, these non-conceptual interpretations are essential for creating and maintaining interpersonal relationships and social cohesion. They are the human equivalent of expressive bonding that occurs in all non-human conscious animal societies. These non-conceptual interpretations also are the unifying meanings of all human aesthetic knowing. Being able to enter into and help maintain interpersonal relationships is essential for psychic stability and further individuation; so also is developing some degree of aesthetic expression. However, producing objective actions by the action mind center and producing objective knowledge by the thinking mind center also are essential for psychic stability and further individuation.

Conceptual, objective representations produced by the action mind center and the thinking mind center oppose the non-conceptual, subjective representations produced by the expressive mind center. This sets the framework for the mind centers' double bond. On the one hand, individuation of the action mind center, or the thinking mind center represses the expressive mind center or subordinates its function. (Some action mind center personality types repress the expressive mind center and subordinate the thinking mind center to accomplishing some action. Other action mind center personality types repress the thinking mind center and subordinate the expressive mind center to accomplishing some action. The same pattern applies to the thinking mind center personality types. Some types repress the expressive mind center and subordinate the action mind center to accomplishing concrete knowledge-action goals. Other types repress the action mind center and subordinate the expressive mind center to producing objective knowledge.) On the other hand, individuation of the expressive mind center represses the action mind center or represses the thinking mind center. (Some expressive mind center personality types repress the action mind center and subordinate the thinking mind center to producing aesthetic knowing. Other expressive mind center personality types repress the thinking mind center and subordinate the action mind center to producing expressive social relating). The double bind is that all three mind centers are needed for the creativity required by diverse human situations. Therefore, in developing the thinking or action mind center, one represses the expressive mind center and thus prevents creativity in some situations. But then, in developing the expressive mind center, one represses the action and thinking mind centers and in a different way prevents creativity.

The Internal Conflict of Eros Drives

Eros-chaos and Eros-order drives propel the individuating human in opposite directions. Especially at the pre-autonomous, that is, pre-personal, stages of individuation, the two drives complement one another to bring about psychic stability at a particular developmental stage. If some factor – genetic or environmental – interferes with this complementarity, then some type of psychic imbalance becomes manifest in a particular stage of development. The imbalance may be severe enough to prevent any further individuation; or it may be carried along to partially disrupt subsequent stages of individuation.

Thus, the Eros drives involve the participatory union opposed to autonomous separation. In the pre-personal stages of development (birth to early childhood around seven years after birth) this opposition can produce psychological disorders. In the personal phases of development with increasing levels of "self-determination," this opposition becomes a true double bind. In each personal phase of development, the person is confronted with choosing between Eros-chaos and Eros-order. Each tends to negate the other and correspondingly each, as an isolated drive, has positive and negative effects on psychic stability. In each phase the person must create for himself/herself a mutuality between the two drives, thereby transcending the double bind.

Fundamental Ambiguity of Human Individuation: Participatory Subjectivity versus Control Objectivity

Human individuation begins with the baby in participatory oneness with its mother (or mother equivalent). The baby's all-consuming subjectivity is its being totally engaged with whatever object elicits a feeling perception from the baby. Individuation leads to the emergence of a physical self, which gives the baby some degree of objectivity. The baby now is a physical self that has feeling perceptions of itself being different from other objects of the world. With the emergence of a body ego, the child objectifies its own feelings. The child's conscious feelings become objects of its awareness, thus producing feeling self-awareness. This transformation to greater objectivity produces greater individuality, which, in turn, generates "the challenge of the impossible project" described earlier. The emergence of a Mind Self solves the challenge of the impossible project by enabling the child to enter into mutual, polar relationships with mother and others rather than attempting willful control of others, the behavior that characterizes the magic structure. By means of the mind the child represents aspects of the body self to itself, thus producing mental self-consciousness. Some of the represented aspects are incorporated into the mind self's metaphorical representation of the child's experiences of the world. Other aspects are repressed into the Individual Collective Unconscious. Further developmental-evolution of the Mind Self progressively gives the child greater objectification of its experiences; that is, the process produces greater conceptual objective representation of them. This gives the child greater control over his/her experiences by enabling the child to see and choose among diverse options of how to respond to these experiences.

Thus, human individuation drives the baby from total participatory subjectivity to ever greater objectivity with correspondingly greater separation from nature, greater individuality, and greater control of one's behavior. As a result, individualistic, objective control consciousness begins to repress and partially overshadow participatory subjectivity. But this subjectivity still has center stage during the pre-personal phase of individuation. The transition to the personal phase of individuation occurs by the Mind Self becoming identified with a set of personas and simultaneously repressing participatory subjectivity. The social innovation of persona is determined by: (1) metaphorical conceptual knowing organized into a literally "true" religious story describing humans in the world, and (2) social roles and rules of behavior prescribed by the literal religious story. In the personal phase of individuation, objective, control consciousness takes center stage and progressively overshadows participatory subjectivity. Yet, the non-conceptual insights of participatory subjectivity are necessary for even minor as well as major transformations to higher levels of objectivity. At the same time, progressively higher levels of objectivity provide correspondingly greater barriers for humans to be open to the transforming insights of participatory subjectivity. Very often the psychically stabilized person must be thrown into chaos in order to force him/her to be open to these transforming insights. When exposed to chaos, some people embrace it and thus are led to a transformation to a higher level of objectivity. Others attempt to repress the chaos or run away from it in other ways and thereby successfully resist any increase in objectivity. Thus, each new level of objectivity provides the platform for transforming to a still higher level. At the same time, each new level produces a greater barrier to transformation to a next, higher level.

At the highest stage of personal individuation, one comes to psychologically exist in what seems to be continuous chaos. If the person is not continually and totally destabilized by this ongoing chaos in which one may go into deep existential despair, then he/she becomes continuously open to and begins to continually receive new transforming insights. These insights enable the person to integrate the various internal oppositions that produce the double binds of individuation. The person also partially transcends his/her own personality type and takes on the power traits of whatever personality type is most appropriate for a particular situation. This internal psychic integration provides the platform for a transformation to trans-personal individuation. In the process of transforming into the trans-personal realm, one transcends the opposition between participatory subjectivity and control objectivity by incorporating these two types of consciousness into what many authors refer to as "the inner witness." The inner witness is trans-personal, is both subjective and objective, and provides simultaneously participation in and control of nature (see chapter 27). For some people, at least, when the inner witness first emerges, it destabilizes the human "self system," which can and often does fragment back into the subjectivity versus objectivity of the personal realm. If and when the human self-system finally does stabilize with the inner witness as its new, primary center of consciousness, the human undergoes further individuation. Various spiritual/mystical traditions describe these further phases of individuation.

Chapter 8
CHILDHOOD OF CIVILIZATIONS OF THE MIDDLE EAST

CHILDHOOD OF CIVILIZATION

Metaphorical, conceptual thinking regulated by myths thought to be literally true produced "the childhood of civilization," called the Age of Myth, that extended from 3,300 to 750 B.C. These societies were stabilized by control consciousness expressed as personas, patriarchal social institutions, and codified laws. However, participatory subjectivity not yet constrained by logical, conceptual thinking provided for playful creativity in these societies. The experiential aspect of metaphorical thinking disposed one toward participatory, subjective seeing of new possibilities of mythological expression. The conceptual aspect of this thinking disposed one toward participatory, subjective seeing new possibilities of conceptual expression and control. Correspondingly, Eros-order actuated these new possibilities along two paths: aesthetic expression related to myth and conceptual expression related to patriarchal control.

> Between prehistory and history proper [involving linear thinking
> and therefore linear perspective of time] there lies an intermediate

period whose literature... throws light on an older mythological perspective rooted in cyclic time; the annual cycle of the year with its feasts and fasts, its processions and rituals, all arranged according to the phases of the moon, the solstices and equinoxes of the sun, and the cycle of the zodiac.

The starting point of this Age of Myth is the first picture writing in Mesopotamia which remains undeciphered. According to legend, the Sumerians brought the art of writing with them when they joined the Semites to build that very early civilization on the banks of the Euphrates and Tigris [rivers].

The Sumerians also influenced the Egyptians in prehistoric times... [but] it was again the Semites... who [with the Hamites] laid the foundations of the Pyramid Age [in Egypt]....
....
It was also the heyday of the Semites. They were centered in Western Asia, including Mesopotamia, Assyria, Syria, and Old Phoenicia. The partially deciphered Linear A writing suggests that they either formed an older stratum of the population in Greece and Crete, or else had a strong influence there.... The watershed at the end of this era, the Age of Myth, comes around 750B.C.

By then Homer's epics had reached their final form....
....
From about 750 BC there was a new widespread interest in history.[9]

Chelwynd goes on to describe the individuation of the "childhood of civilization" to modern "adult" civilization.

> ... the birth of the conscious ego of modern man [what I call the *rational, patriarchal self*, that individuates via the scientific enlightenment to what I call the scientific, autonomous self] ... [occurred] at around the beginning of the Classical Age of Greece.... Homer's era marks the watershed, when men still experienced life in the old mythological way. Up until that time ... two centers of arbitration were struggling for supremacy: the imagination and the intellect. But from that time the intellectual ego got the upper hand and increasingly became the dominant conscious attitude. When the intellect won, it was a great triumph for the human race, as it is today when a child reaches the age of reason and discernment [which begins to occur during adolescence]....

9. T. Chelwynd, *The Age of Myth* (Hammersmith, London: Mandala, 1991), pp. xi - xii

... The gain for the human psyche – the mind – also involves a loss....

....

What was lost had been expressed in the grandeur of civilization in the Bronze Age. [This was] the childhood of civilized man.

When civilized man came of age, his conscious ego emerged at the expense of its unconscious background [magic, feeling insights, and imagination which had been repressed into the Individual, Collective Unconscious at the first emergence of civilization].

This period was such a playful and inventive time ... [when humans] still experienced life in terms of imagination, feeling, and intuition: the mythical worldview, which has survived only in dreams at night.10

The emergence of the persona self involves the creation of the Individual Collective Unconscious into which the previous centers of self-consciousness are repressed. The metaphorical, conceptual thinking characteristic of this period results in the self of the persona self still being open to these repressed centers. That is, the metaphorical aspect involves the subjective aspect of persona self incorporating the magical, feeling-intuitive, and imaginative perceptions repressed in the unconscious. The objective persona ego does its best to reject those aspects of mythical thinking that go against socially prescribed personas and rules of behavior. The objective self also integrates these personas and rules of behavior with mythical thinking. Thus, in the age of myth there is the self conscious duality of subjective mythical thinking and objective conceptual thinking. In some spheres of self consciousness there is a patriarchal marriage of these two types of thinking, that is, the relationship is dominated by conceptual thinking. In other spheres of self consciousness, there is a truce or a conflict between the two types of thinking.

ANCIENT MIDDLE EASTERN CIVILIZATIONS

The Middle East is bounded by the eastern shore of the Mediterranean Sea, by China in the west, Russia in the north, and India and the Arabian Sea in the south. The major civilizations that set the cultural milieu form which the Persian empire emerged are: Assyrians and Babylonians, Aryans, and Medes. The Babylonians and Assyrians are a Semitic people who are the same ethnographically and linguistically. They have the same religion, language, and literature and therefore may be grouped together as the same civilization. The Aryans were a semi-nomadic Nordic Whites who between 2,000 and 1,500 BC migrated in two waves southward from the steppes of southern Russia and Central Asia. The first wave settled in northern India. The second wave settled in northern Mesopotamia in the 1,000s and 900s BC. The Persians in 843 BC provided definite evidence of Aryan settlements in Iran. The Medes-Assyrian inscriptions dating to 836 BC attested to Aryan settlements in Central Iran. The word, *Aryan*, is the self-

---

10. T. Chelwynd, *The Age of Myth* (Hammersmith, London: Mandala, 1991), pp. 1-6

designation of the people of Ancient India and Ancient Iran. However, this term is primarily a linguistic idea that indicates closely related Indo-Aryan and Iranian languages. For example, the Indo-Aryan language changed into Greek, Latin, Hittite, Sanskrit, French, German, Latvian, English, Spanish, and Russian.

The *Assyrian empire* differentiated in three phases. The first phase lasted from about 2,500 to Samsi-Adad I (1813-1781 BC) whose son became a vassal of king Hammurabi of the Old Babylonian empire. The Middle Assyrian period began when Assur-Uballit (1364-1328 BC) led Assyria to regain its independence. His successor continued the expansion of the empire, which was stable for awhile. The ruler, Tiglath-Peleser (1140-1076 BC) marked the beginning of a century and a half decline of the empire. The third phase is the Neo-Assyrian empire beginning with Shallmaneser III (858-824 BC) who consolidated Assyrian power in the west and founded a new capital at Nineveh. Between 750-700 BC the Assyrian kings conquered the Medes of the Zagros, which is the great chain of mountains that separate the fertile crescent from the upland plateau of Iran. The Khorasan Highway that snakes across these mountains provided the trade route and cultural interaction between Asia and the western border of the Middle East. Assyria's greatest king, Sargon II (721-705 BC) absorbed into his empire the Medes living along the Khorasan Highway. His son, Sennacherib (704-681 BC) sacked Babylon in 689 and deported the Babylonians to Nineveh.

Just before Assyria's decline, King Assurbanipal (668-631 BC) created a great library of 22,000 cuneiform tablets including texts, the Epic of Gilgamesh that contains the account of the Great Flood also described in the Jewish Bible. The beginning of the end occurred when Nabopolassar became king of Babylon in 626 BC and defeated Assyrians near Harran in 616 BC. Then with an alliance with the Medes he conquered Ashur that along with Nimrod and Nineveh were among the "sacred cities" in what is now northern Iraq. In 612 BC the Medes and Babylon sacked Nineveh. The final collapse of the Assyrian empire occurred in 608 BC.

Hammurabi (1792-1750 BC) established the first *Dynasty of Babylon*, and during his reign, he issued a famous code of moral laws. The religion and the cuneiform writing of both Assyria and Babylon were derived from the older culture of Sumer. The writing evolved into an abstract form based on cuneiform symbols. As described in the "Epic of Creation," Marduk, the city-god of Babylon, became the supreme deity, king over all gods of heaven and earth. Beginning in 1,180 BC Babylonia flourished as a subsidiary state of the Assyrian empire, but as described earlier, was destroyed by Sennacherib. In 625 BC Nabopolassar established the New Babylonian empire that evolved to its greatest power under the rule of his son, Nebuchadnezzar (604-562 BC). Nebuchadnezzar sacked Jerusalem in 586 BC and hauled all its inhabitants off into exile in Babylonia. Cyrus, king of Persia, conquered Babylonia in 539 BC.

The *Medes* are the ancient Iranian people who lived in the northwestern portions of Iran along the Khorasan Highway. Around 836 BC the Medes were intermingled with another steppe tribe, the Scythians, who were the dominant group. Deioces (700-647 BC) was called the first Medes king, but he was merely a chieftain who along with his kinsmen were defeated and transported by Sargon II to Hamath (Hannah) in Syria in 715 BC. Phraortes (674-653 BC) took over the rule of the Medes chieftains and gained their independence. After dieing in battle, his son, Cyaxares, renewed the war with the Assyrians, but while besieging Nineveh, he was defeated by Scythians who then ruled Media from 652 to 625 BC. During a great banquet involving Medes and Scythians, Cyaxares (624-585 BC) killed the Scythian chiefs and formed the first Iranian empire. He captured Ashur in 614 BC and then sacked Nineveh in 612 BC. In 609 BC he conquered Urartu and took part in the final defeat of Assyrians in 608 BC. The

Medes swept into Lydia, what is now Turkey, in an attempt to conquer the Lydians. However, a solar eclipse that manifested over the battlefield persuaded the two sides to draw back. In a hurriedly patched-together treaty, the river, Hales, flowing midway between Media and Lydia, was established as the boundary between the rival empires. This led to the balance of power that produced peace in the Meddle East for the next thirty years. Astyages, son of Cyaxares, ruled Media from 585 to 550 BC.

According to Chelwynd, see beginning of this chapter, the Age of Myth ended in 750 BC when Homer's epics, created in the Middle Eastern Ionian shores of the Mediterranean, reached their final form. However, for the rest of the Middle East, the Age of Myth did not end until the final defeat of the Assyrians in 608 BC. It lingered on with the reign of Astyages over Media from 585 to 550 BC. A new kind of consciousness oriented toward a more abstract, mystical-mythical humanism emerged when in 550 BC Cyrus defeated Astyages and founded the first Persian empire. Cyrus' expansion of this empire was in sharp contrast to the previous civilizations in this region. He provided the groundwork for Zoroastrianism during the reign of Darius (522-486 BC) to establish a mystical-mythical humanism. The Assyrians epitomized to an extreme the contrast between the old and the newly emerging consciousness. Their kings spread terror and extermination across the land. They always saw it as their duty to flatten any resistance in the wilds beyond their frontiers. They began moving entire populations around their empire, transplanting one defeated enemy into the lands of another. These kings honed their rights of conquest to a peak of savagery in prescribing unspeakable cruelties for defeated enemies. Not only were cities burned to the ground and its inhabitants moved to the interior of the empire, the aristocracy were flayed, fed to animals or impaled.

UNIT IV EVOLUTION TO THE PERSIAN EMPIRE AND THE GREEK ENLIGHTENMENT
Chapter 9
FORMATION OF THE PERSIAN EMPIRE AND THE THREE STAGES OF THE GREEK ENLIGHTENMENT

CYRUS BECOMES THE FIRST PERSIAN EMPERER

In his book, *Persian Fire*,[1] Tom Holland describes the emergence of the Persian empire in association with the dreams, double treachery and betrayal of Astyages, king of the Medes. Astyages, see end of Chapter 8, was tormented by dreams that seemed to warn him of the ruin of his kingdom. He sought advice from the *Magi*, a whole class of the Medes dedicated to interpret dreams and visions, to acquire occult and sacred knowledge, and to direct mythical-religious rituals. Astyages married off his daughter, Mandane, to a vassal, a Persian prince of a backward and inconsequential kingdom. After Mandane became pregnant, a *Magus* interpreted a dream-vision of Astyages to mean that Mandane's son would imperil the Median throne. Panic-stricken Astyages gave orders to have his grandson, Cyrus, put to death by Harpagus, the commander of the Median army and the most prominent of the clan chiefs. Harpagus claimed to have carried out the murder of the infant Cyrus, but years later the truth came out that Cyrus not only survived but prospered and won the Persian throne. "Astyages was rumored to have wreaked a bloody revenge, butchering Harpagus' son, jointing the corpse, and then serving it dressed as mutton to the unsuspecting father.[2] After consuming his own son, out of fear for his life, he feigned to be chastened and loyal as a servant of his king. His act was so convincing that when war against the Persians broke out, he was appointed to the supreme command.

In 553 BC Astyages heavily outnumbering the Persian army with all the resources of a mighty empire would have triumphed had it not been for the betrayal by Harpagus. In mid-battle he led a rebellion to capture Astyages and deserted to Cyrus. Thus, in 550 BC Cyrus scattered the large armies of the Medes with his small army. He captured Ecbatana, the stronghold of the kings of Media, emptied and carted away its treasury, but he did not destroy the city. It remained the capital of Media and during the hot summer months in effect became the capital of Cyrus' whole empire. Cyrus, being the grandson of Astyages, had a family relationship that mirrored close cultural ties, which bound Persians to the Medes. Both peoples were Aryan and therefore were foreign only to non-Aryan rather than to one another. Thus, Cyrus in defeating the Medes preferred the course of mercy by not treating his fellow countrymen as slaves. He pensioned Astyages off into princely retirement and calculated that it is more pragmatic for him to have Medes as loyal servants rather than as an abject conquered people. They came to feel, though not equals of their conquerors, as associates of the great adventure of their new king's reign.

Cyrus died in battle in the summer of 529 BC, but since his initial victories in 550 BC, he expanded his empire to include Lydia that includes Ionia that is on the border of the coast of the Mediterranean Sea, eastward toward the uplands of Central Asia, what now is Kazakhstan, and south to the flatlands of what is now southern Iraq stretching from Assyria to the Persian Gulf.

---

[1] Tom Holland. *Persian Fire* (New York: Anchor Books, 2007), pp. 8-12.
[2] Tom Holland. *Persian Fire* (New York: Anchor Books, 2007), p. 11.

Cyrus brought the intersection of two, radically different enlightenments: one, Western individualism created by the Greeks in Ionia, and the other, universal peace for allegiance and subservience to an emperor who links his authority and power to ties with a universal, all-powerful God. In 547 BC Cyrus sent Harpagus to take command of the Persian forces in Lydia and Ionia. The scattered Greek city-states there had by now fully developed *Pre-Socratic individualism* of the first stage of the Greek enlightenment, see Chapter 10, which shifted to Athens under the tyranny of Pisistratus, who ruled from 560 to when he died in 527 BC. Harpagus brutally subdued all the coastal city-states there. Many Greeks rather than submit to Persian rule emigrated across the sea to Sicily or the Italian peninsula.

While since 900 BC the Ionian Greeks were developing their Pre-Socratic individualism, a new mystical-mythological vision associated with the teachings of Zoroaster (Zarathustra) began to emerge in the Middle East. He is said to be born between 1,500 – 1,000 BC, but his traditional stated birth and ministry is around 600 BC, just before the rise to power of Astyages. Zoroaster was neither a Mede nor a Persian. His teachings first came to the Zagros (the mountains in Iran associated with the Khorasan Highway) from the East. The Persians worshipped their ancient gods, honored mountains and flowing streams, and sacrificed horses before the tombs of their kings. In contrast, Zoroaster preached that *Ahura Mazda*, the Lord of Life, Wisdom and Light, was the supreme, the all-powerful, the only uncreated god. But at this time, instead of rivalry there was synergy between the two visions. In particular the Magi and the priests of Zoroaster engaged in numerous interactions. Instead of dividing the Medes and the Persians from their Aryan cousins in the East, the expression of a single religious impulse had been evolving over the centuries and became a unifying force. Fire had long been sacred to the Iranians as well as to Zoroaster himself. His teaching that the flames of fire were the very symbol of righteousness and truth became embedded in the imagination of both Persians and Medes alike. Followers of Zoroaster had the sacred duty daily to pray before fire.

Cyrus, partly calculating a way to rule over diverse conquered peoples and partly influenced by the emerging new religious vision, represented himself as an *avatar* (the descent of a deity to the earth in an incarnate form) of peace. While invading an enemy territory, he claimed to be defending it against the irruption of chaos. He claimed to be a model of "righteousness and justice," and his "universal lordship" was a payback from the gods.[3] Once his regime was well established, there were no more organized massacres, executions were kept to a minimum, and his diktats were couched in a moderate and gracious tone. Instead of aligning with Ahura Mazda or to a particular god, he posed as the favorite of all of them. Assorted priesthoods scrambled to hail him as their own as he made himself the patron of ancient cities such as Ur and Uruk. To this end Cyrus adopted certain customs from the ancient heartlands of various Iranian peoples. In the midst of his empire he built three new structures that were fire-holders made of stone, in which white-hot ashes were kept burning continuously. Thereafter, similar sanctuaries emerged all across his empire. Magic guarded the flames, which only were extinguished on the death of the reigning monarch. All this led to Cyrus being "remembered with an almost unqualified admiration for his exceptional nobility of character and as the architect of a universal peace.... [he persuaded] a host of different peoples that he understood them, respected them and desired their love. No empire had ever before [500 BC] been raised on such foundations. No conqueror had ever before displayed such clemency, such restraint."[4]

---

[3] Tom Holland. *Persian Fire* (New York: Anchor Books, 2007), p. 19.
[4] Tom Holland. *Persian Fire* (New York: Anchor Books, 2007), p. 20.

ZOROASTRIANISM

Zoroastrianism is a religion based on the teachings of Zoroaster (Zarathustra) who was born around 600 BC in west or east Iran. The basic beliefs include the following.[5] (1) Ahura Mazda is the one, transcendental and uncreated Creator to whom all worship is directed. (2) The entire universe is the battleground for conflict between *asha* equal to truth and order and *druj* equal to falsehood and disorder (chaos). All humans are drawn into this conflict. (3) Humans have free will that is expressed as active participation in life through good thoughts leading to good words that lead to good deeds. (4) There will be a final resolution of the conflict between good and evil, in which all creation including souls of the dead who were initially banished to "darkness" will be reunited to God. (5) Zoroaster's monotheism nevertheless produces de facto dualism: *Spenta Mainyu*, which is the instrument or "Bounteous Principle" of the act of creation, opposes *Angra Mainyu*, the "Destructive Principle." In creating humans by means of Spenta Mainyu, Ahura Mazda is immanent in them. The Creator through Spenta Mainyu also is immanent in the world and continually brings order out of chaos and thereby ultimately triumphs over Angra Mainyu. (6) Other more concrete, specific beliefs include that fire is the symbol of the energy of the Creator, equality of males and females, and the human soul leaves the body on the fourth day after death.

There are three major interpretations of Zoroastrianism.[6] The first is *ethical dualism*, in which the two principles of Good and Evil are purely psychological, mental, and abstract. Instead of being immanent in all the world, they only are active in the human mind, heart and soul. In the interpretation of Spenta Mainyu opposing Angra Mainyu, Mainyu means *Mind*. Thus, the choice between Good and Evil is made possible by mind self-consciousness. Evil does not exist in itself but only arises in the world through wrong choices and actions of humans. While wrong choices producing evil actions retards the eventual triumph of the Creator God and keeps humans from Him, every morally good action advances God's work on earth and brings humans closer to the Wise Lord.

Cosmic dualism, the second interpretation, became the classic Zoroastrian doctrine. *Mainyu* now is interpreted as *Spirit*. This led to the view that the principles of Good and Evil represent personified, living beings called the "Twin Spirits." Each of them is thought to have active lives with a soul that generates a self, producing thoughts, works, and deeds. In later mythologized versions of Zoroastrianism, the Spenta Mainyu, the Good Spirit, became identified with Ahura Mazda, the one Godhead called *Ohrmazd*. Angra Mainyu, the Evil Spirit, became *Ahriman*, the lord of lies and hater of humans. In this mythological vision the God of Goodness fights against an independent spirit of evil. However, in the midst of this cosmic battle there only is one God, Ahura Mazda; Angra Mainyu was never thought to be divine. Rather the Evil Spirit was considered to be a subordinate entity in rebellion against "the One God and His Truth." The reign of Evil is temporary, not eternal. In the Jewish-Christian tradition the Evil Spirit is Satan.

The third, more abstract interpretation transcends the opposition between ethical dualism and cosmic dualism. Archetypes and the archetypal world "are things which, though perceived

---

[5] http://en.wikipedia.org/wiki/zoroastrianism
[6] http://www.pyracantha.com/Z/dualism.html

by human minds, go beyond the individual and live their own, non-physical but in the very real world of archetypal reality. Zarathustra's Two Spirits exist in this archetypal world and therefore they have both a cosmic existence and a psychological or ethical existence."[7] One way of understanding this cryptic statement is in terms of the Collective Non-conscious as one aspect of soul as presented in Chapter 7, p. 68. On the one hand, the soul is a direct manifestation of SOURCE equal to Ahura Mazda, but, on the other hand, by means of archetypes expressing Spenta Mainyu they guide the individuation of each human to generate dualities that oppose one another. Correspondingly, each human generates and endures the opposition of mind versus body. An archetypal understanding of the Jewish-Christian doctrine of Original Sin is that while the Creator God produced the embodied human soul that is good, the individual is propelled into a spiritual journey. The journey guided by archetypes in the Collective Non-conscious always involves the emergence of disharmony and conflict between mind and body. Also, the soul as Will generates Eros-chaos opposing Eros-order. The SOURCE provides the basis for these two drives to collaborate to produce order out of chaos, creativity, and a limited new harmony. This human individuation mirrors the universal physical individuation involving the mutuality of Order and Chaos as described in Unit I.

A postmodern, relativistic understanding of Zoroastrian dualism points to a human or a group of humans in a position of power determining what is True and Good and what is False and Evil. This is how Darius, who became the emperor of Persia, used Zoroastrian dualism.

DARIUS' VISION: HOLY WAR, UNIVERSAL PEACE

Darius Becomes Emperor of Persia

After Cyrus died in 529 BC, his elder son, Cambyses became the new king and his younger son, Bardiya, became governor of Bactria, the largest and the most important of the eastern provinces. In 522 BC after invading and conquering Egypt, Cambyses was murdered or died of a freak accident. Bardiya was formally invested by the Magi and became the new king of Persia. In that same year, Darius along with six other conspirators and his brother, Artaphernes, murdered Bardiya and took over the Persian empire and expanded it in all directions. He routed the army of Nebuchadnezzar III and then completed the annihilation of Babylonian forces. In 521 BC he suppressed widespread rebellions across his empire, and in 512-511 BC he conquered Thrace, which is in Europe north east of Macedon that is just north of Greece.

Darius Unites Heaven and Earth

Darius had no direct, biological connection to Cyrus, but similar to him he was attuned to the emerging new spiritual vision in the Middle East and intent on exploiting this to enhance his power as emperor. He was not committed to the Zoroastrian religion, but he became convinced of some of its central beliefs. In particular, he embraced the idea that the entire universe is the battleground for the conflict between asha equal to truth and order and druj equal to falsehood – the Lie – and disorder. He conjoined this with the corollary view that Ahura Mazda will bring a final resolution of the conflict between good and evil. After his victories in 521 BC in which he brings universal order and peace in his Middle Eastern empire, he sees himself as the chosen one

---

[7] http://www.pyracantha.com/Z/dualism.html

of God. Everything he had achieved was due to the favor of Ahura Mazda. He brought history to a glorious close. His victory over the liar-kings representing druj had been a great and terrible one that proved the truth that he was indeed the champion of Ahura Mazda. Now that he had redeemed the world of the Lie, the Persian empire might be expected to endure for all eternity. However, there still remained adherents of the Lie, worshippers of "daivas" – false gods and demons. They threaten to upset world harmony as established in Darius' Persian empire.

Correspondingly, Darius claimed that any people who rebelled against his reign or refused to submit to it – as did the Greeks beginning in 500 BC – are also in rebellion against the order of Ahura Mazda and adherents of the Lie. Consequently, those who Darius sent to war against them should expect divine blessings both in their lives and after death. Darius assured his men that they would experience glory on earth and eternal bliss in heaven. Thus, he instituted the world's first holy wars. Foreign foes were to be crushed as infidels. Warriors were promised paradise and conquest in the name of Ahura Mazda became a moral duty. The every concept of universal order was tied to the timeless nature of Persian power. He visioned his empire as a fusion of cosmic, moral, and political order. The Persian rule embodied the covenant: "harmony in exchange for humility; protection for abasement; the blessings of a world for obedience and submission."[8] All this serves to justify global conquest without limit.

THREE STAGES OF THE GREEK ENLIGHTENMENT

First Stage: 900 BC in Greek communities in Asia Minor to degeneration of Athenian democracy that produced a new kind of tyranny (of the masses) under Pericles (died 429 BC) and degeneration of Pre-Socratic philosophy to relativism and skepticism.

Second Stage: Classical Greek rationalism beginning with Socrates (469 – 399 BC) to Alexander the Great (336 – 323 BC) establishing a Greek empire.

Third Stage (post-Classical Greek Enlightenment): consisted of two phases:

First phase: Hellenistic expansion of the second stage of Greek enlightenment: began with the death of Alexander to 31 BC when the kingdom of Ptolemaic Egypt was defeated by the Romans at the battle of Actavian.

Second phase: Greek-Roman Hellenization: began 31 BC to the fall of the Roman Empire, 476 AD.

Chapter 10
FIRST STAGE OF THE GREEK ENLIGHTENMENT: WESTERN INDIVIDUALISM PRODUCED BY IONIANS IN ASIA MINOR

---

[8] Tom Holland. *Persian Fire* (New York: Anchor Books, 2007), p. 60.

MYCENAEAN ANCIENT GREEKS

Around 2,000 BC two civilizations emerged, the Minoan that settled in the island of Crete in the city of Knossos, and the Mycenaean that settled on the Greek mainland in the city of Mycenae. The Mycenaeans spoke Greek derived from the language they brought with them when they moved away from Caucasus. This language, now called Indo-European, is considered to be an ancestor to most European dialects from India through Europe. Mycenaeans brought with them highly developed techniques in architecture, pottery, and metallurgy. By 1,600 BC the Mycenae was firmly established as an important center of the ancient world. They had written language preserved on numerous clay tablets. People lived in Mycenae in northern Argolis, in the Peloponnese of southern Greece as well as in Athens, Pylos, Thebes, and Tiryns. Like other ancient civilizations described in chapters 8 and 9, they were organized as monarchial societies governed by warlord kings and an aristocracy. Homer called Mycenaeans *Achaeans*.

Archaeological finds indicate in agreement with legend as described by Homer that the Mycenaeans' fought a war against one of their trading partners, Troy, a city on the coast of Asia Minor along the sea route between Europe and the Near East. Troy was conquered and burnt to the ground sometime in 1,200 BC.

Emergence of Epic Poetry

In the mythical age, "the childhood of civilization," see chapter 8, societies expanded celebrations and rituals concerning the great cycles of nature to oral remembrance of significant events in the past. This oral remembrance – oral history – became poetic stories sung accompanied by a stringed instrument to an aristocracy of society at ceremonies to commemorate a significant past event. These improvised performances evolved to what we now recognize as epic poetry. The defining features of epic poetry include:

> **A hero**: A central figure of exceptional character and position (though not necessarily exemplary in any moral sense), around whose actions the plot turns. This character's legendary, religious, mythic, or (less commonly) historic significance transcends time and space….
> **Narrative**: A story (as opposed to lyric or dramatic poetry) taking in a grand subject or theme, and broad in scope of its contents and length.
> **Elevated style**: Descriptions and speeches that reflect established rhetorical traditions, elegant meter, conventionalized lexicon.
> **Supernatural elements**: The intervention of gods and their interest in the outcome of events demonstrate transcendent meaning of story.
> **Unity**: One hero, one theme, but numerous episodes, with reversals… and recognition scenes…, leading to reconciliation.
> **Vast Scope**: Narrative of heaven and earth, deeds of great valor, extreme prowess and spirit, cataclysmic violence.
> **Invocation**: Opening invocation of the Muse, together with statement of theme.

In medias res: Beginning in the middle of things, with a narrative that makes use of flashbacks and foreshadowings, rather than a progressive, chronological narrative.

....

Epic Simile [and metaphor]: Simile compares two items explicitly using the connective words "like" or "as" and metaphor omits the explicit comparison: extended comparisons that broaden the plane of action, and move its frame of reference beyond fighting and adventure, connecting heroic exploits with other human action.[9]

Greek Dark Ages

Between 1,200 and 1,100 BC the Mycenaean warrior, tribal civilization abandoned their once mighty cities. This became the beginning of the "dark ages" in that these people lost the ability to read and write. Some speculate that the Dorains may have caused the Mycenaeans to abandon their cities. In any case, the scattered Mycenaeans in 1,100 BC became Achaeans, and Dorians emerged to become Sparta in the Peloponnese. Ionians displaced by Dorians migrated to and settled in Athens. Many of these settlers further migrated to Ionia in Asia Minor and the island of Samos. After the scattering of Mycenaeans, Greeks consisted of four tribes: 1) Achaeans, 2) Dorians, 3) Ionians, and 4) Aeolians.

CLASSICAL GREEK CIVILIZATION

Emerging Humanistic Ideal of Heroism

Instead of celebrating tribal gods and a past event, the epic poems took on the role of differentiating Greek clans from non-Greeks, referred to as barbarians. A major difference between the Greeks and barbarians was an emerging humanism centered around a new type of hero mythology. The heroes of the embellished epic poems asked for and received help from the gods; or sometimes a god or goddess would intervene on behalf of a hero even without being asked. However, the narrative of the poem made it clear that the god or goddess only helped the hero to achieve what he or she decided to do. The gods could not force heroes to do things against their free will. The hero had at least two aspects. The first was consistent with the core idea of the persona self and civilization; namely, the hero attempts extraordinary feats in battle out of a sense of social obligation. The second aspect is the beginning of humanism. He/she seeks glory even, as often was the case, if this meant death. To die with honor is greatly preferred by a hero over living an average, mundane, non-glorious life. But to have a glorious death one must prepare to be very skillful in battle and then very intensely struggle to win. Actually winning was not necessary for obtaining glory, but the preparation and the intense struggle were necessary.

As the Greek epic poem evolved to celebrate this emerging humanistic ideal of heroism, so also the story of Greek tribal gods evolved to a humanistic, patriarchal religion.

---

[9] Mr. Hahn's Eng. 140 Homer And Greek Epic Poetry Handout
http://www.courses.rochester.edu/hahn/eng140/homer.html

In the *theogony*, Hesiod [8th century B.C.] spells out an account of
the creation of the universe and the generations of the gods.
According to Hesiod, the universe begins when Chaos … brings
forth three natural forces: Gaia (Earth); Tartarus (the region
beneath Earth and Sea); and Eros (love, or sexual desire). Gaia,
the female principle, then bears three children without the
participation of a male: Uranus (sky), Oura (mountains); and
Pontus (sea). Next…

…

The continuation of this mythic history of the sources of life is
essentially a story of repressing violent natural impulses, and of
shifting power from the matriarchal first-generation, Gaia to the
patriarchal third-generation Zeus, who is called – not quite
accurately – the Father of the gods.

…

The Homeric epics and the subsequent Greek dramas constantly
refer to the various Olympian gods, so called because the family of
Zeus resides on Mount Olympus, distinguishing them from the
earlier generations of gods who dwell in or below the earth and
consequently are designated Chthonic. (Chthon, or Earth, is
another name for Gaia). By and large, the Chthonic gods personify
forces of nature; the Olympian gods, by contrast, who act more
like human beings, are thus "anthropomorphic," in the shape of
men.[10]

The Greek ideal of humanistic heroism probably influenced their attitude toward the
Olympian gods. This ideal celebrated a defining characteristic of a hero as the intense free
choice to attempt a feat that may lead to one's death. Fate or the gods totally determining what
happens in human society would undermine this ideal of heroism. To be consistent the Greeks in
contrast to the cultures of Asia and Africa, considered the stories of the intervention of gods as
not literally true. The stories only were metaphors for aspects of what happened in the epic
drama. Just as the Muse was thought to be the source of a poet's inspiration, so also the gods or
goddesses were responsible for various turn of events in an epic battle. In the Greek mind-set
then these muses for diverse types of creativity and the gods were unseen *Realities*. These
"realities" were directly experienced but remained unknown in themselves. Mythology was a
way of representing these "experienced mysteries."

From 1,200 B.C to around 700 B.C. stories were told about the fall of Troy. The story of
the Trojan War as told in the *Iliad* in its present form is probably the work of Homer, who lived
in the 8th century B.C. in Ionia (modern Turkey). He inherited from many generations of singers

---

[10] http://newman.baruch.cuny.edu/digital/2000/c_nc/c_01_epic/oral.htm
Note: in the above 2000/c_n_c/c_01_epic/oral.htm While at this site, click on "table of Contents"
and then click on "Background for Reading the Iliad and the Odyssey"

the traditions of versification, diction, and phrase that reach back to the earliest emergence of the Greeks from barbarism:

> The "Odyssey" [also by Homer] probably belongs to a somewhat later era than that in which the "Iliad" took final shape. The wanderings of Odysseus reflect newer experiences of the same Achaean stock which had won success in stirring conflicts in Asia, and was now pushing out in ships over the Mediterranean to compete with the Phoenician trader. The "Odyssey" presupposes the events described in the "Iliad;" unlike the "Iliad," it is not a story of battles and sieges, but of adventure and intrigue which center about a bold sailor.[11] (web site quoting Charles Burton Gulick)

Greek Innovation: the Greek City-State

The Mycenae kingdom tyrants who built grand palaces originated from tribal kings who were military chiefs. They validated their rule by claiming to have inherited it from the gods or God. For example, this is what Darius did with respect to Ahura Mazda. The God-granted ruler is the common feature among all ancient empires, such as Babylonian, Persian, China, Egypt, Israel, and India, as illustrated by the way Darius ruled the Persian Empire. The king possessed absolute power and all conquered people were subject to his will. Homer's epics also described the heroic God-granted kings, but the communities they ruled also had warrior aristocracies. From about 800 BC the Greeks began to emerge from the dark ages wherein they adapted the Phoenician alphabet to Greek and began keeping records. From about 900 BC migration from the mainland and colonization abroad produced independent communities. Greece became a sea-based civilization with the city (polis) as the basic unit of Greek government. The geographic position of Greece was particularly conducive to fostering this development. Greece mainland was a peninsula, and numerous scattered islands linked Greece with Asia Minor and Italy. Sea transportation facilitated large scale migration, trade, and cultural exchange and influence.

When mainland Greeks were still in the "dark ages," Greek communities in Ionia began to evolve into *city-states*. From about 1,100 BC each Greek community was ruled by a king and a warrior aristocracy that as a result of the evolving epic poetry adopted the cultural perspective of hero self-sufficiency and independence. The king came to be regarded only as a member of the aristocracy; so warrior heroes were validated by the gods along with or sometimes instead of the king. Thus, royalty along with the idea of the "God-granted ruler" disappeared without a violent revolution. The city-state developed out of the growing class of small landowners who defined themselves in relation to the region where their farms lay rather than to some kingdom. Initially the aristocratic elite, the heroes of the epic poems, made all the political and legal decisions, fought the battles, and controlled access to religious rituals. Further evolution of the city-state involved the spread of the privileges of the heroes to a larger group called *polites*, who were members of the city-state community. Women, free born men with no land or property and slaves were excluded from being a polite; also excluded were men who were polites in a

---

[11] Web site quoting Charles Burton Gulick

different city-state. Each city-state maintained an army of citizens, the polites, called *hoplites*. They could be called upon to fight at any moment; each provided his own armor. Each city-state had its own religious rituals that were performed by individual citizens rather than a class of priests.

The city-states evolved in two subsequent stages: Dynasteia and Tyranny. *Dynasteia* was a senate system ruled by an aristocracy that developed a democratic custom within the aristocratic class. It gradually evolved into a system with a comprehensive set of internal rules. It was an institutional structure ruled by law that represented the will of the aristocratic class that took precedence over the will of the individual ruler. *Tyranny* involved a *tyrant*, who called himself an *archon*, and who came to power in the context of the struggle between dynasteia and non-aristocratic, economically depressed citizens who wanted more input into political decisions and who were suppressed by the aristocrats. The archon was needed to maintain the social order by force. In order to maintain his power the tyrant enacted policies that benefited the non-aristocratic citizens. Tyranny was a short-lived transition phase to "direct democracy" that first occurred in Athens in 508 BC.

Greek Mythic, Personal Meaning

Homeric poems, besides portraying the heroic ideal, emphasized the hero's self-sufficiency. These poems are anthologies of heroic episodes in which one hero after another exhibits *aristeia* referring to brilliant achievements in battle. The root *aristo* means "the best" or "excellence." The hero in the *Odyssey*, Odysseus, expanded the traits of the hero to include adventure, cunning, and the choice to actively accept the chaos that befalls him, see new possibilities, and conceive a plan to overcome the chaos. Sometimes no plan is possible, so Odysseus must endure the chaos until he is "metaphorically speaking" saved by the gods or by happenstance. His salvation amounts to a personal transformation. Thus, Homer's epics, though only describing humans of the aristocracy of society who become heroes, nevertheless symbolized for later generations the possibility of any civilized human to be a hero. Every civilized human is a possible hero who has two natures: (1). *Earth nature* equal to the Inner Self, which in the epics are symbolized by the Olympian gods, and (2). *Spirit nature* equal to self consciousness that defines itself to some degree in terms of socially prescribed social roles. If the possible hero, when thrown into chaos, chooses to endure the chaos, the *earth nature*, which included the gods, help him/her to develop the traits of the heroic ideal and thus undergo a personal transformation. In Homer's time the hero became more separated not only from nature and barbarians but also from ordinary humans. In the classical period of city-states each civilized individual has the possibility to become extraordinary, not just in battle but in diverse human situations.

As described for Odysseus, the extraordinary human exhibited not just social virtue and excellence of some activity but other traits relating to one's ability to make plans; this ability will be seen by the pre-Socratics as relating to mind. Also, it is the mind that undergoes heroic transformations. The Greek word, are*te* (which perhaps is related to aristo), came to symbolize all that made any human individual extraordinary. Thus, *arete* represented not only static traits such as excellence and the virtues of duty and courage, but also the dynamic creative process of personal transformation. This is the basis for Robert Pirsig's referring to *arete* as Dynamic Quality.[12]

---

[12] . R. Pirsig, *Lili* (New York: Bantam Doubleday Dell Pub. Group, Inc.,1991)

Competitive athletic games provided an artificial circumstance for a human to exhibit and develop *arete*. First, the athlete had to train and hone his skills, just as the hero had to prepare for battle. If one obeys the rules of the game and fights the good fight, then there is dignity, virtue, and honor even in losing. There are no winners unless there are losers, so the willingness of people to compete, knowing full well that they may lose, is what makes the game possible. Therefore, provided one has trained diligently, just entering the contest is an important contribution to society. Independent of whether one wins or loses he contributes just by intensely competing. In so doing he exhibits Dynamic Quality (arete).

Precursors to the Olympic games probably occurred since the 10th or 9th centuries B.C.

> The origin of the Olympic Games is linked with many myths referred to in ancient sources, but in the historic years their founder is said to be Oxylos whose descendant Ifitos later rejuvenated the games. According to tradition, the Olympic Games began in 776 B.C. when Ifitos made a treaty with Lycourgos the king and famous legislator of Sparta and Cleisthenes the king of Pissa. The text of the treaty was written on a disc and in <u>the Heraion.</u>
>
> In this treaty that was the decisive event for the development of the sanctuary as a Panhellenic center, the "sacred truce" was agreed. That is to say the ceasing of fighting in all of the Greek world for as long as the Olympic Games were on.
>
> As a reward for the victors, the cotinus, which was a wreath made from a branch of wild olive tree that was growing next to the opisthodomus of *the temple of Zeus* in the sacred Altis, was established after an order of *the Delphic oracle*.[13]

It was a one-day celebration consisting of a single event, the running of one Stadion, a foot race 600 feet long. The games were held every four years during the month of July or August in Olympia, a rural sanctuary site in the western Peloponnese. Over the years more events were added, so that by the 6th century B.C. there were 10 events (running, jumping, discus, "ekebolon" javelin, wrestling, boxing, the pancration, chariot racing, horse racing). The games were primarily a part of a religious festival honoring Zeus.

Greek mythic, personal meaning, then, is an individual being able to acquire some degree of Dynamic Quality and then increasing it over a lifetime. *Arete*, Dynamic Quality, was celebrated at the Olympic Games, which came to include a religious celebration at which the Homeric epics were sung. These festivals represent what one might call "the Classical Greek enlightenment."

> In the eighth century B.C. ... so far as our remains indicate, there cannot have been much to show that the inhabitants of Attica and Boeotia and the Peloponnese were markedly superior to those of,

---

[13] http://www.culruew.gr/2/211/21107a/0g/games.html

say, Lycia or Phrygia, or even Epirus. By the middle of the fifth century the difference is enormous. On the one side is Hellas, on the other the motley tribes of "barbaroi...."

> If we wish for a central moment as representing this self-realization of Greece, I should be inclined to find it in the reign of Pisistratus (560-527 B.C.) when that monarch made, as it were, the first sketch of an Athenian empire based on alliances and took over to Athens the leadership of the Ionian race.

> In literature the decisive moment is clear. It came when... Homer came to Hellas.... under Pisistratus... the Homeric Poems, in some form or other, came from Ionia to be recited in a fixed order at the Panathenaic Festival, and to find a [written] canonical form and a central home in Athens till the end of the classical period.... [Homer's oral narratives probably were written down in the 7th century B.C during the period when the Greeks were beginning to use the alphabet.]

> In religion the cardinal moment is the same. It consists in the coming of Homer's "Olympian Gods,"[14]

However, the Greeks participated in a revolution unique to the evolution of Western consciousness:

> The Olympian religion, radiating from Homer at the Panathenaea, produced what I will venture to call exactly a religious reformation....

> It was a move away from the "beggarly elements" toward some imagined person behind them. The world was conceived as neither quite without external goverance, nor as merely subject to the incursions of *mana* snakes and bulls and thunder-stones and monsters, but as governed by an organized body of personal and reasoning rulers, wise and bountiful fathers, like man in mind and shape, only unspeakably higher.

> For a type of this Olympian spirit we may take ... the reiterated insistence in the reliefs of the best period on the strife of men against centaurs or of gods against giants.... to the Greek, this battle was full of symbolical meaning. It is the strife, the ultimate victory, of human intelligence, reason, and gentleness, against what seems at first the overwhelming power of passion and unguided strength. It is Hellas against the brute world....

> the Olympian Religion is only ... intelligible and admirable if we realize it as a superb and baffled endeavour, not a *telos* or completion but a movement and effort of life.[15]

---

14 Gilbert Murray, *Five Stages of Greek Religion* (Garden City, N.Y.: Doubleday & Co., Inc., 1955), pp. 38-42.

The original Sparta was a Mycenaean kingdom that before 1,200 BC was ruled by Menelaus whose wife, Helen, was abducted by Paris to Troy. In 1,250 BC after 10 years of war, Troy was burned to the ground, Trojans massacred, and Menelaus and Helen returned to Sparta. Around 1,200 BC Sparta and the surrounding area was sacked and burned to the ground. Between 1,200 and 1,000 BC Dorians migrated into the Peloponnese and the Dorian city of Sparta was founded on the basis of five villages. By 800 BC Sparta evolved to a Greek city-state. Spartans were devoted to the age of heroes as described in the epic poetry of Homer more than other Greek communities. Because of this devotion, Spartans called themselves "Heraclids" (after Heracles described by Homer) and as such claimed dominion over much of Greece. Between 725 and 640 BC Sparta invaded and conquered the surrounding area, which was Messenia. This success, however, became the source of great stress.

On the one hand, in the 700s and 600s BC Sparta had been notorious for their materialism and greed. By 600 BC throughout Greece there was a widening gap between rich and poor citizens. In Sparta this gap was intensified to produce extreme hatred between the two classes of citizens. On the other hand, during the same period, a great transformation in war tactics began to spread across the whole of Greece. Attack units of soldiers were arranged in a V-shaped *phalanx* in which each fighter carried a circular shield, called a *hoopla*, which was a meter high and wide and faced with bronze across its wood. Each fighter was also protected by a bronze helmet and body armor and he carried a spear. Each *hoplon-holder*, called a *hoplite* used the shield not only to protect himself but held it in such a way that it protected the hoplite on each side of him from arrows from enemy archerors and from spears and swords of an advancing enemy. However, in order for the phalanx tactic to succeed, each hoplite had to have the courage and discipline to stay in line.

During the Messenian War, the Spartans further developed this new form of warfare. But the citizen soldiers supplied themselves with weaponry; so those of the poor class began not to be able to afford it. Moreover, intense class hatred and mistrust between rich and poor caused hoplites not to be motivated to "hold the line in a phalanx." What made matters worse, the revolt by Messenians almost destroyed Sparta and now the defeated Messenian population outnumbered the Spartan population by ten to one.

Spartans invented a revolutionary new political system to solve this crisis. Much of Messenia was partitioned into allotments for the impoverished farmer hoplite who now became a full time warrior. Messenians were turned into agricultural slaves called *helots*, who worked small plots of land on estates owned by Spartans. Every citizen, whether aristocrat or peasant, was a member of the military institution. At age seven every male was sent to military and athletic school that taught toughness, discipline, endurance of pain, and survival skills. At twenty the Spartan became a soldier who lived in barracks and ate all his meals with his fellow soldiers. He married but only at thirty did he become an "equal" and was allowed to live in his own house with his own family. At sixty he ended his military service. "The ideology of Sparta was oriented around the state. The individual lived (and died) for the state.... The

---

15 Gilbert Murray, *Five Stages of Greek Religion* (Garden City, N.Y.: Doubleday & Co., Inc., 1955), pp.56-59.

combination of this ideology, the education of Spartan males, and the disciplined maintenance of a standing army gave the Spartans the stability that had been threatened so dramatically in the Messenian revolt."[16] In 520 BC Cleomenes became king of Sparta and in 519 BC he led an army against Athens. <u>Note</u>: In 522 BC Darius took over the Persian Empire.

FURTHER GREEK ENLIGHTENMENT IN IONIA

Olympic Games Rationality.

Humanism is the differentiation of self-consciousness in which the individual asks questions and then wants to know the reasons for the answers he/she intuits or is given by others. By contrast, in order to be accepted into society, the non-humanistic persona self does not ask questions or believes in the truth of the answers that society gives to these questions. The first poetic stories of events related to the fall of Troy were literally true except for some details. Over several generations minstrels began to change the original historical narrative to focus on values implicit in the recorded events. The emerging epic poems indirectly answer the question: "Who are we and what do we believe about the world and the way humans should live?" Embellishments of these stories included interactions between humans and gods and goddesses that symbolized the differentiating Greek cultural attitudes about life.

The OLYMPIC games consisting of competitive sports collectively served as a metaphor for Greek rationality. Any competitive game structures the way we perceive reality. For example, a tennis game consists of the autonomous parts: tennis players, tennis ball, tennis rackets, and a tennis court where each part is specified by characteristics that can be defined operationally, for example, measurable dimensions of the ball, racket, and tennis court and describable procedures for hitting the tennis ball. A set of rules prescribes how all these parts interact. When two tennis players follow these rules, they bring about a synthesis of the interaction of these autonomous parts, which is the completed tennis game. The two players and anyone else who "knows tennis" and watches the game understand this whole game in terms of the interactions among its parts. This is the essence of any rational analysis, also known as mechanistic analysis.

The OLYMPIC games symbolized the belief that Homer's OLYMPIAN gods organized humans' life in the world similar to organizing a game. Each human is a contestant competing against other contestants in the game of life. If an athlete knows a particular sport, trains diligently, becomes skillful, and obeys the rules when he is allowed to enter the game, then he may win. To some extent the relative intensity of one's desire to win and lucky breaks influence the game's outcome, but overall, the game is totally comprehensible by means of rational analysis; there is no mystery to it. The players have some degree of control of their destiny. If they lose, they know that it is because someone else had greater skill or equal skill coupled with some lucky breaks or some winning combination of skill, luck, and intensity. In any event, loosing will not be due to the caprice of blind FATE or of a despot; likewise race, political affiliation, philosophical perspective, or religious belief will not prejudice the outcome.

Thus, by the 6th century B. C. Greeks in a non-self-conscious way saw the world as "rationally knowable" in the same way as competitive games are knowable. Without reflecting and then choosing a mindset,

---

[16] http://www.wsu.edu:8080/~dee/GREECE/SPARTA.htm

> ...the Greeks assumed ... that nature would play fair; that,
> if attacked in the proper manner, it would yield its secrets and
> would not change position or attitude in midplay.... There was
> also the feeling that the natural laws, when found, would be
> comprehensible.[17]

Beginning with Thales (624 – 546 B.C.) several Greeks now referred to as *Pre-Socratic philosophers*, who lived and taught in Asia Minor, in Thrace, in Sicily and in South Italy, began to ask abstract questions: (1) Where does everything come from? (2) What is it, for examle, the world, made out of? (3) How do we explain the diversity of things found in nature? (4) Why are we able to describe diverse things with a singular mathematics? For example:

> Thales seems also to have been the first to go about proving
> mathematical statements by a regular series of arguments,
> marshaling what was already known and proceeding step by step to
> the desired proof as inevitable consequence. In other words, he
> invented deductive mathematics, which was to be systematized
> and brought to a high polish two and a half centuries later by
> Euclid.[18]

To answer these questions the Pre-Socratics invented a new kind of game which we may call the game of rational discovery.

Game of Rational Discovery

The purpose of the rational discovery game is to create stories using ordinary language that provide experientially understandable answers to abstract questions about nature. Because the stories use ordinary language involving metaphorical concepts, these explanatory answers link subjective experiences with the objective, conceptual questions. That is, the metaphorical aspect of ordinary language "pulls in" subjective experiences, and the conceptual aspect relates to the objective, conceptual questions. One is free to propose any story, even if it appears to be absurd, for exaple, Thales proposed that the fundamental "stuff" of the universe is water even though one's experience of rocks is radically different from one's experience of the "stuff" that flows in rivers. The freedom of story-telling means that one does not have to satisfy the dictates of any human institution or ruler; nor does one's story have to be consistent with a socially prescribed myth.

However, just as in any competitive sport, there are rules that impose limits that define the nature of the game. First of all, the stories must never refer to or depend upon any epic myth or the presumed existence of any of the Olympian gods. These myths that referred to gods and goddesses were not thought to be literally true. The myths symbolized and prescribed human

---

[17]. Fred L. Wilson, http://www.rt.edu/~flwstv/presocratic.html

[18] Fred L. Wilson, http://www.rt.edu~flwtv/presocratic.html

values and attitudes rather than representing explanations of nature. Second, in rejecting myths, the stories also must exclude anthropomorphic explanations. That is, the stories are about Nature that includes humans merely as one type of "part" making up Nature. Therefore, human values and attitudes are irrelevant to these abstract explanations. Third, the stories had to be reasonable, that is, logical. This was so even though the pre-Socratics did not reflect on and spell-out what "being logical" means – Aristotle is the "after Socrates" philosopher who did this. Nevertheless, the pre-Socratics had a presumed logic, which meant that they attempted to show how a particular story explanation comes from and explains empirical observations; we now call this induction. The story also had to be logically consistent, that is, people without any formal training in deductive logic can recognize glaring contradictions.

Pre-Socratic Arete

As described in chapter 7, the first stage of differentiation of the persona self corresponds to the *Age of Myth*, which extended from 3300 to 750 B.C. In the age of myth there is the ego conscious duality of subjective, mythical thinking versus objective, conceptual thinking. The subjective, mythical thinking is a non-conceptual perspective, that is, an existential perspective that is an experiential dialogue with what Heidegger calls Being. Being "speaks" to humans, who respond by interpreting what they "heard" and then representing the interpretations in some form, such as works of art, myths, and rules of behavior. Usually the myths were considered to be literally true. This aspect of the collective consciousness of the age divided people into conflicting civilizations that made war on one another. In the Western culture the ancient Greeks were the first to begin to reconnect with the direct experience of Being by viewing their epic myths and Olympian religion as fanciful stories that only symbolized the culture's anthropomorphic attitudes and values. Greek culture structured by Homer's epic poems and the Olympian religion influenced the pre-Socratics to create what Heidegger called the great beginning, that is, "poetic philosophy."

Poetic philosophy is a metaphorical, conceptual objectification of a passionate insight after a struggle; that is, one expresses the insight in ordinary language. I was privileged to observe this happening to students in a sophomore high school biology class I taught in 1958; they struggled in writing essays on "What is Life" and on scientific thinking. I watched a similar phenomenon in some 10th grade high school algebra students I taught in 1959 when they wrote essays on "What is Mathematics" and on the concept of number. My favorite story of passionate insight concerns a college senior, biology major taking my cell physiology course taught in 1971. The course involved open book take-home exams emphasizing gut understanding of bio- and physical-chemical theories of cell function/structure. Over 60 percent of this class (of pre-med majors very intent on getting into medical or dental school) received a D or F on my first open-book exam. By the end of the course most got an A or B. Toward the end of the course one dental student from New York City told this story of himself:

> "I woke up one morning so excited that I shook my roommate and shouted at him:
> 'Do you realize that all living things are made up of cells! If you understand how
> cells function and how they interact, you understand what life is!'"

Of course he had heard this sort of statement since grade school, but he had not really understood it until his senior year in college! How sad that it took him this long to have this kind of experience of knowing, when in fact people can be brought to such passionate involvement as early as the 10th grade and perhaps even earlier. The point here is that instead of having a static, book-knowledge about life, he experienced a personal engagement in a knowing process relating to an aspect of life. The knowing process put him in relation to an aspect of being, in this case, an aspect of life. The "energy" of this subjective kind of knowing is real but incomprehensible. We are confronted with a tautology: *some aspect of nature is knowable because we know it.* What is it that we know is inseparable from the subjective process of knowing. We experience the mutuality between an objective something that is known and a subject that knows, but we neither comprehend the subject nor the object.

The pre-Socratics who struggled to understand some aspect of Being were like heroes of Homer's epic poems or contestants in the Olympic Games. When such a thinker experienced this engagement in the knowing process and created a story to represent empirical observations, he (almost never a she) exhibited *subjective, proto-rational arete*. It was "proto-rational" because the explanatory story consisted of ordinary language, which is imprecise. This imprecision allows for diverse interpretations that could contradict one another, and there was no way of deciding which interpretation is the correct one. Personal preference turned out to be the criterion of truth for the early pre-Socratics. In spite of this imprecision the explanatory stories could be beneficial to the individual and/or to society. At the very least, such stories pointed individuals toward further empirical observations about the world, rather than rely on tradition or myths. Also the stories directed a person to be mindful of the spiritual qualities of *choosing* to *create stories* rather than passively accepting what tradition dictates to him. The pre-Socratic *arete* associated with creating explanatory stories had greater worth than "heroic battle or competitive game arete" because it transcended concrete situations. Instead of *arete* being tied to a specific battle or competition, it now was associated with a transformation of the mind. As such it could be "transmitted" to other humans who could be guided to create for themselves an understanding of an explanatory story. Now, not only individuals exhibit *arete*, but shared explanatory stories exhibit it. But this kind of subjective knowledge exhibits *arete* only for those individuals who create their individual interpretation/understanding of it rather than merely "learning" the story in a non-creative way; that is, accepting the explanatory story in the exact form it is taught to them. *Arete* of shared proto-rational stories led to the development of pre-Socratic humanism.

Pre-Socratic Humanism

Creating explanatory stories led the Greeks to believe that they exhibited the highest, most perfect level of being human. All Greeks – mostly aristocratic males – who participated in this story knowledge began to live in the realm of spiritual mind that transcended concrete events embedded in space and time. Using modern terminology one could say that the Greek seers individuated to a higher level of self-consciousness. By contrast, Barbarians, like ants carrying out genetically determined behavior patterns, conformed to personas and rules dictated by literally true myths. This Greek higher level of self-consciousness shifted the focus from provincial, anthropomorphic heroism to non-provincial, non-anthropomorphic spiritual heroism. The pre-Socratic philosopher is like the sea captain who departs from the safe, known regions of the seashore to the dangerous, unknown vastness of the open sea. Metaphorically speaking, the open sea is Being. The spiritual adventurer both then and now is a person who leaves the safety of conventional wisdom, such as, literally true myths, in order to gain new wisdom that engenders greater power. This new kind of hero enters a spiritual battle to conquer an aspect of Being by comprehending it with an explanatory story. That is, the spiritual hero attempts to master "the overpowering of Being" by reducing it to an ordinary language representation. In effect the hero is saying: "If I can create a true story representation of Being, I know Being and have power over it."

Many pre-Socratics created mechanistic story representations that proved to be unsatisfactory. However, according to Heidegger, Parmenides and Heraclitus participated in Being in order to create some valuable though inadequate representations of it. That is, they gave up total mastery of Being and committed to *openness to Being* that implies a mutuality between thinking and Being. Heidegger translates a maxim from Parmenides as: "There is a reciprocal bond between apprehension and being"[19] By this he means that Being forces one to conceive what Being is, which is to say that Being in manifesting becoming and appearances "speaks" to humans who perceive these changes and appearances. Humans take in and remember (after the invention of writing, they record) these perceptions as empirical observations. Then humans respond by creating story explanations of these observations. In this way the explanatory stories indirectly represent Being.

However, both Parmenides and Heraclitus persist in their openness to Being. They acknowledged that Being manifests becoming and appearances but Being in itself is not becoming or appearances. Using modern terminology we may express this conviction as: Becoming and appearances are "real" only as manifestations of Being. They are dependently real. If there were no Being, there would not be any becoming or appearances. But conversely, if there were no becoming or appearances, there still would be Being. Becoming and appearances imply that there must be Being, but Being does not imply that there must be becoming and appearances. (This interpretation is consistent with Thomistic metaphysical theology and with Samkara'school of Advaita-Vedanta. It is, I believe, inconsistent with Zen Buddhism for which absolute Emptiness manifests Being and becoming as well as Being and non-Being; also there is a total mutuality wherein absolute Emptiness implies becoming and becoming implies absolute Emptiness.) Thus, any story explanation of becoming or appearances does not represent Being in itself. The explanations only represent an aspect of Being, which itself remains entirely mysterious.

This realization led Parmenides to interpret Being as the mutuality: Existence implies Logos, and Logos implies Existence. Being as Logos manifests humans who individuate to a persona self capable of metaphorical, conceptual thinking. The Eros drives emanating from Logos lead some humans to freely choose to be open to Being, to participate in some aspects of It, such as, becoming and appearances, and then to create story explanations of these aspects. My understanding of Heidegger's translation of a maxim of Parmenides is that Being determines thinking, that is, drives humans to metaphorical conceptual thinking which, in turn, creates explanatory stories of Being, that is, interpretations of Being. Heraclitus expresses the same insight in terms of a paradoxical understanding of Logos. My theory of creativity makes the paradox explicit and then transcends it. Logos via participatory subjectivity associated with Eros-chaos drives civilized humans to be engaged in concrete events that have no intrinsic order; that is, all is change – "everything flows." Logos via control objectivity associated with Eros-order drives civilized humans to create story explanations of the continuous flow of events in

---

19. M. Heidegger, *An Introduction to Metaphysics*. Trans. By Ralph Manheim (New Haven: Yale Uni. Press, 1959), p. 145

nature. Creativity is the mutuality of Eros-chaos and Eros-order so there is a mutuality between becoming and permanence. Thus, Being as Logos is Creativity (see figure 10.1). Logos of the universe is hierarchal individuation as described in chapters 1 and 2. Logos of humans is individuation to the persona self and then hierarchal individuation of the mind to hierarchal levels of conceptual thinking.

LOGOS▶   **PARTICIPATORY SUBJECTIVITY**

Eros-Chaos ━▶ Drives

Humans to <u>engage</u> in concrete events

that have no Intrinsic Order

All is change

Everything flows

opposition

LOGOS▶   **CONTROL OBJECTIVITY**

Eros-Order ━▶ Drives

Civilized Humans to create story

explanations of the continuous

flow of events in nature

BEING as LOGOS
is
CREATIVITY

1. Hierarchal Individuation in the Universe
2. Hierarchal Individuation of Self-Consciousness in Humans

FIGURE 10.1 Heraclitus' Concept of Logos

Heidegger expanded his interpretation of the core idea common to Heraclitus and Parmenides. This became the foundation for his radically new approach to philosophy. I take Heidegger's insight and expand it as follows. Being is ultimate Reality that is totally mysterious, that is, Being cannot be comprehended or adequately represented by any explanatory story. At the same time, Being manifests becoming and appearances, such as qualities of things that stimulate sensations in animals. Being as Logos draws humans to subjectively participate in their sensations of motion and appearances and then create objective, story explanations of them. The explanations only interpret sensations of the manifestations of Being. In this manner unknowable Being is contrasted with three aspects of Being: (1) motion, (2) appearances, and (3) metaphorical, conceptual thinking about motion and appearances. Heidegger proposes four aspects of Being, the first three are the ones listed above and the fourth aspect is *ought* equal to right action that is guided by one's explanations of any situation. I propose to add one more aspect to the four that Heidegger lists: Being as Logos draws in humans to create an *aesthetic appreciation* of any situation; this is the fifth aspect. The aesthetic appreciation, though conditioned by one's thinking and story explanations of nature, will draw one to participate in the situation for the sake of pleasure or spiritual joy.

This five-fold list of unknowable aspects of unknowable Being plus the idea of Logos as Creativity provide a framework for specifying pre-Socratic *arête*. Pre-Socratic arete is the mutuality of Being as Creativity and human creativity that results in metaphorical, conceptual thinking that produces: (1) explanatory stories of each of the five aspects of Being, (2) right action in any situation, and (3) aesthetic engagement in concrete situations All of these have great practical value in that they, for example, can be applied to achieve excellence in combat and in sports in competitive games. A Greek exhibiting *arete* was radically different not only from barbarians but also from most other Greeks. Most Greeks, similar to the barbarians, believed that the epic poems, though not literally true, nevertheless dictated laws of right action and therefore, the "good life," producing happiness consisting of conforming to these laws accepted as divine and eternal. In radical contrast, pre-Socratic *arete* dictated a radically new version of Greek humanism. An individual who exhibits his humanness to the greatest extent is one who also rejects this traditional view of the good life. Rather he creates for himself explanatory stories that explicitly exclude all myths. As a result, having no tradition as a guide, he creates the intent for right action, that is, orientation toward the Good and aesthetic engagement, that is, orientation toward the Beautiful.

Greek Eros Stress: Individualism in Opposition to Community

Greek humanism led to opposition between Greek and non-Greeks, opposition among Greek city-states, especially Sparta versus Athens, opposition within a city-state between aristocracy and the ordinary citizen. Survival of Greek civilization and culture depended on collaboration among Greek city-states. Survival of an individual city-state depended on collaboration between aristocracy and ordinary citizens.

GREEK TRANSCENDENCE OF EROS STRESS: DEMOCRACY

Spartan Democracy

Instead of some among the aristocracy becoming heroes, all Spartan citizens were raised to be warrior heroes. This eliminated the opposition between aristocracy and ordinary citizens. The ideology of Sparta became oriented around the state; the individual lived and died for the state. Thus, Spartan ideology eliminated opposition between Greek individualism and Greek community. "The Spartans, who were the masters of their own bodies and appetites as well as a vast population of slaves, were the freest men of all precisely because they were the subjects of the hardest and most unyielding code."[20]

Athenian Revolution of Direct Democracy

Build up to democracy. By 600 BC Athens was confronted with two crises: their agrarian economy was failing and their military endeavors failed or at least were not accomplishing their stated goals in spite of the fact that their population and land holdings gave them the potential to be a great power. In 594 BC Solon, a descendent of an ancient Attic king, was made archon of Athens. He realized that both crises were due to the same root cause: rural impoverishment enfeebled the reserves of Attic manpower. More and more farmers were sinking into ever deeper debts and thereby into serfdom. Solon ordered a general pardoning of debts to landlords thus setting the peasantry free. He created a constitution that put forth the "middle way." Wealth rather than membership in the aristocracy was made the prerequisite of office. But the poor could be members of a *citizens' assembly* though they were not allowed to speak in it. Solon's watchword was *eunomia*, the personification of order. Laws guaranteed to the poor freedom and legal recourse against the abuses of the powerful. At the same time, laws gave to the rich exclusive right to magistracies and the running of the city. Now, thanks to Solon, even the poorest peasant could look upon himself with a new sense of self-worth: he could know himself to be as free as any of the hereditary aristocrats of Athens or other Greek city-states.

Sequence of events leading to Athenian democracy. Pisistratus seized military power and became tyrant of Athens, but he left in tact a Solon-based democracy, though he ran the city by a dictatorship of sorts. As a result of his enthusiasm and support of the tradition stemming from Homer's epics, during his reign from 546 to his death in 527 BC, the center of Greek culture moved from the Ionians in Asia Minor to Athens. When he died, his sons Hippias and Hipparchus became joint tyrants of Athens. In 513 BC Hipparchus was murdered and panic-stricken Hippias began to rule to an ever greater extent by imposing naked terror. In 512 BC Cleisthenes attacked Hippias but was defeated by him. But then in 510 BC, upon the request of Cleisthenes, Cleomenes, king of Sparta, defeated Hippias; so the tyranny was temporarily finished. Athens dramatically and unexpectedly was free.

However, the conspiratory collaboration between Cleisthenes and Cleomenes broke down. In 508 BC Isagoras, a rival nobleman was elected to the archonship. In the mean time, Cleomenes now aligned himself with Isagoras against Cleisthenes. But as this was happening, the collective social consciousness in Athens had radically changed. Citizens found themselves no longer able to commit to the authority of tyrants. Thus, in <u>508 BC</u> Cleisthenes, intuiting this

---

[20] Tom Holland. *Persian Fire* (New York: Anchor Books, 2007), p. 89.

change of attitude, proclaimed at an assembly of citizens that all citizens debate policy, vote on it, and implement it. Citizens should do this without regard to qualifications of class or wealth. Now power became invested in the *demos*; Athens became a *demokratia*. Everyone agreed that Athens was sick. Cleisthenes and his associates decided that the only hope for Athens was: "To break the mold: to harness the ambitions not only of the elite but of all the Athenian people; to create, from their energy, a future for Athens that would at last match the full measure of her potential."[21]

Then the first challenge to democracy occurred in the summer of 507 BC. A herald from Sparta, citing an ancient curse, demanded that all the aristocratic families that included Cleisthenes leave town; Cleisthenes left town. With a small bodyguard of soldiers Cleomenes came to Athens and ordered 700 families, deemed to be anti-Spartan, to be purged. Isagoras and Cleomenes began to dictate a new, non-democratic constitution. The consciousness-transformed Athenians rebelled and forced a truce. The Spartans were allowed to return to Sparta and Isagoras managed to slip away into exile, but his fellow conspirators were rounded up and put to death. The democracy was reestablished when Cleisthenes returned in triumph. Athenians celebrated a new *eunomia*, which is equality before the law and equality of participation in the running of Athens. All citizens, no matter how poor or uneducated, were free to speakin the assembly of citizens and to openly debate policy. All the Athenian people now voted to pass or not all measures and all laws. "The sponsors of the Athenian revolution were no giddy visionaries moved by shimmering notions of brotherhood with the poor, but rather hard-nosed pragmatists whose goal, quite simply, was to profit as Athenian noblemen by making their city strong."[22]

PERSIAN ATTACK TO ANNIHILATE THE GREEK ENLIGHTENMENT

Fate of Democracy in Ionia

The Persian mythical-mystical vision of universal peace in stark contrast to Greek individualism culminating in direct democracy became a violent, destructive force in Ionia. As described in Chapter 9, in 547 BC with the defeat of Croesus, King of Lydia, Cyrus sent Harpagus to Lydia who thereupon brutally subdued all the city-states of Ionia. The city-states kept their relative autonomy, but the tyrants that ruled them were under the surveillance and restriction of Persian rule. Between 512 and 511 BC Darius conquered Thrace, which is in Europe north east of Macedon that is just north of Greece. Just a few years later when the spiritual virus of direct democracy emerged in Athens in 508-507 BC, it quickly infected most Greek city-states. Between 500 – 494 BC 500 tyrannies throughout Ionia were toppled and replaced by democracies similar to the Athenian model. However, by the end of 494 BC the Persians under Darius' brother, Artaphernes leadership, totally ended Ionian resistance to the Persian vision of peace. Cities were burnt to the ground; populations annihilated. Thus ended the core ideals of the Greek Enlightenment for Ionia.

Key Dates and Events of the Persian Wars in Greece

---

[21] Tom Holland. *Persian Fire* (New York: Anchor Books, 2007), p. 132.
[22] Tom Holland. *Persian Fire* (New York: Anchor Books, 2007), p.134.

Darius' holy war. After learning of the Athenian revolution to direct democracy, Darius resolved to destroy it wherever it began to flourish. Any democracy was a stronghold of the Lie, but Athens, in particular, stood as the home of demons, *daiva*, false gods who had chosen the path of rebellion against the Lord Mazda. Not only had Ahura Mazda delivered the world into his hands, Lord Mazda had laid the sacred duty to storm and to destroy these demons. Only by burning Athens to the ground could he redeem her and her temples from the lie. For the spiritual good of the universe, the Persians must expand its empire by transforming the entire Aegean into a Persian lake. This rage against Athens rose to an even greater intensity that spilled over toward Sparta as a result of Greek sacrilege. In 491 BC, a year after the Persian conquest of Macedonia, Darius sent ambassadors on an exploratory tour of Greece with demands for earth and water. Giving them simply meant that a particular community was open to collaboration with the Persian empire. Most cities scurried to oblige. But in Athens the assembly of citizens not only dismissed Darius' demands out of hand, in blatant defiance of international law, put the ambassadors on trial, convicted them and then put them to death. The Spartans committed an even greater act of sacrilege. Instead of a trial the ambassadors were flung down a well where they were told before they drowned that "if they wanted earth and water, they could find it there."[23] If the Persians won their holy war, the fate of Athens and Sparta was preordained. Not only would the cities be burnt to the ground, women raped, Greek citizens taken as slaves, all male children would be castrated thus preventing future generations of Athenians or Spartans. Moreover, the population of the two cities would be transported to the interior of Persia thus aborting the emergence of the Greek enlightenment.

Battle of Marathon. In 490 BC the Persian generals, Datis and Artaphernes, led an expedition across the Aegean, sacked Eretria in Euboea, which is near Athens, and landed on the beach of the Bay of Marathon 26 miles north of Athens. The whole hoplite army marched away from Athens to set up camp on high ground to block the only two roads leading from the Marathon beach to Athens. In the open plains, such as that which separated the Athenians on the high grounds and the Persian's camp, no Greek army had ever succeeded in fifty years in defeating the Persians in open combat – this was because of the military efficiency of the Persian cavalry. However, the Persians aware that the Athenians had left their city unguarded by any army decided to take most of their cavalry in their ships to sail out to the Sardonic Gulf to attack and sack Athens. Once spies made Athenians aware of this, the generals voted to venture out in the open plane and attack the Persians, even though the Athenians had no archerers or cavalry and were greatly outnumbered. However, the military tactic of the phalanx combined with the social cohesion brought about by the commitment to direct democracy totally routed the Persians. Now each Greek hoplite was fighting not just for some aristocratic leader but for himself as a free person. Now it was all for one and one for all. After the battle ended before ten in the morning, 192 Greeks were killed in contrast to 6,400 Persians that perished. Realizing that the Persians fleet was sailing off to attack Athens, the Athenian army quick-marched 26 miles south to protect their families left behind. In spite of each hoplite carrying about 70 lbs of armor, shield and weapons, the army arrived in late afternoon just as the Persian ships could be seen on the horizon approaching the city. The great victory at Marathon announced to the world as well as each hoplite that all mainland Greeks motivated by the spiritual vision of individual freedom could defeat the Persians despite their superior manpower and wealth.

---

[23] Tom Holland. *Persian Fire* (New York: Anchor Books, 2007), pp. 178-179.

Other significant events and battles. Darius died in 486 BC and his son, Xerxes, became the king of Persia. Athenians discovered a rich vein of silver in 483 BC that enabled them to build a navy in 482 BC. In 481 BC Xerxes arrived in Sardis in Ionia and prepared to invade Athens. Citizens at the congress of Greek city-states committed to resist the Persian invasion. In 480 BC Xerxes with an army of about 240,000 soldiers crossed the Hellespont into Europe. Athens, after being totally evacuated by the Athenians, was occupied and burned by Xerxes. However, the Greeks achieved a great sea victory at the Battle of Salamis. In 479 BC as Xerxes leisurely retreated to Sardis, Thrace, Macedonia, Thessaly, Thebes and central Greece stayed loyal to him. Athens was occupied a second time, but at the Battle of Plataea, the Persian general, Mardonius, was killed and all but about three thousand of his soldiers were massacred. Throughout the 470s BC the Greeks conquered Persian garrisons in Thrace and around the Hellespont. The Greeks achieved even more spectacular successes during the 460s BC, all of which led to Peace being signed between Athens and Persia in 449 BC.

GREEK DEMOCRATIC ARETE

Beginning with Pisistratus (560 –527 B.C.) Athens became the center of Greek culture. By the time of Pericles (495 – 429 B.C.) the Athenian empire became a Greek citystate with a democratic constitution. By virtue of this constitution, any Greek, instead of just aristocrats, who developed pre-Socratic *arete* could aspire to some position in public administration and thereby achieve status, power, and material possessions. However, this "metaphorical, conceptual *arete*" generated insurmountable problems. Any explanatory story is imprecise and often could be interpreted in diverse ways some of which contradicted one another. Moreover, it often was possible to entertain two contradictory stories wherein each gave a "reasonable explanation" of the same set of empirical observations. There were no objective criteria by which one could decide which story is "true." This led students of the pre-Socratics, later known as sophists, to claim that all metaphorical, subjective thinking is purely a matter of subjective opinion.

The first great sophist, Protagoras (490-420 B.C.) taught that: "Man is the measure of what exists." This is quite consistent with Parmenides and Heraclitus who asserted that civilized Greeks exhibiting *arete* only *interpret* the manifestations of Being. Being in itself is unknowable and the manifestations of Being only have meaning in relation to a particular human's interpretation of them. The barbarians have subjective experiences that have meaning by virtue of some dogmatic tradition imposed on them. The Greek citizen of a city state is free to "give meaning" to his experiences via his personal *arete*. This experiential, subjective relativism of Protagoras readily leads to the Skepticism of Gorgias (485-380 B.C.). Any human does not know a world that exists; he only knows his or someone else's interpretations of sensations in response to stimuli *supposedly* generated by an objective world. Gorgias' dictum was: "Nothing exists." If the world "out there" does really exist, a human cannot know it. But, even if a human could know an aspect of the world, he could not teach this to anyone else. His knowledge is via a metaphorical, conceptual story that is open to many different interpretations.

Democratic Athens provided the context for a social solution to relativism and skepticism. The solution was the game of competitive argumentation, which was one aspect of rhetoric. When there are many different explanatory stories for the same experienced events, there are two mutually dependent conditions for selecting the "right story." The first is *arete* that formulates an explanatory story in the most reasonable, sometimes most dramatic, and most pleasing way. The second condition is one person or a group of people empowered by society to collectively decide, such as vote, for what they regard as the right story. This is the way Greek democracies dictated explanatory stories, morals, and aesthetics.

The sophists were itinerant tutors who guided students to develop personal *arete* that would enable them to be successful in city state politics, especially in Athens. When this *arete* of argumentation is guided by "reasonableness," there is some semblance of "fair play" to it, but not the same kind of fairness as in a competitive sport. In sports there are non-ambiguous rules; whereas the ambiguity of explanatory stories would allow two or more people to sincerely, that is, in a fair-minded way, disagree. What began to happen is that the Sophists subordinated *arete* to techniques that could make a story seem more reasonable than others or very reasonable, when, in fact, it was less reasonable or not at all reasonable. This degeneration of rhetoric as *arete* to rhetoric as technique led to "might of the argument makes right," amoral utilitarianism, and hedonism.

Chapter 11
SECOND STAGE OF THE GREEK ENLIGHTENMENT: RATIONAL INDIVIDUALISM

SOCRATES AND THE EMERGENCE OF LOGICAL, CONCEPTUAL KNOWING

Since 700 B.C., some members of Greek aristocracy had been evolving toward a shift in their metaphorical, conceptual thinking from an emphasis on metaphors to an emphasis on concepts. The Olympic Games was a concrete enactment of conceptual thinking that led to pre-Socratic rationalism. This rationalism rejected the moral restraint provided by Homer's Olympian gods and within about 100 years led to the relativism and skepticism taught by the sophists. Sophists from scattered Greek communities relocated to Athens where Pericles (495 – 429 B.C.) led a democratic party that was successful in overriding an aristocratic party led by Cimon and later by Thucydides (not the historian). Pericles, born to a wealthy family, was educated and guided by Sophists, e.g., Zeno, student of Parmenides, and Protagoras, to embrace the sophist's rationalism. According to Plutarch, "he was the best orator of the day, both for style and content" and thus practiced the more noble rhetoric as arete rather than rhetoric as technique.[24] Though Pericles was never elected to any public office, he led the populace to overrule the elected politicians of the aristocratic party by his power as an orator (Plato in Laws, III, 701 referred to this policy-making steered by public opinion as theatrocracy). In time Pericles "… made use of the masses against his political opponents so that he became a king disguised as a champion of the people."[25] As a result of a personal conviction to traditional virtues and to reasonableness, Pericles always strove to make Athens prosperous and powerful. He succeeded in this as well as making Athens the center of classical Greek culture. He used money collected from Greek allies, on the pretense to finance the War against the Persians, to build the Parthenon and other now famous statues and buildings. Pericles showed moderation in the use of his power as "king." Though there was some corruption during his "reign," corruption in Athens got much worse after his death.

Thus, during the reign of Pericles, there were many Athenians who rejected the tradition of Homer's Olympian gods but also saw the need for a definition of *arete* restricted and specific,

---

[24] www.e-classsics.com/pericles.htm

[25] www.e-classsics.com/pericles.htm

rather than the relative and ambiguous perspective taught by the sophists. In terms of our present vantage point, these Athenians were psychologically prepared for and saw the need for a transformation from metaphorical, conceptual thinking to exclusively logical, conceptual thinking. As so often happens in history, it took an enlightened person to show the way for making this transformation. Socrates (469-399 B.C.) created a "yoga" for such a transformation now called *Socratic dialogue*. Plato, Socrates' student, describes this "Socratic Yoga" as the process of guiding a group of students to "see" a concept latent in the mind. This latent concept would then be seen as the eternally true definition of an idea under discussion, for example, the idea of virtue.

From my postmodern perspective analogous to that of Protagoras and Gorgias, Socratic dialogue is a creative process of converting metaphorical, conceptual ideas into exclusively conceptual ideas. The metaphorical subjective aspects of an idea are filtered out or "boiled off" until all that is left is "pure concept." The definition using this concept is not absolutely or approximately true. Likewise, the concept is not an eternal idea latent in one's mind (Plato); nor is it an eternal form, nature, or essence that defines the "whatness" of a being (Aristotle). Rather Socratic dialogue is the beginning of the creative process of producing a system of categories where each category is only *nominally true*. The process of creating these categories may be carried out by a single person; for example, this is what Plato and Aristotle did. But it is essential that a group of people forming a society accept the set of categories as valid, i.e., "true" for them. This agreement requires that all members of the society will suspend their subjective perceptions whenever they apply the set of categories to understand the world and to solve concrete problems. Thus, the set of categories are said to be *objectively true*, though their "truth value" depends totally on a consensus among humans.

In any case, whether one takes the classical Greek view of eternal truths or the postmodern view of nominal truths, this transition was an enormous advance for humanity. Now, at least, the aristocracy of a society could create a consensus on a set of categories that would solve many problems. Science and mathematics depend on this approach and, more importantly, so do morality and politics. This conceptual rationalism protects society from the dogmatism of mythical-religious tradition and from the tyranny of some ruler or ruling class or the tyranny of "policy-making steered by public opinion." Socrates' enlightenment was passed on to Athenian aristocracy through Plato and Aristotle and became *moral intellectualism* present in all subsequent Greek thought in which practice was completely dependent upon theory; "knowledge in and of itself is virtue." This moral intellectualism should have saved Athens from the above described dogmatism and tyrannies. However, history quickly proved the inadequacy of this approach. Alexander the Great from Macedonia and student of Aristotle for one year conquered Athens and other Greek city-states and colonies. This ending of the classical Greek period (700 B.C. to fall of Athens) was the beginning of the Hellenic period that profoundly influenced Western and Middle Eastern cultures.

In one sense Greek *arete* that evolved from Homer's epics to the rhetoric of the sophists was replaced by conceptual thinking. However, in another sense this *arete* was transformed to what I call *validated, vision narrative*. In this new kind of *arete* an individual or a dialogue group: (1) creates a shared vision or an explanatory story, (2) converts the story into a conceptual model, and (3) creates or uses consensual criteria by which the model can be judged to be valid – it works. From the time of Socrates until today most people, including creative geniuses, do not acknowledge this kind of *arete*. Prejudice in favor of absolutely true dogmas or for exclusively conceptual thinking caused the creative process of producing rational models to be suppressed into the Individual, Collective Unconscious. Nevertheless, *arete* still is manifest in creative geniuses throughout history even though some do not acknowledge it.

GREEK, RATIONAL THINKING

Plato and Aristotle

As indicated in the description of the pre-Socratics, for Parmenides and Heraclitus Logos was a paradoxical bringing together of one's apprehension of Being in terms of its appearance and Being in itself. The paradox comes from Being as appearance is *motion*, that is, change or becoming, whereas Being in itself is pure existence, which excludes any motion. This was Parmenides' insight. Heraclitus had a similar insight but expressed it in terms of the paradox of eternal order underlying continuous change. In chapter 10, p. 99, I summarized Heraclitus' insight in terms of creativity. Being as Logos is Creativity, which involves the mutuality of chaos and order. This kind of ambiguity of pre-Socratic thought led to the skepticism of the Sophists. Socrates eliminated ambiguities by creating, via Socratic dialogue, conceptual definitions relating to questions concerning human morality. In the same tradition of eliminating ambiguities, Plato and Aristotle created a conceptual understanding of Logos in such a way that it represented Being in itself rather than just representing the appearances of Being. That is, for Plato, Logos becomes the hierarchy of eternal ideas that can be known by direct intuition. Aristotle modifies Plato's insight but retains the conceptual understanding of Logos.

Plato tells us that our experience of individual, concrete reality is an illusion; true reality is the hierarchy of ideas representing the hierarchy of forms. An individual, concrete entity only seems to be real as a result of participating in an eternal, unchanging form or idea. However, Aristotle "corrects" Plato and brings us back to earth in saying that forms are in individual beings. We can know individuals by abstracting these "forms" in concrete things into *universals* (*species equals intentional beings*) that exist in our minds (they are "given" existence by our agent intellect). Furthermore, we can make existential statements about things and their interrelations that are true because of our rational intuition of "self-evident philosophical (ontological) principles," such as the principles of efficient and final causality. Thus, with Aristotle and to a lesser extent with Plato, we still can participate in nature by rational insight, such as, rational contemplation of reality. However, now rational models have totally displaced poetic philosophy, grounded as it was in the passionate insights of metaphorical, conceptual thinking.

Now the I versus not-I (Self versus Other) has become: rational self consciousness dominating the Individual, Collective Unconscious into which feelings are repressed, and correspondingly, masculine patriarchal society dominates females and Nature considered to be feminine. A new psychic homeostasis of this newly emerged Rational, Patriarchal Self is maintained by rational models, which tend to be dogmatic and dualistic; they involve statements that are true or false, good or bad, right or wrong. The self consciousness now is primarily characterized by (1) rational knowledge leading to power over instinctual emotions and feelings and aspects of the external world and (2) repression of participatory subjectivity.

Plato and Aristotle brought closure to the question of the opposition of Being to appearance and to becoming. For Plato becoming that equals motion is the illusion, the not knowing of the ineffable meaning present in everyday language. The essential truth of being is the hierarchy of forms, which we can represent by a hierarchy of ideas. For Aristotle, motion is an aspect of material beings, which we can represent by the formula: motion is the actualization of a potency residing in a material being. We can observe such motions in order to ascertain their patterns. Any particular motion is predetermined by an efficient cause, the motion of some other body, and by a final cause. Motion is analogous to our modern concept of epigenesis. The final cause is not something out there that pulls the body toward it; rather, like epigenetic development, the moving object is predetermined toward some end, the accomplishment of which may be obstructed or otherwise modified by the conditions under which the motion takes place.

Plato acknowledges that being as *physis* is the power to appear, which leads to apprehension via the idea. But then idea has permanence only in relation to being as an eternal presence that continues to appear, as Heraclitus proclaims, governed by Logos: in this situation this way, in that situation that way which conflicts with the first. Plato chose to have Logos give rise to the permanent idea, which is the "sole and decisive interpretation of being."[26] Logos as eternal idea replaces Logos as the ever flowing gathering together and maintaining the tension of opposites. This is analogous to when an individual or society is "urged" to transform to adolescent adulthood, it needs to create dualistic thinking to stabilize self consciousness against the overwhelming archetypes of the unconscious, and rational models must replace metaphors.

Plato retains our experience of Being (as *physis*) as manifesting itself for our contemplation, but he rejects our experience of motion. According to Plato, this is an illusion embedded in the ineffable meaning in everyday speech. Aristotle makes our experience of motion rationally comprehensible by distinguishing between substantial and accidental beings. Substantial beings are substances that have being in and for themselves; whereas accidental beings are traits or phenomena that have being only by participating in the being of a substance. For example, there is no walking as such; there only is a human being who walks. Material substances have potentials for particular motions that remain in potency; or the potentials are in the process of being actualized. These substances express their enduring existence (*esse*) only by motions determined by their essence. Thus, we can contemplate changing beings via the science of physics, and we can contemplate the general principles applicable to all beings whether material or not via the science of meta-physics.

Inadequacy of Greek Rationalism

The conceptual perspective of Socrates-Plato-Aristotle is the highest level of individuation of Greek rationalism. This level of Greek individuation, hereafter called *Socratic rationalism*, had three major flaws. The first flaw has to do with the paradoxical differentiation of *arete*. As described earlier, *arete* evolved from the heroic ideal of Homer's epic poems to the subjective creativity of the Sophists, who guided students to develop rhetoric as *arete*. The paradox is that this individuation had, in the short run, both good and bad consequences for humanity. The good consequences were that this higher level of subjective creativity: (1) was a higher level of individualism with a corresponding higher level of free will and power and (2) was the necessary and sufficient condition in the context of democratic, Greek city-state politics for the transformation from metaphorical, conceptual thinking to conceptual thinking, that is, from proto-rationalism to rationalism. The bad consequences were: (1) rhetoric as *arete* and more

---

[26] M. Heidegger, *An Introduction to metaphysics*. Trans. By Ralph Manheim (New Haven: Yale Uni. Press, 1959), p. 182

especially rhetoric as technique destroyed the traditional moral structure necessary for the stability of any society, and (2) the newly emerged conceptual moral structure was in a form that could not sustain or guide a transformation of classical Greek culture. Conceptual morality rejected and replaced traditional, myth-based morality, and in so doing gave rise to the second and third flaws of Greek rationalism. The second flaw of Socratic rationalism is that rational individualism, control, and hope for creating personal happiness applied only to those who were allowed to play "the game of life," and these only included male citizens of the city-state. The Greek word for *human* also meant a male who attained sufficient development of mind-body and rational morality to be "born again" as a citizen of the polis. Women were thought of as children who lacking developed rational faculties, needed to be taken care of. All others were considered barbarians because they still were "...subject to the incursions of mana snakes and bulls and thunder-stones and monsters." However, the excluded barbarians such as the Macedonians north of Greece would conquer the city-states, build an empire, and then use Greek-enslaved teachers to put together a less provincial Hellenistic culture more suited to empires consisting of diverse conquered tribes.

The third flaw of Socratic rationalism is that it absolutized rationality. By the end of the classical period Greek rationality was grounded on the ethical choice that nature or any aspect of it can be analyzed into autonomous parts each of which is defined by characteristics stemming from its form or essence. Rules such as the principles of efficient and final causality and the rules of logic were thought to determine one's understanding of how the parts interact to produce a unified whole which describes what one can observe in the world. If one cannot analyze a thing into autonomous parts and then think of how the parts interact to produce its properties, then the "non-analyzable thing" was assumed not to exist.

For example, in the 5th century B.C. Parmenides noted that change involves something new coming into existence, but this insight makes our experience of the world paradoxical. Parmenides argued: Being is that which is, and non-being is that which is not. But this absolute dichotomy makes change problematic. If being changes, it exists in a new way and the new way was non-being before the change. Hence, before the change, a thing was both being and non-being at the same time which is a contradiction. One way out of this paradox is to deny the validity of our experience of change. This added a new twist to the evolution of patriarchal consciousness, namely, women concerned themselves with change, a world of illusion; men demanded to be cared for in this world of illusion so they could pursue objective true knowledge of eternal ideas.

All change in this material world of illusion involves motion defined as change in location in space. Zeno, Parmenides' student, set out to show that indeed, motion cannot possibly exist because it cannot be rationally analyzed. For a modified version of Zeno's argument, consider the following rational analysis of a person's motion toward a wall. First he must go half the distance toward the wall, but before he can do this, he must go half the distance toward the first halfway mark, i.e., one fourth the distance toward the wall. Of course in moving toward the one fourth mark, he must first arrive at one half this distance, that is, one eighth the distance toward the wall, and so on: 1/16, 1/32, 1/64,... the number becomes indefinitely small but never becomes any final first step toward the wall. Thus, motion when analyzed in this way can never get started. Our experience of motion is an illusion because it cannot be understood by rational analysis.

Chapter 12
FIRST PHASE OF THE THIRD STAGE OF THE GREEK ENLIGHTENMENT: GREEK-JEWISH PATRIARCHAL IDEOLOGY

EXPANSION OF CLASSICAL GREEK ENLIGHTENMENT

The word, *classical*, refers to the first two stages of the Greek enlightenment. Thus, it consisted of the emergence of Pre-Socratic humanism that evolved to Greek democratic arete that proved to be the necessary and sufficient condition for the emergence of the rationalism of Socrates, Plato, and Aristotle. The first emergence of Greek rationalism was what Heidegger called the great ending and the great beginning. It was the great ending by virtue of not only rejecting the mythological thinking of the "childhood of civilization," but also deemphasizing pre-Socratic arete involving metaphorical, conceptual thinking and correspondingly personal, subjective, participatory engagement with all of nature. At the same time, it was the great beginning of rational individualism that celebrates two types of individual freedom. On the one hand, the rational person is set *free from* domination by monarchies and dogmatic religions. On the other hand, one is *free to* evolve to create new knowledge and thereby gain more control of one's self and of nature.

Relevant Dates and Events

The first phase of the third stage , also called *Hellenism*, began in 323 BC with the death of Alexander the Great and ended in 31 BC when Rome defeated Ptolemaic Egypt at the Battle of Actavian. In 341 BC the Macedonian king, Philip II attacked several cities in northern and central Greece and established Macedonian power there. In 338 BC Philip defeated Athenians and Thebans at Chaeronea. Demosthenes (384 – 322 BC), the leading orator of Athens, negotiated a peace with Philip so that Athens remained free. Athens was not attacked but was forced to give up what was left of her empire. Philip organized the Greek city-states into the Federal League of Corinth thus ending the political independence of these cities. After Philip was assassinated in 336 BC, his son, Alexander, became king in 334 BC. That same year his army crossed the Hellespont into Asia Minor with 30,000 infantry, 5,000 cavalry, no navy, and almost no money, but nevertheless he was able to conquer this outpost of the Persian Empire. In 331 BC Alexander conquered Nineveh in Babylon and then in 330 BC, he burned Persepolis, the capital of the Persian Empire, to punctuate the ending of this empire and in revenge for Xerxes burning Athens in 480 BC. Alexander invaded the Indus Valley in India in 327 BC, but his army demanded to go home; so by spring of 324 BC he was back in the Persian gulf. Alexander caught a fever in 323 BC and died in Babylon at the age of 33. Alexander was a legendary figure for the Greeks in that he never lost a battle or failed in a siege. He conquered a vast empire with a very small army. He established over 70 cities built in the classical Greek style.

After his death, three of Alexander's generals and all of the House of Alexander were either executed or murdered. During 306-305 BC various surviving generals attempted and three succeeded in founding dynasties: 1) Ptolemy I (367 – 283 BC) in Egypt; 2) Seleucids I (358 – 280 BC) in Mesopotamia; 3) Antigonus I (382 – 301 BC) in Asia Minor and Macedon.

Macedonians

These people were of Greek stock and spoke Greek. Macedonian nobility thought themselves as Greeks; their kings admired the culture of Homer and claimed descent from the god, Hercules, and the Homeric hero, Achilles. Macedon was allowed to participate in the Olympic games as a Greek state, but its citizens were considered backward and semi-barbaric. Greco-Macedonian conquers during the Hellenanistic period became a professional class of rulers, soldiers, and merchants. Alexandria became the most prominent center of commerce and learning, but as was true of classical Greece, women played no part in public life. Hellenistic societies became stratified in contrasting urban Greeks, Macedonians, and Hellenized upper class with the bulk of native peasants who lived in poverty, and of course there were a large number of slaves.

Hellenization of Greek Rationalism Was a Great Ending – Great Beginning

Hellenization of Socratic rationalism was a great ending in that it reintroduced the importance of metaphorical, conceptual thinking. The new Hellenistic philosophies that emerged espoused a collaboration between metaphorical, conceptual thinking and logical, conceptual thinking. In effect, pre-Socratic arete involving subjective engagement with reality was resurrected though now as a partner with rational, objective knowing. This new partnership of subjective and objective knowing produced the great beginning, which was transformation of the elite in Hellenistic societies to rational individualism.

The Hellenistic great ending, great beginning occurred in the context of provincial, classical Greek thought becoming more universal correspondingly producing greater individualism. It was linked by a common language – Greek – in the old and the new city-states created by Alexander. Though all the city-states lost their independent political life, the citizens had general freedom of thought and religion. The centers of life changed from assemblies and councils to *gymnasia* (schools). Alexandria, founded by Alexander, became a research center in literature and science. Athens became a center for Greek philosophy. Plato founded the Academy in 385 BC. After the death of Aristotle, his pupil, Theophrastus, founded in 317 BC the Peripatetic school. In 307 BC Epicurus began to teach Epicureanism and Zeno of Citium, who founded Stoicism, came to Athens in 313 BC. Other important centers of Hellenistic culture included Pergamun, which only was second to Alexandria, Antioch, and Rhodes that specialized in rhetoric.

Major Hellenistic philosophies. Aristotle and Plato remained important philosophies, but whereas Aristotelianism continued to emphasize logical, conceptual thinking, Plato's philosophy became *neo-Platonism* that incorporated metaphorical, conceptual thinking. *Epicureanism* proposed that the aim of life was pleasure and the pleasure of the mind was preferable to that of the body. In *Stoicism* God and nature are the same. With respect to humans this means that every human has a divine eternal spark in him that after death would return to the divine eternal spirit. The source of human happiness was a life lived in accordance with reason. *Stoics* strove for *apathaia*, which is freedom from passion, the root cause of unhappiness. *Skepticism* also reemerged. It espoused the idea that humans cannot obtain certain true knowledge about anything. This led to cynicism that denounced conventional behavior and advocated a crude life in accordance with one's unregulated passions.

Whether Yahweh chose His people or whether a Semitic tribe invented this transcendent being to represent their new experience of life, transcendental monotheism greatly contributed to the formulation of the patriarchal ideology to be described later in this chapter. Gerda Lerner notes:

> The development of monotheism in the Book of Genesis was an enormous advance of human beings in the direction of abstract thought and the definition of universally valid symbols. It is a tragic accident of history that this advance occurred in a social setting and under circumstances which strengthened and affirmed patriarchy. Thus, the very process of symbol-making occurred in a form which marginalized women.... Here is the historic moment of the death of the Mother-Goddess and her replacement by God-the-Father...[27]

Hebraic monotheism is the control objectification of subjective participation in nature. Control objectification caused everyone in the believing community to create for himself/herself a *persona* based on the same poetic-mythological understanding of the ultimate meaning of nature and human life. In this religion, humans were understood as radically different from nature, but both were related and had intrinsic value as expressions of Divine creativity. Yahweh chose the people of Abraham rather than the other way around. IT is totally other and one cannot communicate with IT through subjective participation with nature or through man-made warrior gods. Hence all other gods and goddesses, especially the Mother-Goddess, must be rejected. Using modern terminology, all ancient pagan religions were projections of the unconscious, which prevented humans from being open to "being touched by" Yahweh, the ultimate, utterly transcendent Other.

Of course if Yahweh is the ultimate other, IT is not human personhood; IT is not human consciousness, and most certainly IT is neither male nor female. Knowledge of, communication with IT occurs only through direct mystical experience, which is granted only to a selected few. How can these chosen ones describe such experiences and instruct the rest on how to worship IT? Objectified metaphor, that is, creating metaphorical concepts, is the only way this can be done. IT is transcendent, whereas the Mother-Goddess is immanent in nature; so IT cannot be a SHE or some animal. Adult, "civilized" males are separate and have control over other males, women, children, and nature (to some extent). Thus, from a metaphorical conceptual point of view, IT could be thought of as a HE. Furthermore, IT is the creator of all things and therefore has power over all things. At this moment in history when Yahweh chose ITs people, the patriarchal family was well established in which the father had supreme power over his wife and children. Therefore, Yahweh, IT, metaphorically must be God-the-Father. BUT IT not only is all powerful and just; IT also is loving and merciful and in the most intimate union with ITs people. Sexual intercourse is a universal metaphor for mystical union. Therefore, IT also metaphorically must be mother and wife, but of course this metaphor would not be appreciated in an already patriarchal culture.

---

[27] Gerda Lerner, *The Creation of Patriarchy* (New York: Oxford Univ. Press, 1986), p. 198.

Not only who Yahweh is but how IT relates to ITS people must be understood metaphorically. In the patriarchal period of Biblical history, beginning around 1800 B.C., humans already were aware of the male role in reproduction. New life came forth from the coming together of male and female animals. In sexual intercourse males were active and females passive. It could appear that males initiated the creative act in receptive females, who then would nurture this new life to maturity. Since about 3,500 B.C. males began to have physical power over nature (especially via agriculture and more efficient hunting) and over females. With the invention of writing, males began to have spiritual power as well. That is, males gave meaning to things by giving them recorded names; males gave significance to events by writing the story of the events, and males wrote the laws that governed how humans interact in society -- the code of Hammurabi was engraved on a diorite stele in 1750 B.C. Thus, it appeared that adult males were materially and spiritually superior to females and to nature. Also a clear cut hierarchy from lowest to highest was firmly established: nonhuman animals, children, women, fathers in patriarchal families, social elite, priests, king.

Yahweh, the totally other, all-powerful, creator of all things, metaphorically would be understood to interact with humans according to the hierarchy established in societies at this period in history. With the divine breath Yahweh creates all things, but He gives males the power to name them thus putting meaning and order into the experienced chaos of nature. The all- powerful Yahweh makes the laws of nature, but He communicates these laws to male priests who write them down for humans to follow. Before the Old Testament, women, though diminished in social stature, were nevertheless equal to men before the divine mystery immanent in nature. The cult of Yahweh not only rejected the Mother-Goddess, but it also marginalized women at a transcendental level. That is, the transcendent God who created nature, granted female fecundity only through the agency of the male seed and ordained that women should be subordinated to men. On the one hand:

> Procreation, then, is clearly defined as emanating from God, who
> opens the wombs of women and blesses the seed of men. Yet,
> within this patriarchal frame of reference, the procreative role of
> wife and mother is honored.[28]

On the other hand:

> Whatever the causes, the Old Testament male priesthood
> represented a radical break with millennia of tradition and with the
> practices of neighboring peoples. This new order under the all
> powerful God proclaimed to Hebrews and to all those who took the
> Bible as their moral and religious guide that women cannot speak
> to God.[29]

Furthermore:

---

[28] Gerda Lerner, *The Creation of Patriarchy* (New York: Oxford Univ. Press, 1986), p. 188.

[29] Gerda Lerner, *The Creation of Patriarchy* (New York: Oxford Univ. Press, 1986), p. 179.

Jewish monotheism and Christianity, which built upon it, gave man [i.e., humans] a purpose and meaning in life by setting each life within a larger divine plan which unfolded so as to lead man [humans] from the Fall to redemption, from mortality to immortality, from fallen man to the Messiah.... Women's access to the purpose of God's will and to the unfolding of history is possible only through the mediation of men.[30]

Jewish monotheism was a great advance in the expression of both participatory subjectivity and control objectivity. The Mosaic law showed the Hebrews how to live harmoniously with one another and in a way that would fulfill the divine plan for them. In order for the law to be effective it had to be specific, precise, and taken to be literally and absolutely true. The Old Testament religion-mythology gave all who believed in it a sense of meaning and purpose in life, and hope for a glorious future. However, in order for this religion to be effective, the ideas had to be imprecise, vague, and often paradoxical, so as to allow each individual to experientially participate in this transcendent reality that was referred to by metaphors. Humans were admonished to enter into a participatory subjective understanding of God and nature.

In fact the Old Testament is full of stories of individuals having these kinds of personal experiences. Perhaps the most sublime exemplar of this is in the Book of Job. The traditional view at the time of Job was that if one suffered terribly, it was as punishment for transgression of the law by him or a member of his family. Job, according to the biblical story, was a just man; he obeyed the Law down to the finest detail. Yet he was forced to endure the most terrible of sufferings. In line with the literal patriarchal thinking of the time, Job kept demanding that God explain how this is just. At long last God granted Job an answer:

Then the Lord answered Job out of the tempest: Who is this whose ignorant words cloud my design in darkness? Brace yourself and stand up like a man; I will ask questions, and you shall answer. Where were you when I laid the earth's foundations? Tell me, if you know and understand. Who settled its dimensions?... [and so on for several verses]
Then Job answered the Lord: I know that thou canst do all things and that no purpose is beyond thee. But I have spoken of great things which I have not understood, things too wonderful for me to know. I knew of thee then only by report [i.e., via control objectification] but now I see thee with my own eyes. [I have a personal, participatory subjective experience of the Lord.] Therefore, I melt away. I repent in dust and ashes.[31]

Literal, objective thinking is not the way to communicate with God, but if one can give up literal thinking, then he may be open to be "touched by God."

---

[30] Gerda Lerner, *The Creation of Patriarchy* (New York: Oxford Univ. Press, 1986), p.201.

[31] The New English Bible (New York: Oxford Uni. Press., 1976), pp. 563-568.

Ideally Jewish primitive science consisting primarily of Mosaic Law was subordinated to Jewish primitive ecology, which consisted primarily of the religion-mythology of the Old Testament. In fact, however, control became the primary value. As a result many chose to interpret the religion-mythology as literally and objectively true. This gave priests control over the masses and gave men absolute control over women. This control, however, was a total block to mystical experiences of God.

Impact of Hellenization on Jews. As a result of intense dialogue with Hellenistic culture, the Hebraic version of the Old Testament was translated into Greek to become the Septuagint Hebrew scripture (the oldest Greek version of the Old Testament traditionally said to have been translated by 70 or 72 Jewish scholars at the request of Ptolemy II). Jewish scholars began adopting the literary forms of the Greek tradition. All this occurred during the *Hellenistic Diaspora*, which was a voluntary movement of Jews into Hellenistic kingdoms, especially Ptolemaic Egypt, Alexandria in particular. This caused Jews to be caught between two extremes: Mosaic Law and Jewish traditions and Hellenistic values. By in large Jews stayed committed to their tradition, but they blended in to the culture in Alexandria by emphasizing common values and lowalty to the monarch.

Mystery Cults

The reemergence of metaphorical, conceptual thinking in Hellenistic Greek rationalism provided a spiritual, holistic vision about the meaning of life, but this satisfied only a portion of the elite. For most people in the Hellenistic world, Greek philosophy, even when modified by participatory, subjective engagement, produced a great spiritual-religious vacuum. There were three major ancient mystery cults that attempted to fill this spiritual vacuum: 1) Dionysian Mysteries; 2) Eleusinian Mysteries; and Cult of Isis, which was related to the Eleusinian Mysteries in its celebration of the mysteries of death and resurrection of Osiris. These cults set the stage for the emergence to dominance of the Jewish, Jesus Christ Mystery cult.

Dionysian Mysteries. The basic mysteries took form in Minoan Crete between 3,000 and 1,000 BC. The rites were based on the seasonal death-rebirth theme and also were associated with spirit possession. The rites involved a primitive liberation from the constraints of civilization and its rules. The Dionysian devotees celebrated all that was outside civilized society and thus provided an escape from the dictates of personas. In terms of the ideas presented in chapter 7, irruptions from the individual, collective unconscious overwhelmed the individual mind self. Participants in these rituals originally were outsiders of society such as women, slaves, foreigners and other non-citizens in Greek democracies. Everyone in the cult was considered equal. In Greece in the late 500s and 400s BC, the Dionysian rites were almost entirely associated with women. The god, Dionysus, is similar to Shiva, "the destroyer" in Indian culture. Bacchus, the god of wine and exuberant pleasure, was another name for Dionysus. Heraclitus, who proclaimed the paradox of Logos as both Chaos and Order, see chapter 10, figure 10.1. was closely associated with the Mysteries. This cult seeded the emergence of acting and theatre, especially tragedy and comedy. During the classical period in Athens and Attica, the Dionysian rites evolved into a great drama festival. Dionysus became the god of acting, music, and all poetic inspiration, and Mysteries evolved into casting off personas and creating a more individualistic, genuine character.

Eleusinian Mysteries. These mysteries are associated with the myth of Demeter and Persephone as recounted in one of Homer's Hymns (650 BC) in which Hades, the god of death and the underworld seized Persephone, the daughter of Demeter, the goddess of agriculture and fertility. Eventually Zeus allows Demeter to reunite with her daughter, but Persephone could not avoid returning to the underworld for part of the year. The Mysteries pre-dated Homer's poetry in that they began in Mycenaean Greece around 1,500 BC. The ceremonies were held every year at Eleusis where people from Athens could walk in procession along what was called the "Sacred Way." Under Pisistratus, tyrant of Athens from 546 to when he died in 527 BC, the Mysteries became pan-Hellenic. The so-called "Lesser Mysteries" were usually held in March. These Mysteries evolved to occult practices that signified the miseries of the soul in subjection to the body. The Greater Mysteries always took place in late summer. Celebrants professed that when the soul is purified from defilement of matter, the individual person could intuit intellectual, spiritual visions. "According to Plato, 'the ultimate design of the Mysteries ... was to lead us back to principles from which we descended [de-evolution to] ... a perfect enjoyment of intellectual good.'"[32] People who were initiated into the Mysteries were forbidden to talk about them under penalty of death. They thought of themselves as blessed in that they knew the beginning and the end of life and would be happy in this life and in the next.

DIFFERENTIATION OF GREEK RATIONAL INDIVIDUALISM

Two Types of Rational Individualism

Evolution to the subordination of metaphorical, conceptual thinking to logical, conceptual thinking in Judaism and in pagan Hellenistic culture led to the emergence of rational individualism. This new rational individuality presents each human (and each human society) with two new options. Firstly, he/she can choose to exert greater control of self-expression and of nature by differentiating individualistic, control consciousness manifested as control individualism with control objectivity. Secondly, he/she can choose to give himself/herself over to participatory awareness of self and of nature by differentiating individualistic, participatory consciousness manifested as participatory individualism with participatory subjectivity. Control consciousness involving reason opposes participatory consciousness involving feelings.

Greek Adolescent Mentality

Some aristocratic male citizens in Greek city-states during the classical period (before 323 BC) transformed from the "childhood of civilization" to an adolescent mentality. The emergence of Pre-Socratic humanism (see chapter 11) represents the progressive death phase of childhood and the emergence of proto-rational mentality. The emergence of Socratic rationalism is the rebirth to adolescence. In classical Greece, women, slaves, and all non-Greeks were thought to be like children who are not capable of transforming to what we now call adolescence.

---

[32] Thomas Taylor. *Eleusinian and Bacchic Mysteries.* Lightening source Publishers, 1997, p. 49; http://en.wikipedia.org/wiki/EleusianMysteries

Transition from Childhood to Adolescence

Emerging guiding role of mind. The transition from Childhood to Adolescence exemplifies the general feature of all creativity, that is that is, transformation as a Life, Death, Rebirth process, but now the MIND plays a major role in guiding this transformation.

In modern Western societies adolescence begins between eleven and fifteen years of age, when the mental-ego dissociates from literal, metaphorical, conceptual thinking (defined by rules and social roles) and begins what Piaget calls formal operational thinking. One can operate on one's own thought, that is, mentally manipulate verbal concepts, e.g., learn geometry or algebra, as well as work on physical objects.

> By the time of adolescence, the formal-reflexive mind begins to emerge. This level is ... the first structure [of consciousness] clearly capable of sustained self-reflection and introspection. The self identified with the reflexive mind would thus be involved in conscientious and self-inquiring modes of awareness and behavior. It would have the capacity to question conventional mores... No longer bound to conformity needs, the self would have to rely more on its own conscience, or its inner capacity to formally reflect and establish rationally what might be the good, the true, and the beautiful, at least in its own case.[33]

Thus, this is the period when rational, psychological factors begin to eclipse biological determinants of behavior.

Transition As a Life, Death, Rebirth Process.

Life. The child is the center of his literal mythical world he has created to represent the real world. Also the child is "god-like" by virtue of his relative innocence and naiveté and his ties with nature. Self consciousness is not yet strong enough to bring total separation of the child from its inner world of fantasy and imagination and from some sense of belonging to nature. Probably because of this, children usually get along well with animals and can communicate with one another in ways closed to adults. The child's self-centeredness is further enhanced by the nurturing of mother and the solicitude of others. The child tends to draw forth unconditional love, that is, the child is lovable just by being a child rather than for doing anything. Also, though greatly separated from *participation mystique,* the child still can take delight in the inner world of creative imagination and fantasy.

Death. The mythical world of childhood begins to crumble when the child is drawn into the "objective" world in which he is not the center. The child is forced to acknowledge the awesome power of his inner world that not only can delight but also overwhelm and terrify the evolving self consciousness of the mental self busy at creating a rational representation of the outer world. From time to time, intense feelings degenerate into raw emotions that overwhelm the self. This, in effect, draws the mind self back into the psychic energy field of the body ego

---

[33] K. Wilber, *Eye to Eye The Quest for the New Paradigm* (Garden City, N.Y.: Anchor Press/Doubleday, 1983), p. 286.

and brings about a kind of death to the mind self. Also, while some fantasy creatures are delightful and friendly, others are terrible and frightening. The child's increased independence and power puts him into situations where inanimate objects frustrate him by not being the way he thought they would be or should be. More importantly, other children as well as adults, such as parents, teachers, older brothers and sisters, make demands that go against the child's desires. The child must begin to give up his free access to the inner world of fantasy in order to begin to control the expression of feelings and at least prevent them from degenerating into raw emotions. Figuratively speaking, the child must "kill" his/her creative imagination in order to adapt to the demands of the objective world. The child's greater cognitive abilities and abstract knowledge leads to a loss of innocence and naiveté. The child must acknowledge good/bad, right/wrong, true/false. The rational code of behavior that he/she learns to understand and accept produces a sense of guilt and inadequacy. This in turn tends to make the child more self-conscious. The child may become shy and in need of reassurance of his intrinsic worth, like what he/she had when he was loved unconditionally for his "god-likeness."

Rebirth. "Killing" creative imagination and shutting out the world of fantasy is a definitive victory for the evolving mind self. By this act the mind self manifesting as self consciousness has taken center stage and its light has eclipsed the dark forces of the Individual, Collective Unconscious. The reborn adolescent has gained some power as a result of increased mental self-consciousness and more control over the expression of feelings. This enables the adolescent to be more aware of the demands of the external world, especially his/her social world, and to adapt to these demands. As the mental self internalizes more abstract knowledge, the young boy or girl attains some ability to recognize and choose good over evil, right over wrong, and truth over falsehood. This increased mental power produces greater self-confidence. The brightness of the ascending mental self reduces fear projected onto the outer world and deflates the emotional energies of the inner fantasy world. The mythical order of childhood is replaced by mental Ego-self's acceptance of the dogmatic rational order taught to the adolescent by human society.

Greek Adolescent Adulthood

Some educated males in cities and city-states of the Hellenistic world gradually transformed from an adolescent mentality (involving Socratic rationalism, see chapter 11 and previous section in this chapter) to an adolescent adulthood mentality.

Transition from Adolescence to Adolescent Adulthood

Throughout most of the history of civilization, puberty was the dark tunnel that led to full adult status in society. In modern scientific, industrial societies, puberty overlaps the hero battle of adolescents and the struggle of the victorious adolescent coming to a preliminary adulthood, which I call *adolescent adulthood.* Full adulthood requires many more years of preparation, and especially since the 1960s, it has become very diverse. Adulthood now depends on one's life situation, and in general, maturity comes after a long series of transformations rather than just one major transformation as in former times.

Puberty. Biological evolution from simple to complex organisms shows a corresponding increase in diversity. The same trend also is present in individual human developmental-evolution. When biological factors dominate human evolution, there are many similarities among all individuals because each young proto-individual is a variation on the biological theme of *Homo sapiens*. Thus, all fetus development is pretty much alike, and all infants and even children have many similar characteristics and similar problems. However, in the adolescence stage when rational psychic factors begin to predominate, there begins a diffraction of the evolutionary process, like white light passing through a prism to become many different colors. The self is the new co-creator of the emerging human psyche, which is the prism that differentiates the evolutionary impulse into diverse personality types. Still, adolescence is manageably stereotyped to allow descriptions of and guidelines for what should occur during this period. These developmental guidelines are particularly standardized in societies that have a clear idea of what it means to be an adult. Like any developmental process, one stage, adolescence, should prepare the individual to enter the next stage, in this case, adulthood.

Despite the beginning of mental self ascendancy during adolescence, Mother Nature still dominates the time of onset and early phases of the transition to adulthood. Puberty is Mother Nature's last fling, as it were, at orchestrating individual human evolution. Again SHE appears to be paradoxically ambivalent in her goals. On the one hand, SHE plays such havoc in the body of the adolescent that the mental self is catapulted into the situation of either taking full charge of further psychic growth or suffering some form of breakdown or regression. In a manner of speaking Mother Nature challenges the individual: "Grow up or else die." Out of the desperate human need for order and stability in life, that is, psychic homeostasis, most individuals in stable societies choose to suffer the anguish and to do the hard work necessary to fashion some sort of adult personality that is acceptable to society. Thus, by forcing the issue, Mother Nature propels the adolescent into adulthood. On the other hand, SHE steadfastly resists and rebels against the ascendancy of the mental self displacing the body ego that metaphorically represents Mother nature. Even after giving way to the reign of reason, Mother Nature in the form of intense feelings rises up to overwhelm mental self consciousness from time to time. Alternatively, SHE becomes the unacknowledged demonic forces in each human being that if not progressively integrated into the mental self will drive the individual insane or to despair and/or to psychosomatic illness.

Normal chaos. The skirmishes of childhood which became periodic battles of adolescence now escalate into all out war during puberty. The emergence/creation of a persona during late childhood produces opposition between the demands of metaphorical, conceptual thinking and the demands of feeling. The child wants to do one activity and must learn to do just the opposite, for example, it wants to yell and scream but must learn to be quiet. These skirmishes between reason and feelings intensify during adolescence, but society helps reason win over feelings by imposing external law and order. Lack of this externally imposed regulation produces the disaster of a spoiled child. At the time of puberty, though social structure may be there to help, the individual to a significant degree is thrown on his own. Human feeling awareness becomes more closely associated with the repressed body ego and individual consciousness becomes associated with the mental self, which now becomes preoccupied with rational learning. The opposition between feelings and rationality is expressed by the integrated triune brain of the body ego becoming alienated from the triune mind. That is, the mind self now has: (1) action mind center analogue to the *emotional brain*, (2) expressive mind center analogue to the *feeling brain*, and (3) thinking mind center analogue to the *perceptual learning brain*. The triune mind of the mind self functions as a result of psychic energy coming up from the triune brain of the repressed body ego, but the triune mind is not in harmony with the functioning of the triune brain. Moreover, each of the three minds of the mind self is alienated from one another; there is no integrated triune mind. The integrated triune brain of the repressed body ego is in opposition to the non-integrated triune mind of the mind self, and each attempts to subdue the other, in order to rule the whole animal system. The trouble is, this opposition that becomes war during puberty is a stalemate; neither can totally vanquish the other. Sometimes perceptual feelings brought up by the triune brain reign supreme, while at other times rational learning of the triune mind prevails. The polar self expressing the mental process of imagination sometimes is co-opted by perceptual feelings and at other times is co-opted by metaphorical conceptual or rational knowing. All too often both perceptual systems fight each other to a standoff, leaving the individual paralyzed in indecision. At these times the adolescent bursting with emotional energy, says he's tired and wants to sleep all the time. Girls in the United States culture may temporarily relieve the inner tension by fits of crying alternating with deep depression that may last all day or days.

War is cruel and one of nature's supreme laws still applies: Only the fit will survive. Puberty always produces some casualties. Everyone must endure the suffering of this war to some extent. In the end, most will come through as transformed persons, but there always will be war casualties. Some adolescents will go insane; others will regress to childhood, and some will commit suicide.

Transition as a Life, Death, Rebirth Process

Life. The adolescent, though still pressured not to be egocentric, nevertheless usually is taken care of by parents or parent equivalents. Usually the adolescent does not have the responsibilities of adult life. He/she does not have to earn a living, develop a career, or formulate an adult sex role and parent role. Therefore, the adolescent still is somewhat tied to the participation mystique carried over from childhood.

Death. Puberty for many adolescents finalizes the split from the participation mystique. As a result of this split, the mental self associated with rationality is set up to dominate feeling awareness associated with the participation mystique. This creates a war within each individual, which produces chaos in overall human consciousness until some higher collaboration is established. The mental self represses the feeling aspect of human consciousness so that it can control how and when particular feelings will be expressed. Thus, participatory, feeling awareness is almost totally shut out from the mental self. Therefore, the usual mode of human consciousness is identical to the mental self consciousness, and all the other modes of human consciousness are said to be in the Individual Collective Unconscious. The unconscious is not the same as non-consciousness of lower animals; rather it contains all modes of human awareness, for example, pure feeling awareness, which are excluded from the mental self consciousness. Thus, the inner war of puberty is the conflict between the mental self and the Individual, Collective Unconscious, including most aspects of the magic, feeling-mythical, and imaginative self-consciousness. This Unconscious now also includes the experiential aspects of metaphorical, conceptual thinking. The rational person often is unaware that he/she is using metaphorical concepts. At first, this inner war is a stalemate. The mental self pervades human consciousness but from time to time is overwhelmed by outbursts from the Unconscious. From a behavioral point of view, the proto-adult behaves rationally and is in total control of himself/herself, but he/she sometimes behaves irrationally and is out of control. Alternatively, the individual withdraws from life for fear of behaving irrationally and out of control.

Rebirth. In being reborn to a higher psychic homeostasis, the adolescent adult fragments aspects of the Individual, Collective Unconscious and then integrates some of these "unconscious pieces" into the mental self. One of the major fragmentations has to do with sexual feelings. Each person divides all things, behaviors, ideas associated with sexual feelings in the Individual, Collective Unconscious into two sets called masculine and feminine traits, respectively. Males attempt to integrate masculine traits with the mental self, thus producing a masculine gender commitment. Female traits, designated as feminine, are held in the Unconscious. Likewise, females develop a feminine gender commitment and hold masculine traits in the Unconscious. In like manner, both males and females integrate socially acceptable behaviors, feelings, and ideas with the mental self thus producing a *mental persona* (mask). The culturally defined persona is what enables diverse personalities to live together and get along with one another in society. Socially unacceptable behaviors, feelings, and ideas are held in the Unconscious. All these unconscious traits are sometimes referred to as one's *shadow*.

Thus, the mental self occupies center stage as a result of a partial victory over and a partial truce with the Individual Collective Unconscious. The victory comes from fragmenting some aspects of the Unconscious, thus forestalling the full force of feeling awareness overwhelming the mental self. Without such fragmentation, the individual would remain in the chaos of feelings that eventually would lead to insanity and/or suicide. The truce comes from allowing some aspects of the Unconscious to be expressed in the mental self. Even though the rationality of the mental self sets up a new psychic homeostasis, feelings color all aspects of this mental state. No human ever is merely rational. Both the victory and the truce are partial because the adolescent adult still experiences inner war between reason and feelings expressing the more fundamental battle between the mental self and the Individual Collective Unconscious. From time to time feelings overwhelm reason; elements of the Unconscious invade the mental self and produce temporary chaos.

The adolescent adult also may fully participate in cultural rituals. These rituals use symbols that connect rational knowledge with feeling values of the Individual Collective Unconscious. For example, for believing Christians, the cross is a symbol that connects a rational theology, such as, Thomistic theology, with religious feelings about the meaning of life. Competitive sports are rituals to the competitors and symbols to the observers. Sports connect rationally controlled behavior with intense aggressive feelings. All these symbols help buffer the inherent opposition between feelings and reason. The emergence and evolution of the mind self produces a suppression of one's earth nature (the Inner Self, which is SOURCE manifesting soul and the Individual, Collective Unconscious), but this culturally determined synthesis of adulthood prevents an all out war between autonomous reason and autonomous imaginative feeling.

Emergence of the Greek-Jewish Patriarchal Ideology

The "Great Ending" equal to adolescent rationality led to a retreat from engagement with the world as expressed in nihilistic philosophies that included cynicism. This corresponds to the death phase of the transition to adolescent adulthood. The rebirth phase was expressed by the Hellenistic philosophies: Epicureanism, Stoicism and neo-Platonism. The rebirth also was expressed by the *Greek-Jewish patriarchal ideology.*

One may obtain an understanding this patriarchal ideology by looking at control objectivity associated with masculine persona versus. participatory subjectivity associated with feminine persona.

Table 12.1 Masculinity versus Femininity

MASCULINE PERSONA CONTROL OBJECTIVITY	FEMININE PERSONA PARTICIPATORY SUBJECTIVITY
1. Control objectivity: objective knowing that gives one control over self and nature	1. Participatory subjectivity: subjective knowing involving insights, feelings that may lead to engagement with another person or with nature
2a. Possibility of approximately or absolutely true knowledge	2a.No absolutely true knowledge, rather all knowing depends on context and is either valid or not valid based on agreed upon criteria
2b. Mechanistic knowing in which any whole system is literally equal to the sum of its autonomous parts.	2b. Any system is a whole which must be understood in terms of a subjective intuition. Then, the parts are not autonomous but are interconnected to one another. Also a knowledge mutuality: the system is understood in terms of its parts and each part is understood in terms of an intuitive understanding of the whole
3.Human relationships are maintained by abstract principles of morality that define justice and duty, for example, in the relationship between a man and a woman, the man should dominate at all times and in marriage, the man should be ruler of the family.	3. Human relationships are sustained by feeling bonding, love, and mutuality, such as., mutual understanding and agreement about the role and duties of each person in the relationship.
4. Structure and order are viewed as good; chaos is viewed as bad. If a system goes into partial chaos, every attempt should be made to reestablish the old order. Thus commitment to maintained order prevents creative change.	4. In some circumstances maintained structure and order is bad and chaos is good. Chaos provides the opportunity to transform to a new order, i.e., undergo a creative change. For example, subordination to England in the 1700s was viewed as bad and the Declaration of Independence which brought the chaos of the revolutionary war was good.

5.Human maturity is identified with developing rationality and learning rational knowledge. Alternatively, one learns how to solve problems and achieve utilitarian goals based on learned rational knowledge.	5. Human maturity is identified with developing subjective knowing that can be the basis of creativity. Alternatively, one via introspection, understands his/her feelings and express one's inner – "so-called true" – self.
6. One chooses rational, abstract spirituality, for example, dogmatic religions that emphasize the rule of law.	6. One chooses subjective, feeling spirituality, for example, religions that accommodate subjective, feeling insights.
7. Because of a patriarchal culture, humans view themselves as autonomous and radically different from all other life forms. Also humans feel justified in exploiting Nature; they fail to realize how interdependent humans and the rest of Nature is.	7. Humans view themselves as not autonomous but interdependent with all other life forms. As a result, humans participate in mother Nature and revere her.
8 Human maturity is defined as one understanding and then obeying the dictates of masculine and feminine persona.	8. Human maturity is defined as conforming to personas in various circumstances but not identifying with any persona. This means that sometimes a mature person will choose not to conform to a persona.
9. Civilized males or masculinized females are viewed as superior to and rightfully dominate uncivilized males, feminine females, and children. The terms, civilized, masculine, and feminine are defined by one's culture.	9.There is an ethic of caring where all humans are viewed as members of a community.

Expanded Patriarchal Perspective

An expanded version of the patriarchal perspective indicates how it resolves the conflicts of civilized individualism. The patriarchal perspective is a socially constructed perspective in which: (1) the mind self represses/dominates the Individual, Collection Unconscious that includes the magic consciousness of the action self, the subjective feelings of the body ego, and the imagination of the polar self, (2) the collective social consciousness represses/dominates individual self-consciousness, (3) within the mind self, the thinking and action mind centers which produce conceptual, objective thinking & action repress/dominate the expressive mind center which produces non-conceptual, subjective knowing, and (4) Eros-order associated with Will-to-control represses/dominates Eros-chaos associated with Will-to-openness as well as sometimes dominating/repressing Eros-chaos associated with Will-to-Egoness.

Chapter 13
SECOND PHASE OF THE THIRD STAGE OF THE GREEK ENLIGHTENMENT: EVOLUTION TO JUDEO-CHRISTIAN RATIONALISM

This evolutionary aspect of Hellenization occurred after the founding of the Roman Empire in 27 BC and lasted until the fall of Rome in 476 AD. It involved evolution to adult rational individualism that paradoxically opposes mysticism but provides the opportunity for the emergence of mystical, heroic creativity.

ADULT RATIONALITY AND MYSTICISM IN THE HELLENISTIC WORLD

Roman History

Events leading to the founding of the Roman Empire. During the 900s BC, a small cluster of huts developed into the city-state of Rome that at first was ruled by kings. In 750 BC Greeks began to establish colonies in southern Italy and Sicily many of which became city-states. The Corinthians established a colony in 743 BC that eventually became a city-state at Syracuse in Sicily that rivaled Athens as the largest and most beautiful city in the Greek world. Rome became a Republic in 509 BC ruled by *patricians* who were a privileged class of Roman citizens whose status was a birthright. Note: Athens became a direct democracy in 508 – 507 BC. The poorer Roman citizens known as *plebeians* withdrew from the city-state, formed their own assembly, elected their own officers, and set up their own religious cults. By 287 BC the wealthier land-rich plebeians achieved political equality with the patricians. This produced a noble ruling class in which there was a unique power-sharing partnership between plebeians and patricians. As a result of piecemeal conflict and diplomacy, the Republic of Rome acquired all of Italy, Sicily, Sardinia Spain, Africa, Macedonia, Achaea, Asia, Cilicin, Gaul, Cyrene, Bethynia, Crete, Pontus, Syria, and Cyprus. This extensive expansion generated social breakdown and political turmoil. In 146 BC Romans sacked Corinth and dissolved the Achaean Confederacy so that now Greece was ruled by Rome. The Roman general, Sulla, sacked Athens in 86 BC. After a series of long and bloody civil wars, Caesar became in 49 BC the dictator of Rome. In 44 BC the Republicans, Cassius and Brutus murdered Caesar. Mark Anthony, Octavian, and Lepidus formed a triumvirate in 43 – 42 BC and defeated the Republicans led by Cassias and Brutus at Philippi in eastern Macedonia. Then Augustus Octavian in 32 – 31 BC defeated Mark Anthony in a naval battle at Actium. Mark Anthony committed suicide in 30 BC just before Augustus entered Alexandria. Representing Rome he annexed Egypt and in 27 BC was established as emperor of Rome; he died in 14 AD.

Pax Romana. Augustus reestablished political and social stability that launched about 200 years of prosperity called *Pax Romana*. The empire flourished and added new territories, in particular, ancient Britain, Arabia, and present-day Romania. Rome developed into the social, economic, and cultural capital of the Mediterranean world. Hellenistic civilization was transplanted to Italy, Gaul, and Spain. Latin replaced Greek as the prevailing language: Lucretius (55 BC), Cicero (43 BC), and Virgil (19 BC) developed Latin into a language capable of expressing Greek philosophy, rhetoric, and poetry. The emerging new Roman culture helped perpetuate art, literature, and philosophy of the Greeks, the religious and ethical systems of the

Jews, the new religion of the Christians, Babylonian astronomy and astrology, and cultural elements from Persia, Egypt and other eastern civilizations. At the same time, it allowed people of many different cultures to retain their heritage. The provinces of the empire were organized into a network of city-states, which came to be dominated by local land-owners. This eventually produced a great gap between the wealthy few educated in Greek-Roman culture and the peasants who titled the soil and paid land-owners rent. In 193 AD civil wars and barbarian invasions began to ravage the empire.

   Major events leading to the downfall of the Roman Empire. Before 67 AD St. Paul preached in several cities. Members of the Jesus cult as well as others who believed in the Christ-event began referring to themselves as *Christians.* The future emperor, Titus (79 – 81 AD) captured Jerusalem in 70 AD and destroyed the Temple there. During the 100s AD, traditional polytheism began to decline. The newly emerged Jesus cult as well as Eastern cults involving death-rebirth mysteries for individual believers became more popular. The new religions stimulated philosophical speculation, especially Neo-Platonism, about monotheism. In 212 AD the Roman emperor, Caracalla (211 – 217 AD) granted full citizenship to all inhabitants of the Roman empire. In the years 257 – 263 AD, the Goths raided Greece and Asia Minor, and in 267 AD sacked Athens, Corinth, Sparta, and Argos. The emperor, Constantine, signed in 313 AD an edict that permitted freedom of worship thus facilitating the spread of Christianity. In 330 AD Constantine established Constantinople as the "New Rome" that became the capital of the Roman empire. The Visigoths defeated the Eastern emperor, Valens, in 378 AD. This battle was seen as a watershed in the decline of the Roman army and its ability to stem barbarian invasions. In 399 AD emperor Theodosius (379 – 395 AD) ordered the closing of pagan temples. After he died in 395 AD, the Roman Empire was again divided into East and West, each ruled by one of his two sons. Alaric, the Visigoth, sacked Rome in 410 AD. At about 426 AD the Olympic games closed when the Temple of Zeus at Olympia was destroyed by fire. The year, 476 AD, marks the fall of the Roman Empire when the German, Odovacer, deposed Romulus Augustus, the last Roman emperor in the West, and became the first barbarian king of Italy. In 493 AD Theodaric the Great (493 – 526 AD) overthrew Odovacer and established the Ostrogothic kingdom in Italy.

Greek-Roman Full Adulthood.

As indicated earlier, in Greek-Roman culture the elite educated minority developed Greek philosophy, especially the Neo-Platonism of Plotinus, and its application to religion and ethics. This greatly intensified the Greek-Jewish patriarchal ideology that in turn influenced all aspects of Hellenistic culture. The expansion of the patriarchal ideology produced the transformation from adolescent adulthood to full, adult, control individualism.

Transition to Full, Adult, Control Individualism.

Attaining full adulthood in modern societies equated with adult control individualism is usually associated with some kind of formal training, though education by no means guarantees adult maturity. It still is unfortunately true that most people in patriarchal societies and in postmodern societies do not acknowledge adult participatory individualism. *The rest of this section will focus on control individualism.* The essence of full adulthood is that the individual has a well-developed mental self and corresponding mental power. In modern societies this development comes from learning rational knowledge and mental skills and applying these knowledges and skills to solving problems. Theoretically and often but not always, the full adult is acknowledged by society by having a high status job that pays enough money to live comfortably. Adulthood in modern scientific, technological societies has positive and negative aspects. On the positive side adulthood implies: (1) well developed rational powers, (2) mind self-control, (3) partial control of one's life situation, (4) radical individualism subordinated to rational rules of various institutions of society, such as formal religion, political system, family, work. On the negative side, adulthood implies a radical separation of reason and feelings and a radical suppression of participatory consciousness. This separation and repression would lead to an internal conflict much more terrible than the puberty war if it were not for the buffering effect of various cultural symbols. Alternatively, some self-destructive behaviors, for example , workaholism and other addictions may keep one from acknowledging an inner war. Psychosomatic illnesses also are manifestations of this unacknowledged inner war. Another negative aspect is that the adult does not have personal individualized wisdom. Either he accepts the predigested transcendental truths that society provides him, or the person avoids thinking about these realities altogether.

Wilber describes what I call mature control individualism as *late formop* (late formal operational thinking). He is worth quoting, since he emphasizes some aspects I have referred to only indirectly.

> At ... late formop, the person becomes capable of hypothetico-deductive awareness (what if, as if), and reality is conceived in terms of *relativity* and *interrelationships* (ecology and relativity, in the broadest sense,...). The self [mental self identity] is viewed as a postulate lending unity and integrity to personality, experience, and behavior (this is the mature self).
>
> But,... development can take a cynical turn at this stage. Instead of being the principle lending unity and integrity to experience and behavior, the self is simply *identified with experience and behavior*. In the cynical behavioristic turn of this stage, the person is a cybernetic system guided to fulfillment of its material wants. At this level, radical emphasis on seeing everything within a relativistic or subjective frame of reference leaves the person close to a solipsistic position.
>
> The world is seen as a great relativistic cybernetic system, so relativistic and holistic that it leaves no room for the actual subject in the objective network. The self therefore hovers above reality, disengaged, disenchanted, disembodied.... And this is precisely ... the fundamental Enlightenment paradigm: a perfectly holistic world that leaves a perfectly atomistic self.[34]

Individual maturity reflects the cultural evolutionary stage of the society of which one is a member. Modern maturity in postmodern societies like the United States reflects all the good aspects of a scientific, technological society, but it also reflects societies in deep trouble as

---

[34] K. Wilber, *Sex, Ecology,* Spirituality (Boston: Shambhala, 1995), pp 261-262

signified by the current sociological and ecological crises. Not only do these societies not recognize participatory individualism, they suffer from a drastic decrease in the number of people who choose to develop this type of maturity. The fall-out from this is that our young people are not cared for in our families or in our schools.

The Christ-Event that Transcended the Patriarchal Ideology

The Christ-event is an historical moment related to a man called Jesus. At the age of thirty he began preaching that not only was he the fulfillment of the Old Testament's promise of a redeemer for all humans; he modified the teaching of the Old Testament to be incorporated into a new vision for all humans. This enraged the patriarchal leaders of the Jews who demanded and succeeded in having the Romans crucify him. During his preaching in the last three years of his life, Jesus attracted hundreds of followers and twelve disciples. Three days after the death of Jesus, eleven of the twelve disciples and several of his followers proclaimed that Jesus had resurrected as the Christ who in some manner provided for the salvation of all humans. Jesus did not come back from the dead, but rather he transformed to a new form that manifested God as the Christ that from all eternity lived in the Inner Self of all humans. Yet, for a time until "his ascension into Heaven" Jesus, the Christ, was experienced by some of his followers as present in the external world. Some time after his death the followers of Jesus became energized to form a Jesus cult that evolved to the world religion known as Christianity. Then and now there are diverse interpretations of the sermons of Jesus, but the core "truth" of this religion is that the life, death, and resurrection of Jesus as the Christ provides for the salvation of all humans. This central theme of the Christ-event transcends for each and all humans the alienation between the external world and the world of the Inner Self. Then and now there are diverse interpretations of the way "salvation" and corresponding transformation occurs.

St. Paul's Enlightenment.

Saul, born to a wealthy Jewish family of the tribe of Benjamin in the city of Tarsus in Cilicia (Asia Minor) was therefore also a citizen of the Roman Empire by birth. Saul moved to Jerusalem to study Old Testament law and become a Pharisee. The Middle East in general and Jerusalem in particular was the cross road for interaction between Hellenic and Jewish cultures. Thus, Saul incorporated aspects of these traditions into his personal perspective, which initially was very anti-Jesus. From his Hellenistic perspective the Christ-story was irrational and anti-intellectual. Athenian Greeks would laugh at the notion of a mere carpenter rising from the dead, talking to people on this earth, and then ascending into Heaven to be the God-Son reunited to God the Father. From his Jewish perspective the Christ-story is blasphemy. There is only one God; Christ is not God. Moreover, Jesus is not the promised redeemer. The real redeemer who is yet to come will not condemn the Pharisees as hypocrites, as Jesus often did. Therefore, Saul, with the aid of leaders of the Synagogues in and near Damascus, sought to rout out and imprison the followers of Jesus.

But as Saul road on horseback toward Damascus, he had a "conversion experience," (48 A.D.). In one of his epistles (letters) he described his experience: As Saul with a group of travelers approached the city of Damascus,

...there was a flash and he was knocked to the earth. Christ appeared before him asking: "Saul, Saul, why do you persecute me?" Baffled, Saul asked, "Who are you sir?" "I am Jesus, whom you are persecuting. Get up and enter the city and you will be told what to do." Though Saul's companions heard a voice speaking, they did not see anything. Now, Saul was blind. His companions took him to an inn where he ate nothing nor drank for three days.[35]

...

Later he interpreted this conversion experience as:

I have been crucified with Christ; it is no longer I who live, but Christ who lives in me; and the life I now live in the flesh I live by faith in the Son of God, who loved me and gave himself to me. I do not nullify the grace of God; for if justification were through the law, then Christ died to no purpose. (Acts 20: 24)[36]

After this experience, Saul was baptized "Paul." This ceremony involving a name change symbolized the internal, spiritual transformation from the natural order to the supernatural order. Paul became like another prominent member of the Jesus cult, and as such he established Christian Churches consisting of non-Jews in Greece, Macedonia, Asia Minor, and even in Rome.

Rather than the Greek dualistic view of humans being spiritual souls each expressing mind locked in a non-spiritual body, Paul espoused the Jewish, holistic view of each human being the all good unity of soul, mind, and body that became polluted by *Original Sin* associated with a real spiritual being called the *Devil*. According to Paul, original sin is due to the Devil tricking the first human , Adam, to disobey God and thereby bring internal conflict to all descendents of Adam – all humans – and chaos to all of nature. Because of original sin, humans and especially Jews by means of the Mosaic Law could know to some extent the morally right things to do but could not do them. Original sin may be understood as the internal conflict in each human between the external mind self and the repressed, dominated body ego that includes subjective feelings. It also produces the conflict between the external mind self and the inner self that includes the individual, collective unconscious and the soul. Developing rational individualism intensifies this conflict but also produces two major benefits. One has greater control over the body ego and one moves to a higher level of spirituality and self-consciousness that may facilitate mystical, heroic creativity.

Core idea of St. Paul' preaching to the gentiles. In a metaphorical sense the unknowable God the Father "spoke" Christ – Christ is God's Word – who, in turn, manifests all of nature and each individual human soul. The transcendent God the Father by means of Christ is imminent in nature and in the inner self of each human. Jesus is the promised messiah and the "second

---

[35] http://campus.northpark.edu/WebChron/Mediterranean/Paul
[36] http://campus.northpark.edu/WebChron/Mediterranean/Paul

Adam" and in some manner is Christ, "the Word of God." The historical Jesus was crucified, died, and resurrected as Jesus, the Christ. The death and resurrection to Jesus, the Christ, provided the basis for salvation for all humans. In order to be saved, each human must consciously express faith in the death and saving resurrection of Jesus, and he/she must be baptized. One's subjective Faith, which is a received Grace from God, also produces objective consequences: to the extent that one is saved he/she overcomes the internal conflict in all humans and is able to accomplish good works. Therefore, one is saved and will be judged according to one's Faith and Hope in Jesus, the Christ, and the Love generated by the Holy Spirit (the third person in the God Trinity) to accomplish good works.

Core ideas of St. Paul's vision of individual, personal salvation. Each must choose to negate total commitment to the mind self defining itself in terms of personas or in terms of rational individualism. This negation produces great chaos that one can endure only by looking to the Inner Self and abandoning one's mind self to Faith in Christ manifesting soul. Negation of the mind self means that Faith in Christ is <u>not</u> belief, which is an intellectual commitment of the mind self. Rather, Faith in Christ produces death of the mind self and Hope for a resurrection. Faith, death of mind self, and Hope produce new possibilities that always eventually produce resurrection to a new mind self expressing new harmony, which is Love and power (Grace) to do good works.

Mystical Heroic Creativity

Genera features. The emergence of self-consciousness and the progressive transitions to the persona self are guided primarily by individual soul in union with ultimate SOURCE. At one extreme, hominids or the proto-human fetus before birth or a baby after birth is impelled by Eros drives from the Will aspect of soul to express self-consciousness. This emergence of self-consciousness, like the emergence of life from non-life, is ultimate SOURCE in nature manifesting spontaneous individuation. Each higher level of self-consciousness generates a corresponding greater individual free will that can oppose further individuation. In spite of this, the Eros drive toward individuation almost always overcomes free will resistance to it until the persona stage. In nature and in pre-personal human individuation, there is the spontaneous mutuality of Eros-chaos and Eros-order that generates all creativity. In the persona stage each human has an objective self (external self) associated with Eros-order that to some extent dominates a subjective self (internal self) associated with Eros-chaos. As a result, the persona self, via the objective self, can repress the subjective self thereby preventing any mutuality of Eros-order and Eros-chaos that produces creativity. However, in the "childhood of civilization" and in late childhood, Eros-chaos breaks through mind self control to produce some anguish and destabilization but also some spontaneous creativity. Beginning in 700 B.C. in an emerging Greek culture, the mind in some individuals guided the differentiation of this spontaneous creativity to become sophist *arete*. Sophist *arete* is what Robert Pirsig calls Dynamic Quality, which, according to Nietzsche, produces the "true" aestheticism and dynamism of classical Greek art. But, as described earlier, it also produced relativism, skepticism and chaos in Greek city-state society.

The great Socrates guided the transition of classical Greek culture to logical, conceptual knowing, which repressed sophist *arete*. In like manner, as described in chapter 10, the young adolescent pressured by formal education of rational societies must "kill" the last vestiges of spontaneous creativity of late childhood. All creativity is mysterious, i.e., it cannot be rationally controlled or explained. Thus, all humans exhibiting spontaneous creativity are also spontaneously mystical. But once a person individuates to logical, conceptual thinking, he is cut off from the mystery of nature and from spontaneous creativity and mystical experiences. Such individuals feel safe in their socially constructed perspective of reality structured by conceptual knowing. Because of the dominance of Eros-order in such a perspective, the only way a person can be creative is to break from conceptual thinking. Instead of looking outward for guidance from social consensus, one must look inward to the Inner Self. The Inner Self, consisting of soul in union with ultimate SOURCE and of repressed centers of self-consciousness, resonates with SOURCE in the of nature. In contrast, the external self sees itself as an autonomous entity that can maintain its separateness from nature and its internal stability, and can exert control over nature, including control over other people. Thus, a look inward confronts the mind self with the overwhelming mystery of Reality both within and without. One can neither understand nor begin to control this mystery. This inward vision of "no knowledge" and of loss of control and self-sufficiency is terrifying; it can drive a person mad.

Thus, the key feature of mystical, heroic creativity" is that one takes the risk of mind self destabilization by looking inward. One gives himself over to participatory subjectivity associated with Eros-chaos. Eros-chaos at first destroys conceptual order and confronts one with nothingness and possible madness. Understandably, most people individuated to the conceptual stage of self-consciousness would avoid such an inward glance. However, ultimate SOURCE may drive some individuals to experience this inward vision. If they survive the temporary vision, they gain experiential faith in whatever process that brought them to this inward glance. Sustained by this experiential faith, they allow themselves to be drawn inward again and again, each time gaining more experiential faith that is synergistic with the increasing courageous will to confront nothingness. These selected few people become the great spiritual innovators of human history. The experiential faith, courage, and created vision can guide others to participate in this mystical, heroic creativity. In this manner the repressed *arete* of the sophists is reborn as mystical *arete*. The transition from metaphorical, conceptual thinking to logical, conceptual thinking produced the "death" of arete implicit in the pre-Socratic philosophers. As noted by Robert Pirsig, Nietzsche, and Heidegger, this was a great loss. But the differentiation of logical, conceptual knowing provided the possibility of a greater gain. Paradoxically, conceptual knowing that destroys spontaneous creativity and one's child-like sense of wonder and awe, simultaneously creates the entrance to mystical experiences. Retaining metaphorical vision, plus breaking from total conceptual control of this vision leads to the source of vision that is soul in union with ultimate Reality. Even a tangential glance at the Inner Self as manifested by ultimate Reality produces a mystical experience.

Producing personal transformations. Mystical, heroic creativity may be summarized by the statement: "no-knowledge that leads to no-self." The meaning of this statement follows from the general features of mystical, heroic creativity described above. Choosing "no-knowledge" means that one commits to transcendental skepticism. That is, a metaphorical conceptual representation of direct perceptions of reality is a distortion of those perceptions. Logical, conceptual knowledge is a distortion of metaphorical, conceptual knowing and therefore is an even greater distortion of perceptions of reality. Thus, the mind self that tries to understand reality only can produce distortions. If these distortions are taken to be true, then the mind self produces illusions. However, in acknowledging its distorted knowing, the mind self experiences reality as mysterious. This implies that the self's understanding of itself, and more fundamentally its understanding of self-consciousness, are distorted. At the same time, the self experiences itself as creating knowledge that, even though it is distorted, gives meaning and value to life and provides valid ways of maintaining and enhancing one's material existence. The processes of self-consciousness and creating knowledge are experienced as mysterious. At the same time, these processes are experienced as coming from or generated by an aspect of the Inner Self. As a result of focusing on the consciousness aspect of self-consciousness or on one's creating knowledge, a person experiences an aspect of the Inner Self as the ultimate SOURCE of self-consciousness, of human creativity, and by extension of creativity in all reality. This vague experience of the ultimate SOURCE that humans call God or Christ or Allah or Void occurs by the self subordinating itself to the Inner Self. But this experience of subordination involving the commitment to "no-knowledge" is simultaneously an experience of mind self death. Often a great existential fear of mind self death and existential anguish of experiencing mind self death come before the joy of experiencing the ultimate SOURCE. It takes great courage to continually choose "no-knowledge" that leads to ever greater mind self death that produces "no-self." But in experiencing the enduring of mind self death, one also, eventually experiences the fruits of human creativity such as new knowledge or a rebirth to a higher level of self-consciousness. Thus, the choice of "no-knowledge" leading to "no-self" is appropriately called mystical, heroic creativity.

The continual experience of mystical, heroic, personal transformations may eventually lead to the experience of a loss of mind self identity. The mind remains but no longer defines the mind in a particular way. The mind self defines itself in a way appropriate to a particular set of circumstances. The mind self sees itself as manifested by soul that, in turn, is directly and at all times manifested by SOURCE. Therefore, the mind self sees itself as united to SOURCE and this connection represents its "true self" that supersedes all self-definitions. However, the human person still has individual identity due to the body ego. Eventually one may experience the loss of the body ego identity. Though the mind self, the body ego, and the body of the individual still exist and are perceived by others, the individual human is dead. This is what St. Paul meant when he claimed that not me – Paul – but Christ in me that preaches to you.

Patriarchal Ideology Produced Rational "Religions"

Serapis. Serapis was an invented god that was a composite of several Egyptian and Hellenistic deities introduced to the world at the beginning of the Ptolemaic Period during the reign of Ptolemy I. Serapis was meant to form a bridge between the Greek and Egyptian religions. The chief center of worship of Serapis was a temple in Alexandria until it was ordered to be destroyed by Emperor Theodosius in 389 AD.
Mithraism. Mithraism is a Roman mystery cult of the god Mithras that began sometime in the 1st century AD. It became the most influential and one the largest religions in the Roman Empire until all the pagan religions were ordered to shut down in 391 AD. Members of this mystery cult kept the liturgy and activities of the cult secret, and they had to participate in an initiation ceremony to become a member of the cult. This religion was restricted to males and was popular with soldiers. Mithra was a divinity of fidelity, manliness, and bravery, and "He" stressed good fellowship and brotherliness.
Manichaeanism (Manichaeism). On March 20, 242 AD Mani (206 or 210 to 276 AD) proclaimed his gospel to the crowd gathered at Gundesapor in Sassanid Persia on the coronation of Sapor I: "As once Buddha (563 – 483 BC) came to India, Zoroaster to Persia, and Jesus to the lands of the West, so came in the present, this prophecy through me, the Mani, to the land of

Babylonia."[37] After many years of travel and founding Manichaean communities in Turkistan and India and writing epistles or encyclical letters that explained his doctrine, Mani was crucified on the command of king Braham I, his corpse flayed, the skin stuffed and hung up at the city gate, as a terrifying spectacle to his followers, who Braham persecuted with relentless severity.[38] Though there is no influence of Iranian mythology or Zoroastrian dualism on Mani's formulation of his gospel, the adaptation of Manichaeanism to the dualism of Zoroastrianism began in Mani's lifetime.

The theology based on Mani's cosmology proclaimed metaphysical dualism: there are two equal principles, the realm of light that lived in peace and the realm of darkness that was in constant conflict with itself. After the realm of darkness attacked the realm of light, the Living Spirit, an emanation of the realm of light, created the universe out of the mixture of light and darkness. Thus, there is no omnipotent good power; rather, there is equal and opposite powers of light and darkness. Each human is the battle ground of the conflict between light and darkness. The human soul is light in opposition to the body that is darkness. The body (body ego) tends to overwhelm the soul expressing mind, which defines the person. Thus, humans may be saved by the mind self knowing two things: 1) that it is an expression of soul, which is light and is eternal; 2) that it is one's true identity that must break from the darkness of the body and matter. This approach to "spiritual salvation" is called *Gnosticism* that despised the simplicity of the non-educated. This intellectual religion professed to bring salvation through knowledge; sin is ignorance. "At its height, Manichaeism was one of the most widespread religions in the world, with Manichaean churches and scriptures being found as far east as China, and as far west as the Roman Empire …. The spread and success of Manichaeism was seen as a threat to other religions, and it was widely persecuted by Christianity, Zoroastrianism, and later Islam."[39]

HUMAN SOUL IN RELATION TO LOGOS

Philo of Alexandria (20 BC to 50 AD)

Philo developed a speculative and philosophical justification for Judaism in terms of Greek philosophy, especially Neo-Platonism. He regarded Jewish religious truths as fixed and determinate. He used Greek rational philosophy as an aid to understanding these truths and to teaching them to others. According to Marian Hellar writing in the Internet Encyclopedia of Philosophy[40], Philo laid the foundations for the development of Christianity in the West and in the East. He may have influenced St. Paul and the authors of the Gospel of John. Philo's works were enthusiastically received by early Christians, some of whom saw in him a cryptic Christian.[41] Philo's synthesis was a radical break from the Platonic and Aristotelian approach to spiritual wisdom. Classical Greek philosophers after Socrates rejected the metaphorical, conceptual knowing of the Pre-Socratics and of the Epics of Homer. In radical contrast, Stoic thinkers sought in Homer the basis for their philosophical insights. Philo sought to translate the

---

[37] Catholic Encyclopedia: Manichaeanism; www.newadvent.org/cathen/o9591a.atm

[38] Catholic Encyclopedia: Manichaeanism; www.newadvent.org/cathen/o9591a.atm

[39] http://en.wikipedia.org/wiki/manichaeanism

[40] http://www.iep.utm.edu/p/philo.htm

[41] http://en.wikidedia.org/wiki/Philo

Jewish religious truths in the Old Testament into these philosophical ideas. In particular, he attempted to show that Platonic-Stoic philosophical ideas can be made to be deductions from the verses of Moses and other Old Testament authors. On the one hand, Biblical truths could be understood as literally true, such as: the One unknowable God is "God the Father," who as a father loves his creatures; thus, the literal meaning is adapted to human needs. On the other hand, Biblical truths contain the deeper truths expressed as metaphorical concepts and stories that each individual could understand as objective truths that all believers hold in the same way. The objective Biblical story often could be translated into an objective philosophical story using Neo-Platonic and Stoic terminology and ideas. At the same time, these truths are subjective in that they are interpreted according to one's personal experiences. The metaphorical, conceptual understanding of philosophical ideas allows for these subjective interpretations.

The core doctrine in Philo's synthesis is his doctrine of the Logos. He fused Neo-Platonic and Aristotelian concepts of Logos with Hebrew religious thought thereby providing the foundation for Christianity. This may have led to the Christian-Pauline gospel and to the ideas of Christ as Logos in the Gospel of John. In Philo's synthesis Logos contemplates that from which it was generated, the One, thereby producing in itself the eternal forms or ideas postulated by Plato; they are the thoughts within Intelligence. Logos also is the Demiurge who is the craftsman of the cosmos in Plato's myth of creation of the cosmos. Philosophically speaking, Logos is the true first principle, but it is not the creator of all beings but rather in generating Soul, another eternal Being, it through the Soul generates all existing individual beings. The power of the One is Logos that generates Soul that is the location in which the cosmos made up of individual beings takes on an objective shape and determinate physical form. The Soul contemplates "its anterior," which is Logos, and through this contemplation it generates individual beings. In like manner the human mind as an extension of the Divine Logos exerts its intellect to divide reality into autonomous parts. Other designations of Logos by Philo include: immanent Mediator of the physical world, revealer of God, Manna and Wisdom.

Philo's fundamental doctrine of Logos is that IT is an intermediary power, a messenger and mediator between God and the world. When speaking of the transcendent God, Philo describes the Logos as God's son who procures forgiveness of sins and blessings. Thus, Philo transforms the Stoic impersonal and immanent Logos into a being who was neither eternal like God nor created like creatures but rather was begotten from eternity. As mediator who was sent down to earth, Logos is the basis of hope for sinful humanity. The most universal of all things is God but in the second place is the Logos of God. Humans receive manna, which Christians call Grace for salvation, from God through Logos. Also Jews receive Wisdom represented by the teaching of the Old Testament from God through Logos. In three passages Philo describes Logos as God through which humans share with God the Father, and are able to perceive Him.

Plotinus (204 – 270 AD)

Plotinus contributed to the great ending of Socratic rationalism by incorporating the Stoic idea of using metaphorical, conceptual interpretations of Plato's *Dialogues*. He expanded the great beginning of the Greek rationalism by producing an approach to knowing Nature that is analogous to that of the scientific revolution. According to Plotinus, reality is simultaneously intelligible and mysterious. It is intelligible because *Intelligence* containing Plato's Forms is *Logos* that generates *Soul* that divides itself into *Higher Soul* that contemplates Logos and *lower soul* that metaphorically speaking, is the single, pure light ray of the Higher Soul that passes through the prism of matter. Matter is *pure possibility* that is totally passive so that when the Higher Soul ray of light passes through matter, the pure light is diffracted into infinitely many colored rays of light, each of which is an individual being that is sustained in existence by an image of an aspect of Logos. This image of Logos is what makes each material being *intelligible*. The material cosmos is mysterious because discursive reasoning generated by Logos as a capacity of mind in each human being only can come to true knowledge by turning inward to the Higher Soul that contemplates the eternal Ideas in Logos. By so doing the individual soul through the Higher Soul in him comes to "mystical knowledge" of Logos that leads one back to the One.

In a manner similar to Plotinus, the scientific enlightenment proclaimed that reality is mysterious but *interpretable*. Modern science rejects the Aristotelian-Medieval view that one can directly obtain absolutely true knowledge of Nature. Rather, one can discover empirical patterns in Nature and then form opinions about them wherein there are many different interpretations of any particular pattern. Then, one can *construct* a mathematical representation of a pattern wherein the mathematical theory is analogous to one of Plato's spiritual, eternal ideas. Finally, one must empirically validate the mathematical interpretation thus producing valid but not absolutely true knowledge. In Plotinus' theory of knowing one constructs diverse opinions along with reasons for holding these opinions, but alternatively, one may choose to commit to dogmatic, mystical introspective philosophy. In so doing, one creates an objective mystical, metaphysical theory that is thought to be absolutely true.

Individuality and personality. Personality involving individuality is something that develops as a result of pure soul (Higher Soul) assimilating alien elements through its assimilative contact with matter. Thus, personality is a by-product of the soul's putting order into matter. The soul receives from matter certain unavoidable impulses, which come to limit or bind the soul in such a way as to make it a "particular being." The individual soul tends to construct and then maintain the illusion of being distinct from its source, which is the Higher Soul. Each particular human being is the duality of the "true self" that is the essence of the individual soul that is sustained in existence by Logos. The body ego in the civilized adult is dominated by the mind self. The peculiar qualities of each individual come from the body ego's contact with matter. According to Plotinus, these qualities should be discarded in that they distort the true nature of the soul. The process of discarding these qualities may be understood as mystical, heroic creativity, described earlier in this chapter, producing personal transformations. One must evolve to no mind that leads to no mind self and eventually no body ego thus arriving at one's true self. Plotinus' spirituality is similar to Advaida-Vedanta of Indian mysticism.

Individuality plays no part in Plotinus' metaphysical theology. The sole purpose of the individual soul is to puts Order into the fluctuating representations of the material realm, that is, one constructs opinions, and put Order in the world by proper exercise of sense perceptions. While "Ordering" matter in this way, one should remain as far as possible in contact with Logos, which readily occurs when one has evolved to no mind self and no body ego. The lower part of Soul, which is the seat of personality, is an unfortunate but necessary supplement of the Higher Soul's actualization of the ideas in Logos that it contemplates. The individual soul "gives" determinate order to the pure passivity that is matter. Matter comes to exist in a state of ever changing receptivity equal to "Chaotic malleability," that is, new possibilities for actualization emerge when a system goes into chaos. The developing personality of the individual soul mirrors the malleability of matter. Individuation of the personality mirrors the individuation of the cosmos. In modern terminology this means that each individual human is a microcosm similar to the macrocosm. Each individual human evolves according to the same principles that guide the evolution of the cosmos: Eros-chaos collaborates with Eros-order to produce a new level of individuation, see chapter 2. When personality is taken to be good in itself, then it becomes a surrogate (a substitute) to the authentic existence provided by and through contemplation of the Higher Soul. The personality, as surrogate, stands in the way of one seeing one's authentic existence.

Ethics. The highest attainment of the individual soul is likeness to Logos as far as possible. One achieves this likeness by contemplation of the Higher Soul, which may be thought of as the pure Inner Self or as the *Inner Witness*, which is the individual soul in its own purified state. Because the Higher Soul, the pure Inner Self, does not come into direct contact with matter, it will remain aloof from the disturbances of the realm of the senses. The individual soul no longer will directly govern the world. Rather, it, as guru, will provide governance of the world to those souls that still remain enmeshed in matter. Thus, "Plotinus was unable to develop a rigorous ethical system that would account for the responsibilities and moral codes of an individual living a life amidst the fluctuating realm of the senses."[42] Plotinus does not deal with the problem of human suffering. For him the soul that has descended too far in matter merely needs to turn inward, thereby through intellectual contemplation become reunited to the Higher Soul. Only the lowest part of the soul suffers and is subject to passions and vices. But since the individual soul has free will, it has the ability to look inward and free itself from the bonds of matter. Thus, all particular questions concerning ethics and morality are subsumed under Plotinus' doctrine of the soul's "essential imperturbability." Therefore, the problems plaguing the lower soul are not serious issues for philosophy. Rather than any mythology or mystery cult, Neo-Platonic philosophy became the highest pursuit of the soul. This pursuit was thought to be able to produce a divine act capable of purifying every soul from the stain of its contact with matter.

Matter and the problem of evil. Matter is an eternally receptive substratum by which all determinate beings receive their forms. In being capable of receiving any and all forms, matter is the principle of differentiation of all being. Plotinus distinguishes between two types of matter: *intelligible matter* that takes on various aspects of intelligible Being and "fecund darkness" equal to the "depth of indeterminacy." With respect to ideas presented in this book, fecund darkness is chaos that presents new possibilities of new collaborations to produce a new order. Intelligible matter is the new order that emerges out of chaos in the process of individuation of a system, that is, self-organization of a system, see chapter 2. Thus, matter as chaos is the necessary though not sufficient condition for any individuation to occur, and matter as emerging new order is the other half of the principle (collaboration of Eros-chaos and Eros-order) governing any individuation (evolutionary) process. This means that matter is not inherently evil. It only becomes evil in relation to the individual human soul that becomes bound to matter through its chosen acts. The individual human experiencing chaos is tempted to think of matter, and indeed his material life, as evil, but this is a confusion stemming from there being two types of matter. The confusion can be overcome when chaos becomes the occasion for one to look inward and eventually experience mystical, heroic creativity.

The individual human that has individuated to a state of consciousness that is continually in touch with the Higher Soul (Inner Self) will spontaneously be creative; he/she will spontaneously bring collaboration between Eros-chaos and Eros-order. The individual human

---

[42] Edward Moore; http://www.utm.edu/research/iep/plotinus.hem

that is disconnected from Higher Soul through "forgetfulness" continually chooses not to look inward. Such an individual begins to live on a purely physical basis. Not being aware of the Inner Self (Higher Soul) one loses its power to be creative. The soul becomes affected by what it produces in physical generation. This is the source of unhappiness and hatred. The person becomes the slave to that over which it should rule. That is, one becomes subject to addictions. One also is affected by any and every feeling or event that comes its way. With respect to self-consciousness, the human soul has lost its divinity. Thus, for Plotinus, Evil is subjective in that "one chooses to forget Higher Soul;" that is, one chooses not to look inward to the Inner Self.

Evil also is a metaphysical condition. Each human is an embodied soul that generates the duality of mind in opposition to body. Paradoxically, as one individuates to higher levels of mind self-consciousness, the mind tends to focus only on looking outward to control nature and thereby forget that it is generated by Higher Soul. Thus, only humans, especially after becoming civilized, are capable of evil. But evil is not a meaningless plague upon the soul. The possibility of Evil is metaphysically associated with the possibility of return to the One. When the One (Goodness) descends into the pure passivity of matter, it produces the duality of mind versus body in humans. This, in turn, produces the possibility of Evil. At the same time, by means of Logos in each human one can by introspection turn to its Inner Self, which is the Higher Soul. This contemplative introspection leads the individual soul back to the One.

St. Augustine of Hippo (354 – 430 AD)

Christian great ending. Augustine's legacy is a large body of work – more than 100 separate titles – that encompasses a "movement from a largely Hellenistic eudaemonism [a system of ethics that considers the moral value of actions in terms of their ability to produce personal happiness] to the increasingly somber eschatology [any system of doctrines concerning such matters as death, judgment after death, the future of humans] of his later works…. [Thus] the diversity contained in this body of work defies any easy or succinct synopses."[43] Augustine contributed to the Christian "great ending" of Hellenistic rational individualism by extensively modifying Neo-Platonic ideas. In his earlier works he sought for a common denominator of conflicting views of Hellenistic philosophies, which primarily were Epicureanism, Stoicism, Skepticism, and Neo-Platonism. Also in his earliest writings (386 AD) he believed in the compatibility of Christian doctrine and Neo-Platonism. But toward the end of his life when he was writing *Two Cities of God* (416 AD), he saw that there are significant points of divergence, though Neo-Platonism still provided a philosophical framework for describing many ideas of Christian doctrine. Augustine's "poetic imagery," leading to the diversity of his ideas and superseded by dogma based on Scriptures, is a partial return to Pre-Socratic arete.

As an aspect of the Christian great ending, Augustine modifies as he incorporates Neo-Platonic ideas. (1) *Creator God versus eternal One.* In contrast to Plotinus who regarded the One as manifesting multiplicity, Intelligence, and Soul, Augustine, as a Christine, regarded the Creator God as the ultimate source and point of origin for all that comes into existence. God is Being, Goodness, and Truth. According to the Old Testament, He chooses to create all things "from nothing" and thereby is the unifying principle that itself remains unchanging. (2) *Metaphysical dualism.* Augustine accepts Plotinus' dualism situated within a unified hierarchy that begins with the unity of One, Intelligence, and Soul that progressively unfolds to increasing

---

[43] Michael Mendelson. http://plato.stanford.edu/entries/augustine/

plurality and multiplicity. This duality is the sensible world versus the intelligible realm. The sensible world implies subjective knowing and personal engagement. It involves transitory objects and produces the anxiety of losing what and whom one loves. The intelligible realm is public and open to all in that it involves objective knowing. According to Plotinus, it contains Plato's eternal Forms and the possibility of objective, absolute truths gotten by introspection. Augustine brings in God as the source of the intelligible realm and as the Being that promises the only lasting relief from the anxiety prompted by the transitory nature of the sensible realm.

(3) *Problem of evil.* Evil is not a metaphysical entity or relationship. The sensible world is not evil, and the human soul in a body is not intrinsically evil. According to "Greek intellectualism," evil is due to a faulty will directing one to false knowledge. Humans as embodied souls tend to focus only on the sensible, thereby failing to perceive the larger unity of which the sensible world is only a part. As a result of this lack of knowledge, the individual's will attaches itself to transitory objects that lead to anxiety. But because of the rational nature of the human soul and the Neo-Platonic idea of Logos creating a rational hierarchy, there is a possible solution to the problem of evil. "The human soul has the capacity to perceive its own limited status as a being embodied partly in the sensible world while connected to the intelligible realm."[44] Therefore, it is possible for one to reorientate one's moral world, appreciating it for its liminal goodness but seeing it as an instrument for obtaining knowledge of the intellectual, spiritual world. Later in his life Augustine rejected this Greek intellectualism.

(4) *Linear Creativity versus cyclical generation.* Augustine, inspired by Judeo-Christian scriptures, provided a linear account of the dramatic unfolding of a morally decisive set of non-repeatable events. The major unique, non-repeatable events that define the drama are: the fall equal to Original Sin recounted in Genesis; the incarnation, passion, and resurrection of Jesus, the Christ, described in the synoptic and Johanine gospels, and the final judgment foretold in Revelations. Augustine was keenly aware that what Christians believe falls outside the austere standards of the Platonic conception of knowledge and understanding. As a result of Augustine's increasing familiarity with scripture, he began to focus more and more on the historical dimension of this tradition that is alien to the intellectualism of Neo-Platonism. In contrast to the Judeo-Christian narrative, the One of the Neo-Platonism is like a self-generating energy coupler, see chapter 2 and figure 2.1, wherein the One couples energy from itself to itself that eternally manifests the cycle: Yang process that is de-evolution to the many and the Yin process that is evolution back to the One.

(5) *Human dualism: body versus soul.* Augustine oscillated between four hypotheses of the origin of soul, but then by 419 AD he writes as if there are only two competing hypotheses: Either all souls are propagated from Adam's soul analogous to bodily reproduction; OR God creates a new soul for each body. Later he implies an unofficial fifth hypothesis that (in my interpretation) is more suitable to his later teachings on Original Sin. Namely, God only created the soul of Adam so that the soul of each new individual human is identical to Adam's soul. As this embodied Adam's soul is played out in a particular life history, it takes on personality traits as a result of its interactions with matter. This, of course, is similar to Plotinus' view of human individuality and personality. This view also allows for the possibility that the sensible realm is alien to soul and thus Augustine's view is not altogether different from the Manichaean ideology that embodiment is a kind of entrapment. The soul is immaterial and immortal in contrast to the body, which is material and mortal. Correspondingly, the soul is superior to the

---

[44] Michael Mendelson. http://plato.stanford.edu/entries/augustine/

body and should dominate the non-rational such as feelings. He also viewed reason as the mind's ability to engage in deductive processes. Reason is the cognitive apex of the human soul. The mind can recognize logical necessity, and therefore, if one starts with true first principles, one can obtain absolute true, objective knowledge.

Christian great beginning. Augustine was the major contributor to the "Christian great beginning," which was the emergence of a *Christian rational individualism* associated with what might be called a "Christian Pre-Socratic arete." This involved a metaphorical, conceptual understanding of the scriptural story of the human spiritual journey starting with the creation of all things. (1) *Original Sin.* Based on his study of scriptures and the views of St. Paul on original sin and Grace, Augustine retained some aspects of Neo-Platonism. Each human has an individual soul that expresses mind that is constricted by a material body over which the mind must control as well as controlling all of matter. As a result of each human acknowledging being in the unifying hierarchy as described by Plotinus, one may overcome the dualities of mind versus body and the intelligible realm versus the material realm. In general, philosophy can help the individual soul to transcend the fragmentation and unhappiness produced by the spirit in conflict with matter. At first, Augustine adopted Plotinus' optimistic view that after many reincarnations all humans will return to the One and thus be saved.

In the last half of his adult life Augustine gradually moved to the pessimistic view that most humans are predestined to suffer eternal damnation in Hell. The wisdom of philosophy cannot bring happiness. Therefore, the rational life will not bring happiness to any individual or to human communities. The hope provided by a life of rational enquiry is an illusion. For many, rational individualism has evolved to existential despair. The "Good news" associated with Christian rational individualism only partially transcends each human's internal, spiritual conflict. Augustine's somber view of the Christian message consists of the following four ideas: One, human duality is that the soul generates mind that can know absolute truth and will that has freedom of choice, but matter, as described by Plotinus, causes the soul to forget its true self and place in the unified hierarchy. Two, because of Original Sin, *the human will is free only to choose sin*, which is the rational choice to pursue what is temporal, changing, and perishable rather than what is eternal, immutable, and incapable of being lost. Three, the death and resurrection of Jesus, the Christ, is retroactive as well as proactive in that it provides Grace to each human soul to turn toward God and be saved. Four, each human soul that freely chooses to turn to matter rather than to God is, by virtue of its freedom, justly damned to eternally suffer in Hell.

Augustine's somber vision goes on to proclaim that God has predestined only a small number of humans to receive Grace to turn to God; thus, only a small number of humans will be saved. Most humans will be damned to eternal Hell including: all those humans who do not convert to Christianity; some, possibly many, members of the Catholic Church will be damned – but there is no sure way of predicting who is saved and who is damned; babies who are not baptized will be damned.

In writing his book, *City of God*, Augustine is certain that the linear movement of history leads to the separation of two groups of people where each is reunited with his/her resurrected body. Each of the small minority of people that are saved will have a vision of God. Each member of the second group of people containing the overwhelming mass of humanity will undergo a second death wherein their resurrected bodies will be subject to eternal torment by flames that will inflict pain without consuming the body. The degree of torment will be proportional to the extent of sin in one's life. "All must suffer without end, for to suffer any less would be to contradict scripture and undermine our confidence in the eternal blessedness of the small number God has saved."[45]

(2) *Christian Pre-Socratic arete*. As a result of the profound impact that St. Paul's teachings had on Augustine, he came to believe, like St. Paul, that Christ present in each individual human manifests the soul. Thus, when any of the educated elite develop a life of intellectual contemplation, they continually turn inward and encounter not only soul but Christ (Logos) manifesting one's individual soul. The Pauline doctrine plus metaphorical, conceptual knowing described in chapter 6, and mystical, heroic creativity, described earlier in this chapter, provide a framework for describing Augustine's "Christian Pre-Socratic arete." In metaphorical, conceptual knowing, one's subjective, sensual perceptions include a "seeing" of similar structures in two systems that previously were understood not to be similar or in any way related. As described in chapter 6, the rational, conceptual pattern of a hormone fitting into a receptor in a membrane of a cell is seen to be similar to one's direct, concrete experience of a key fitting into a lock. This "seeing" is associated with a concrete sense experience, but it is not merely a sense perception. At the same time, this "seeing" is associated with objective knowing of a logical, conceptual pattern, but it is not merely objective, rational knowing. Using Neo-Platonic terminology we may say that the "similar structures in the two systems" (lock-key and receptor-hormone) are intelligible realities embedded in the sensible realm, but "seeing these realities" is due to something other than mere sensual perception or rational analysis. What Augustine realized – I presume based on his personal experiences – is that this seeing is a concrete, direct intuition equal to an intellectual illumination.

Often times these illuminations occur spontaneously in children and as such are merely accepted as part of a child developing into an adult. Sometimes when Adults experience these types of illuminations such as $1 + 1 = 2$ or two straight lines will never meet, they are viewed as a normal way humans know some things. However, when one is in the midst of confusion of struggling to understand some pattern and then has an intellectual illumination, the experience is appreciated as a pure gift. Because of the struggle, one realizes that this spiritual gift does not come from the mind self or from the body ego that organizes sense perceptions. Rather, according to Augustine, it comes from the mind self looking inward to the equivalent of the Inner Self, which is Christ manifesting the individual soul. Christ is always there sending forth light that mostly is reflected back by the barriers the mind self sets up. However, when the mind self in some manner breaks down one of these barriers, it becomes open to be illuminated. When a mystic, such as St. Paul, evolves to be totally open to the Inner Self, such as described by mystical, heroic creativity, then he or she may rightly proclaim as St. Paul did, it is not me but Christ in me that speaks to you.

However, this divine illumination is available to all humans. It is the basis, for example, for Christians learning from pagans, which is what Thomas Aquinas said about Christians learning from Aristotle. If we generalized the Pauline-Augustine doctrine to the idea that the ultimate SOURCE generates individual soul in each human, then we have a universal, secular basis for divine illumination. Humans in diverse religions or atheistic spiritual traditions, such as Hinduism or Buddhism, may experience profound divine illuminations, but how can these insights be shared? The illuminations always come as ideas expressed in ordinary language representing metaphorical, conceptual knowing. As such, they are ambiguous and open to diverse interpretations, but because of this, like music, they can unite people coming from very

---

[45] Michael Mendelson. http://plato.stanford.edu/entries/augustine/

different cultural traditions. The unifying aspect remains so long as no one group of people attempt to superimpose their literal, dogmatic interpretation on every one else. At the same time, each of many different groups of people may reduce the narrative to a logical, conceptual model thought to be "true" according to a consensually agreed upon "truth criteria." The narrative unites people and also becomes contextually useful when it is reduced to a validated, logical, conceptual model.

(3) *Will.* Early in Augustine's career when he wrote *Confessions*, he emphasized the importance of introspection and the central importance of free will. The importance of Will was validated by his experience of intellectual certainty about the rational truths of Christianity that nevertheless was not sufficient to enable him to convert to Christianity. Rather, his conversion, though influenced by his Neo-Platonic understanding of Christianity, especially in relation to the writings of Plotinus, was finalized and energized by continuous free choices of the Will that have an essentially non-intellectual component. From the very first emergence of Pre-Socratic arete to the full differentiation of Neo-Plotonic rationalism, Greek intellectualism always insisted on the importance of free will, but it maintained that the hierarchy of knowledge determines what one should do. Free will has the power to carryout the dictates of knowledge, but in the face of not having sufficient knowledge, free will has no power. Moreover, even when one apparently has sufficient knowledge, free will still does not have the power to do right actions. First of all, under the influence of Plotinus Augustine saw that the "intellectual religions," such as Manichaeanism, not only were "dead ends" but produced great unhappiness. Secondly, under the influence of St. Paul and Augustine's study of the scriptures, he began to realize that all knowing by itself also is a dead end. As preached by St. Paul over and over again, the Jews of the Old Testament knew what they should do and yet could not do it. Moreover, St. Paul's doctrine of Original Sin convinced Augustine that, as stated earlier, *the human will is free only to choose sin.*

Augustine began to realize that the Greek hierarchy of mind determining the expression of Will must be turned upside down. Will transcends mind, which means that "right knowing" does not predetermine "right action." This means that there is a non-rational component in human free choice. Moreover, Will is prior to mind self knowing in that Will determines what one is open to learn or even determines what sort of knowing one will attempt to achieve. The diversity of opinions even among the most brilliant thinkers who pursue a life of philosophical contemplation produces the despair of spiritual ignorance and the difficulty of the human condition. This new hierarchy of Will transcending Mind, which is called *voluntarism*, produces the crisis of how a person should organize his life when there is no definitive knowledge as a guide for action. Augustine's response was to produce two radically different ideologies, one suitable for a few mystics who become believers in the Christ-event. The mystical ideology may be summarized by Augustine's statement: "Love God and do what you will." I surmise – based on my own experiences – that the synergism among Augustine's ideas about original sin, Christian Pre-Socratic arete that includes the idea of divine illumination, and voluntarism leads to the vision as described earlier in this chapter and called mystical, heroic creativity. Such a vision probably was not appropriate for most people during the Hellenic period. Augustine's second, radically different vision provides the framework for the emergence of the Medieval Enlightenment to be described in Chapter 14.

(4) *Inner Self.* In the course of his life Augustine's idea of soul shifted from a focus on spiritual mind that produces rational knowledge to a focus on voluntarism. The doctrine of original sin, stripped of the mythological interactions among God, Satan, Adam and Eve,

proclaims that each human is created with an embodied soul that produces a "living hell" that on its own propels the individual away from rather than toward reunion with God. Initially the embodied soul generates a body ego that expresses ego-centric feeling willfulness, see chapter 4. Later, a polar mind self emerges that partially regulates the body ego but also expresses mind selfish willfulness, see chapter 5. As humans become civilized, see chapters 6 and 7, the mind self differentiates into an objective self (external self) and subjective self (internal self). The external mind self not only is in continual conflict with the body ego, it pursues its own selfish willfulness that leads it away from God. Augustine, in agreement with Plotinus, believed that the subjective mind self could continually choose introspective, intellectual contemplation that would lead one back to God. Later Augustine insisted that God's Grace was absolutely essential to enable one to choose a life of introspection involving understanding scriptures in terms of Neo-Platonic philosophy. Thus, Augustine was committed to developing Christian, rational individualism. But by the time he began writing *City of God* he realized that rational individualism, even with Christian modifications was a dead end. Because of original sin, the willfulness of the body ego or of the external, objective mind self or even of the internal, subjective mind self leads one away from God. However, the Grace of Faith, Hope, and Love from God channeled through the Catholic Church representing the Mystical Body of Christ, can turn the individual soul toward God. In particular, each embodied soul is a direct manifestation of Christ. If one freely chooses to be a member of "Christ's Mystical Body," which is the Church, then he/she may receive Grace and be saved. What many now call the *Inner Self* of each human manifests willfulness in various forms as described above. But Grace transforms this willfulness into progressive, willful union with Christ. Thus, the human embodied soul is *pure Will* that without Grace manifests ego-centric or selfish willfulness separating one from Christ. When one puts himself in the stream of Grace from God, through Christ and His Church, then and only then, one is carried back to God.

Thinkers reflecting on Augustine's later writings about personal salvation produced a secular view of the Inner Self. The *soul has an objective aspect*, which is seen as a direct manifestation of divine Logos, the ultimate SOURCE (Christ). The *soul also has a subjective aspect*, which is will that has some degree of freedom of choice. (According to Augustine, this freedom is the basis for the justice of eternal suffering of those individual souls predestined to damnation.) This freedom is the site where non-merited divine Grace enables the individual soul to choose to turn away from the realm of senses and turn toward God. This secular vision is the basis of the US democracy; namely, because each human has an embodied soul that has an *objective mind aspect* and a *subjective free will aspect*, all humans are equal with respect to their spiritual potentials. The government empowered by the people, that is, a democracy, must not only acknowledge this potential, spiritual equality, it must facilitate individualism that promotes developing these spiritual potentials.

UNIT V MEDIEVAL ENLIGHTENMENT

Chapter 14

PREPARATION FOR ENLIGHTENMENT

GREAT BEGINNING OF HELLENISTIC CHRISTIAN RATIONALISM

From the very beginning of the Jesus cult evolving into a world religion, the Christ-event became the Christ-story that was passed down from one generation to the next. The apostles of Jesus and other leaders in the Jesus movement became like the Greek epic poets who produced the myths of the Olympic gods and goddesses intertwined with the celebration of the deeds of Greek heroes. As each generation of Christian preachers retold the Christ-story, they embellished it, along with new interpretations of it. Soon after the death of Jesus preachers wrote letters called epistles, for example, the epistles of St. Paul, that gave more specific interpretations of the Christ-event appropriate for answering specific questions and solving problems of a particular community of believers in Jesus. Also, several gospels were written that retold the Christ-story from a particular perspective. Four of these gospels eventually came to be recognized as legitimate, true representations of the Christ-event, see chapter 13, pp. 128-129. Thus, there arose a set of metaphorical, conceptual narratives and interpretations of the Christ-event that formed an oral and written Christian doctrine. Ideas of this tradition contradicted one another. For example, some preachers claimed that Jesus was God who only appears to take the form of a human; in fact, He was not a human. Others claimed that Jesus only was a human who resurrected to a direct union with God; in fact, he was not God. Ideas in this tradition were expressed as metaphorical concepts that were vague and therefore open to diverse, contradictory interpretations. For example, what does it mean to say there are three persons in one God? What does the idea of "person" mean? In what way can this idea be applied to humans and to God? According to Plotinus the idea of personality, equivalent to the idea of person, only applies to humans, and it is secondary to one's true self, which is soul. Personality is a by-product of the soul's putting order into matter.

From the very beginning all these conflicting narratives and interpretations set in motion the individuation of a class of preachers that eventually became a priestly hierarchy of the Catholic Church. This hierarchy gave structure to the Jesus cult in several ways that produced potential internal conflicts. First of all, the Jesus preachers transcended patriarchy by prescribing a Christian persona that superseded the distinction between masculine and feminine roles in society. In this formulation a person is obliged to conform first and foremost to a Christian persona and then to the masculine or feminine personas respectively. However, the Christian persona included the idea of males dominating females, which idea is not contained in or even is totally against what Jesus preached. Secondly, influenced by the Hellenic culture and by the Greek-Jewish patriarchal ideology described in chapter 12, the class of priests converted some narrative interpretations of the Christ-event into logical conceptual models. This then focused on the precise ways in which ideas in an emerging Christian doctrine conflicted with one another. The co-emerging hierarchy of priests could decide which rational doctrines are true. Then all other doctrines that contradicted them are false. When the power of the class of priests joined the power of state leaders, true and false doctrines could be established and maintained. Individuals who extensively reflected on these doctrines could sincerely decide for themselves which of them are true and which ones are false. But in all civilizations committed to the need for absolute certainty, the men in power determine the truth or falsity of any doctrine. This emergence of ruthless dogmatism also facilitated the differentiation and adoption of logical, conceptual thinking by Christian societies. A certain small elite felt compelled to learn to read and write, but the large majority of illiterate Christians were forced to various degrees to think logically. In order to avoid excommunication that later led to torture or death, the non-educated humans had to logically distinguish between true and false doctrine and then act in accordance with the "true" doctrine.

This individuation of Christian societies to various degrees of logical, conceptual thinking produced the related possibilities of internal social conflict and mystical, heroic creativity. One of the factors that greatly enhanced these possibilities was the aspect of the Christian persona that forbids any form of participation in pagan religions or in magical practices. As a result, the Christian had to "kill" the spontaneous creativity associated with the childhood of civilization, as described in chapter 8. Most Christians conformed to the prescriptions of their personas. However, there always are individuals who embrace the drive to be creative. Under these specifications creativity was channeled into a new form. The creative person had to look inward to the Inner Self. Those who did not go mad or commit suicide as a result of an intense inward orientation experienced mystical, heroic creativity as described in chapter 13. This creativity produced ideas that sometimes contradicted the traditional doctrine or were not covered by this doctrine. The priest hierarchy incorporated some of these new ideas into the traditional doctrine while proclaiming other ideas as heresies. For example, the religious-philosophical ideas of Manichaeanism, see chapter 13, were thought to be heretical. The religious-philosophy of Meister Eckhart (1260 - 1327) was more akin to the non-dualism of Eastern thought, even though he was familiar with Aquinas's thought. Eckhart died just before the Church was about to judge and condemn him as a heretic.

The early individuation of the Jesus cult to a Catholic dogmatic religion greatly facilitated the individuation of Western culture. It did this by providing the circumstances for the continual emergence of mystical, heroic creativity. Creativity only emerges when the Order of a system goes into partial Chaos. Under these circumstances the partial Chaos allows for new potentials, embedded but not available in the original system, to emerge. Thus, Order plus appropriate Chao is the basis for the emergence of new possibilities. The acceptance and differentiation of some of these new possibilities leads to the emergence of a new order. The Catholic Church hierarchy selected what new possible interpretations of the Christ event would be accepted and further differentiated. This continual creativity produced a vigorous though internally conflicted culture.

Also part of the "great beginning" was St. Paul's preaching that the Old and New Testaments no longer were meant only for Hebrews. God's revelations now was meant for all humans, both Jews and Gentiles. St. Augustine as well as many other thinkers participated in this great beginning by claiming that scriptural revelations are not necessarily literally true and should no longer only be understood as a true story using metaphors and analogies. Rather the true stories of both the Old and New Testaments can be understood by objective Greek philosophy, especially by Neo-Platonism as represented by Plotinus.

THE GREEAT ENDING OF HELLENISTIC, CHRISTIAN RATIONALISM

According to the modern historian Thomas Cahill, Augustine simultaneously was the last Hellenistic man and the first medieval man. Ken Wilber credits Augustine with expanding much of the mystical vision of Plato and Plotinus by converting Plato's *internal spectator* into the *ever-present wakefulness* in Plotinus, which Augustine transformed into the idea of the *interior Witness*. For Augustine this Witness is an individual's soul. Anyone who reflects on the questions: "Am I conscious?" or "do I exist?" would most likely answer "Yes" to both of these questions. The yes answer affirms the ever-present wakefulness. The soul expressing this wakefulness does not see how or why it has consciousness or how or why it now exists. Acknowledging this spiritual ignorance points one to a transcendental "higher power," God, that thereby is the ground of one's immediate and primordial awareness and therefore cannot under any circumstances be doubted. Therefore, the path to acknowledging an ultimate SOURCE, such as God, involves developing the mental habit of looking inward. Wilber goes on to point out that in becoming a mature person one develops reason that allows one to look inward and go beyond reason to the ground of the reasoning self. Then one can see that this interior ground also is the ground of the universe.[1]

Augustine in the Hellenistic phase of his life influenced the Church's rejection of some new interpretations such as Manichaeanism, and the Church's acceptance of other interpretations of scriptures in terms of Neo-Platonic philosophy. However, as indicated in chapter 13, toward the end of his life Augustine came to believe that focusing on Neo-Platonic understanding of the scriptures leads one away from rather than toward union with Christ. In this profound shift of belief away from the adequacy of Neo-Platonism, Augustine participated in the "great ending" of Hellenistic Christian rationalism.

One may understand Augustine's shift in belief in terms of Wilber's comment about reason allowing a person to look inward and *then* go beyond reason. The key question here is whether the mind self that expresses reason can go beyond reason to the ground of reason. Augustine's answer was a "partial yes" and "a more fundamental no." In describing divine illumination as a part of some thinkers ordinary knowing (see chapter 13, "Christian Pre-Socratic arete"), Augustine proclaimed that the mind self can go beyond reason. However, even with divine intellectual illumination one still ends up with a mental understanding of spiritual "truths" that may include aspects of understanding God (or the ultimate SOURCE). Augustine eventually came to believe via his own experiences that the mental understanding of even the most lofty spiritual truths about God is not the same as any kind of union with God. This, for example, is the fundamental lesson for humans in the story of Job, see chapter 12. One does not "see God" via any mental process; rather one "sees" God "face-to-face" by some means other than a mind process. I do not know whether Augustine interpreted the Book of Job in this way, but I presume – based on my own experiences – that Augustine came to "see" that one does not

---

[1] Ken Wilber. *Sex, Ecology, Spirituality* (Boston: Shambhala Pub. , Inc., 1995), pp.357-359.

directly experience God by any mental process, even including the most lofty mental, intellectual introspection, such as that described by Plotinus.

In the second half of his adult life Augustine gradually came to the view that each human is not a soul trapped in a body. Rather, each human is an embodied soul that after birth starts its unique individual life's story toward salvation or damnation. Furthermore, each individual life story is part of a grand story for the whole universe as described by the Old and New Testaments. This shift amounts to changing emphasis on intelligible matter by what Plotinus called Logos to individual human narratives determined not by what one knows leading to choices of what to do but by dispositions of the soul as pure will that generates a sequence of free choices culminating in one moving toward or away from union with Christ. Of course the mind self having appropriate knowledge is not only helpful but necessary for a civilized adult to be saved. How can one choose to believe the Christ-event and the morality associated with it if one does not know about them in the first place? But such knowledge is not sufficient. Of course the willfulness of the body ego expressing feelings or the willfulness of the objective, external mind self exerting control over itself and its environment leads one away from union with the ultimate SOURCE (e.g., Christ). Likewise, the willfulness of the subjective, internal mind self causes one to cling to intellectual insights that become – as it did for Job – a barrier for "seeing God face-to-face," or a barrier to even moving in that direction. This shift from Greek, Neo-Platonic intellectualism to a "Christian voluntarism," see chapter 13, led Augustine to facilitate the "great ending" of Hellenic, rational individualism. Most members of the Catholic Church during Augustine's lifetime were not educated enough or had the necessary leisure to engage in introspective, intellectual contemplation. Moreover, such intellectual activity for those who did have the education and time for contemplation often led to a dead end or worse yet to heresy.

Thus, Augustine began to promote what may be called a patriarchal vision for personal salvation. He anticipated the Inquisition by advocating use of force against heretics. He "advocated the use of force against the Donatests [a heresy], asking why … should not the Church use force in compelling her lost sons to return if the lost sons compelled others to their destruction."[2] He believed in Papal supremacy and promoted obedience to authority and the dictates of the Church. He put less emphasis on rational analysis and logical argumentation and before he died he concluded that introspective rationalism will not lead to happiness or one's salvation. Correspondingly, he put more emphasis on: pledged community membership, trans-generational authority, obedience to divinely-sanctioned standards and *overt suspicion of intellectualism*. He preached the necessity of divine aid for moral transformation. This view became a *patriarchal way to salvation* and may be summarized as follows. One must believe that the Catholic Church is God's representative and therefore, it is the only institution that can truthfully interpret the Christ story and all the scriptures. Therefore, it propagates dogmatic

---

[2] Michael Mendelson. http://plato.stanford.edu/entries/augustine/

doctrine. The Church is the only institution that receives God's saving Grace that is dispensed via the sacraments, such as baptism, to members of the Church. Finally, it has the ability and the right to decide who is worthy to receive Grace. One must give himself/herself over to be ruled by God's Church.

DARK NIGHT OF WESTERN CIVILIZATION'S SOUL

The phrase, "dark night" is a metaphor for the loss of literacy, rational thinking, and personal, interior experiences. The developmental evolution of individual and social consciousness provides a framework for understanding how these three losses are interrelated and what conditions might provoke the losses to occur. As described in chapter 7, the persona self that co-emerged with metaphorical, conceptual thinking divided the mind self into an *external self* and an *internal self* and created in each human an *individual collective unconscious* as an aspect of the Inner Self. The external self expresses personas appropriate to particular situations as dictated by society. For example, males develop a dominating masculine persona and females develop a complementary, subordinating feminine persona. The internal self gains some vague awareness of an Inner Self by repressing inappropriate attitudes and intentions into the individual, collective unconscious that is an aspect of the Inner Self. Civilized humans evolve to an awareness that can be represented conceptually as a conflict not only between the external self and the individual, collective unconscious but also between a self-consciousness constantly buffeted by external events and a spiritual aspect of the Inner Self called the soul. The Greek enlightenment that began with Pre-Socratic arete, see chapter 10, led to the emergence of a *rational, mental self* for elite members of society as described in chapter 12. The rational self emerged in the successive transformations from childhood to adolescence and from there to adolescent adulthood. The rational mental self in the Hellenic world differentiated into *control objectivity* associated with the masculine persona and *participatory subjectivity* associated with the feminine persona.

This third stage of the Greek enlightenment that included the differentiation of the *patriarchal ideology*, see chapter 12, table 12.1, generated new conflicts and paradoxes in individuals and in society. In a process I call *mystical, heroic creativity*, see chapter 13, the rational mind self can look to the Inner Self and create new ways of seeing the world or new ways of seeing one's self. The new vision will depend on whether the control objective mind self or the participatory subjective mind self does the introspection. However, the patriarchal ideology of the Hellenic culture greatly complicated this creative process in several ways. This culture – and virtually all of Western cultures up until the late 20th century – did not recognize participatory subjectivity associated with the feminine persona as a legitimate type of adult maturity. But most creativity involves the collaboration between Eros-chaos, participatory subjectivity and Eros-order, control objectivity. Hellenic thinkers made the situation even more confusing by claiming that the soul is feminine even while looking down upon individual females in their societies. Furthermore, the mystery cults originally peopled mainly by women, rejected inequality of males and females even before the differentiation of the patriarchal ideology. In the first few decades of the Jesus mystery cult, men and women treated one another as equals wherein in some communities men took on leadership roles but in other communities women took on these roles. All this was in accord with Jesus as a model who both in his behavior and his preaching, as told by oral tradition and later by the gospels, was feminine as well as masculine. Moreover, individual and collective mysticism that energized the early Christian Church was more in accord with the "so-called" feminine participatory subjectivity.

The conflict between the patriarchal ideology that advocated control objectivity and Christian mysticism involving participatory subjectivity was intensified by the reemergence of metaphorical, conceptual thinking in Hellenic philosophies. This type of thinking involves subjective insights and provides the language for describing creativity, which is neither exclusively subjective nor exclusively objective. Rather, a new vision comes from the collaboration between the subjective insight embodied in metaphors and the objective form contained in conceptual knowing. The dilemma for Hellenistic thinkers was how to reconcile their strong commitment to the patriarchal ideology and the "feminine" subjectivity absolutely necessary for creativity in general and mysticism in particular. Augustine's development of "Christian, rational individualism" provided an apparent resolution of this dilemma. Christian rationality proclaims that the Inner Self manifests soul that produces subjective insights that are objectified into rational, conceptual knowing. Intellectual contemplation brings the internal mind self to look inward to the soul. Since the soul is a direct manifestation of God (Christ) that produces divine illumination to a receptive mind self, this spiritual reflection brings one to direct, intellectual knowledge of God. Intellectual knowing is consistent with the patriarchal ideology and the idea of divine illumination covers over the idea of feminine, subjective insight. Thus, this patriarchal vision encourages one to develop Christian rationality that enables one to pursue intellectual contemplation that eventually unites the person with God.

The patriarchal ideology combined with Hellenistic rationality to convert the Jesus mystery cult into a patriarchal, institutional Church. Before this could occur the Roman pagan persecution of the Christians in the 1st century AD had to give way to the acceptance of Christianity and to factors that facilitated its spread. As listed in chapter 13: 1) during the 100s AD, traditional polytheism began to decline along with diverse mystery cults becoming more popular; 2) in 212 AD the Roman emperor, Caracalla, granted full citizenship to all inhabitants of the Roman empire thus eliminating persecution or prejudice against anyone who joined the Catholic Church; 3) in 313 AD Constantine signed an edict that permitted freedom of worship thus facilitating the spread of Christianity, and 4) in 399 AD emperor Theodosius ordered the closing of pagan temples. A parallel set of events converted the scattered Christian communities into a unified institutional Church defined by a hierarchal priesthood and a set of religious dogmas. Almost all of the early Christians were non-educated, lower class citizens or even slaves who were only individuated to metaphorical, conceptual thinking. They could believe the stories about Jesus in the same way as citizens of the Mythical Age (3,300 - 700 BC, see chapter 8). But by the time of the Roman emperor Decius (249-252 AD), Christians had become a cross-section of Roman society including members on all levels of its social scale. Because of the influence of Hellenic philosophies, Christians, especially the clergy, who were educated in Greek thought, felt the need to obtain a logical, conceptual understanding of these stories. Clement of Alexandria (150-230 AD) began to integrate belief in these stories with a study of scriptures and Neo-Platonism. When Constantine, who became emperor of Rome in 306 AD, converted to belief in the Christ-event, Christianity became a legitimate religion along side of the other religions in the empire.

The evolution to the "great beginning" of Christian rationalism, as described in the beginning of this chapter, generated a conflict between orthodoxy and individual, mystical experiences. Conversions to the Jesus cult in the 1st century came via personal vision and trance or a communal mystical experience. By rationally conceptualizing these personal and shared visions, Christians created diverse doctrines. As this was occurring, a male priest hierarchy consolidated into a Church began to designate some doctrines as orthodoxy. All doctrines that contradicted the orthodox view were designated as heresy. Because "heretical views" were created as a result of personal introspection, the Church began to discourage one developing a life of "*interiority*," which many people call a *Gnostic* response to life. Berman relying on the views of many scholars, argues that Christianity was a Gnostic/magical (possibly Essene) cult within Judaism; that is, it was a Jewish heresy.[3] As this cult, under the influence of St. Paul, became a universal religion, it also increased its stability and the number of converts by eliminating diversity of doctrines by outlawing heresy. It increased its members by encouraging people simply to believe the orthodox doctrines, obey the rules, and participate in the Church rituals, but not bother with personal reflection on the core ideas of Christianity. As described earlier in this chapter, Augustine developed a possibility for interiority via Christian, intellectual contemplation. Initially this approach seemed to overcome the conflicts between orthodoxy and Gnosticism and between the patriarchal ideology and feminine subjectivity. This intellectual vision marked the highpoint of Hellenistic, Christian rationality. Augustine's rejection of this vision in the years toward the end of his life marked the great ending of Hellenistic, Christian rationalism.

Based on my experiences of mystical, heroic creativity, I presume that Augustine had mystical experiences that caused him to see that direct union with God does not occur via intellectual contemplation. In fact, over time he came to realize that intellectual contemplation produces mental activity that is a barrier to communicating directly with Christ. Moreover, it often enough produces heresies that destabilize the Church thereby endangering some of its members. Therefore, Augustine preached that most Catholics, even and perhaps especially educated persons, must reject intellectual contemplation and embrace the patriarchal way to salvation described earlier. Even though levels of literacy were never high in the Hellenistic world, there was a rise in literacy among Roman citizens between 250 and 100 BC but then a

---

[3] Morris Berman. *Coming to Our Senses* (New York: Bantam Books, 1990), pp. 153-154.

decline of it between 200 and 400 AD. Augustine's vision collaborated with this decline in literacy to reinforce the attitude that began to emerge in the 200s AD that education was not relevant to salvation and that ignorance had a positive spiritual value. Neo-Platonism and the intellectual religions such as Manichaeanism, the last manifestation of Gnosticism in the ancient world, began to fade away from Hellenistic culture. The Church became insulated from new ideas by the authoritarian tradition of Augustine, whose views sounded the death knell for the voice of interiority.[4] Not only did most of the educated elite begin to reject introspective intellectualism, most operated in their daily activities as if they did not have an inner self.[5] Berman sites Charles Radding ("Evolution of Medieval Mentalities: A Cognitive-Structural Approach," *American Historical Review*, 83, 3, pp.577-597) claiming that what happened after Augustine was the disappearance of interiority – not the repression of interiority, but the actual disappearance of it – that lasted down to about 1050 AD.[6]

Monasticism is a retreat by religious men and women from society to pursue spiritual salvation via an ascetics of denial of materialistic desires. It began with hermits, such as Anthony (252-356 AD), but by the 400s AD several sites of communal monasticism rapidly emerged across Europe. St. Benedict of Nursia (480-543) founded a community of disciples at Monte Cassino in 529. Toward the end of his life he constructed a rule for his community that served as a constitution for other monastic communities. In the 500s and 600s AD hundreds of monasteries were founded. Many, especially the Irish ones in 700 AD became learning centers in which books written in the classical Greek and Hellenistic periods were copied in order to be preserved for later generations. Some of the clergy proclaimed books other than the Bible as heathen, pernicious or dangerous works of the devil and so were burned. In contrast, monks in the learning centers thought the ancient classics should be preserved; however, the monks involved in preserving these works had no understanding of what it was they were transcribing. By this time all interiority and intellectual reflection vanished from medieval culture. Monasticism, which followed Augustine's prohibition of intellectual contemplation, opposed much of the tradition of classical antiquity while preserving it in written form. In agreement with Augustine, monks and clerics regarded philosophy as a source of heresy and by 500 AD believed that secular learning was incompatible with Christian culture. Copying manuscripts was manual training as an aspect of ascetic contemplation rather than an intellectual pursuit. The rule of St. Benedict included this approach to monasticism so that abbots in monasteries across Europe led the way for monks to cut their ties to the world of antiquity.[7]

Many thinkers interpret Augustine's patriarchal way to salvation as a very negative legacy for Western civilization. However, from the point of view of mystical, heroic creativity,

---

[4] Morris Berman. *Coming to Our Senses* (New York: Bantam Books, 1990), p.177.

[5] Morris Berman. *Coming to Our Senses* (New York: Bantam Books, 1990), pp.176-177.

[6] Morris Berman. *Coming to Our Senses* (New York: Bantam Books, 1990), p.179.

[7] Morris Berman. *The Twilight of American Culture* (New York: W.W. Norton & Co., 2000), p.81.

it was a great positive legacy. Hellenistic rationalism, tied as it was to the Greek-Roman culture of the Roman Empire, was about to fade away with the fall of the Roman Empire. Because of the efforts of many thinkers culminating with that of Augustine, the teachings of the early Church became intertwined with Hellenistic rationalism. Augustine's legacy severed many Church leaders as well as educated lay people from this form of rationalism. This disconnect allowed the Church to become an important influence on the emerging new culture of the medieval dark ages. Thus, the Church's future viability was no longer tied to the fate of Hellenistic culture that was in rapid decline. Aspects of this culture would emerge later in the 1200s AD and again in the Renaissance (1400-1500), but they would be partially transformed into aspects of new ideologies. Just as appropriate chaos is essential for any transformation to a new ideology, so also the "dark ages" was essential for the transformation to a new Western culture. Augustine as the "first medieval man" was instrumental in bringing about the appropriate chaos for the birth of a new Western culture. Using modern terminology, I would say he initiated *destructive creativity* for Western societies. In terms of medieval mysticism, for example, St. John of the Cross, Augustine was instrumental in bringing a collective "dark night of the soul" to European communities. This purging was absolutely necessary for an awakening that was to lead to the medieval enlightenment.

THE MEDIEVAL AWAKENING

After over 500 years of medieval Europe being cocooned within the authoritarian tradition of Augustine, troubadours (a class of medieval lyric poets who flourished in southern France from 1000s to 1200s AD) and monks began to awaken to an interior life. This renewed interiority was expressed as intellectual reflection that produced heretics, such as the Cathars, and a movement celebrating romantic love. The rediscovery of the inner self spread to the larger populations where the experience of *intent* became more common place. People acknowledged guilt and priests hearing confessions of guilt took on the role of spiritual advisors. Murder with intent became a crime singled out for special punishment. A new sense of spiritual fervor sparked the religious fanaticism that energized the Crusades, but often enough this new religious urgency was coupled with the rising world of laissez-faire that produced rapid economic growth in craft, trade, and business. This phenomenon collaborated with suddenly great numbers of individuals interested in new skills and knowledge for personal advancement. Also, urban schools began to challenge the supremacy of monastic schools wherein this challenge finally led to forming universities. The "free market" mentality led to family-owned private property. This produced the need for individual learning leading to lay professionals, such as secular scribes, notaries, and lawyers. Correspondingly, this revived Latin studies because charters, court decisions, and records were written in Latin.[8]

---

[8] Morris Berman. *The Twilight of American Culture* (New York: W.W. Norton & Co., 2000), pp.85-86.

What were the causes for this radical shift in self-consciousness? The end of chapter 2 describing physical individuation provides a general answer. The radical shift is individuation that involves an order, chaos, new order process in which an emerging new order incorporates a modified old order to produce a new, higher level of individualism. This principle applied to mind self transformations is manifested as a transformation to the persona self as described in chapter 7 and as a sequence of individuations to adolescence and then to adolescent adulthood as described in chapter 12. The specific factors driving the radical reawakening from the medieval dark ages may be described in terms of Sufism and the dynamics of romantic love.

Sufism

According to Robert Graves, Sufis believe Sufism to be the secret teaching within all religions. They are not bound to any religious dogma; they use no regular place of worship, have no sacred city; nor do they have monastic organizations or religious instruments.[9] Its universality is because its foundations are in every human mind so that Sufi development must inevitably find its expression everywhere.[10] I propose that Table 14.1 (see below) that lists 28 contrasts between participatory subjectivity and control objectivity provides a way to characterize Sufism. The fully individuated Sufi, like Jesus, himself, emphasizes participatory subjectivity but also has developed control objectivity. As a result, he or she can be in the world but not of it. Because modern societies subscribe to the patriarchal ideology that represses participatory subjectivity, the average citizen will not understand Sufism. The person who seeks to understand it must first develop some of the perceptions, attitudes, and skills associated with participatory subjectivity as listed in Table 14.1. Some of the more important of these, listed in bold face, are: 1) subjective knowing involving direct experience, 2) transformational-evolutionary development, 3) empathetic engagement, 4) non-verbal, metaphorical or analogical intuition, 5) mind self and body ego subordinated to soul, 6) empathetic love, and 7) love others and want to be loved. Table 12.1 provides a more abbreviated and somewhat different perspective of a Sufi who is seen as both masculine and feminine.

---

[9] Idries Shah. *The Sufis* (New York: Anchor Books, 1971), p.vii.
[10] Idries Shah. *The Sufis* (New York: Anchor Books, 1971), p. 27 and p. 34.

Table 14.1

PARTICIPATORY SUBJECTIVITY VERSUS CONTROL OBJECTIVITY

SUBJECTIVITY Via The primacy of metaphorical, conceptual knowing leads to the primacy of **PARTICIPATORY CONSCIOUSNESS**	**OBJECTIVITY** Via Logical-conceptual knowing that leads to the primacy of **CONTROL CONSCIOUSNESS**
1. **Subjective Knowing, direct experience**	1. Objective Knowing, book knowledge
2. Experiential Feelings	2. Rationalized Feelings
3. Open to new ways to approach a "solution" to a problem	3 Problem Solving via pre-established methods and guidelines
4. Situational Morality	4. Rule-Defined Morality
5. **Transformational Development**	5. Adaptation Development
6. Expressive Individualism	6. Utilitarian Individualism
7. Receptive, Open	7. Socially determined perspective, Conservative
8. Oceanic—Explorative	8. Structured—focused

9. Gestalt-Integrative-Visionary	9. Mechanistic-Analytical
10. Perceptive-Engaged-Empathetic	10. Judgmental-Separate-Critical
11. Sharing Dialogue; Intimate Conversation	11. Discussion-Debate-Argumentative; Didactic Explanation
12. Unstructured Playful	12. Goal-oriented Play
13. Meaning determined by context	13. Meaning = Information independent of context
14. Non-verbal or Metaphorical or Analogical Intuition	14. Conceptual or Literal
15. Harmony	15. Order
16. Social cohesiveness via an "ethic of care"	16. Social cohesiveness via rule of law & order
17. Non-cultural-social but individual and personal	17. Non-individual and non-personal but cultural-social
18. *Zen creativity* (dreaming, seeing creativity); the metaphor for this is Virgin birth of a god, e.g., the Christmas story and *passion creativity* the metaphor for this is: Life, Death, Rebirth process , that is, Order, Chaos, New Order that includes modified aspects of the old order.	18. Maintain homeostasis rather than be creative or be open to a transformation.

19. Communal—Interconnected	19. Autonomous within Hierarchal Community Structure
20. Unconditional love & feeling bonding	20. Respect based on merit & bonding via a social contract
21. **Mind self and Body ego are subordinated to soul and therefore love is primary,** but in an unhealthy person self-centeredness (ego centeredness) causes feeling love to become self-centered love, possessive love, narcissistic love	21. Self-centeredness (ego centeredness) causes the control oriented person to be (1) addicted to progress, (2) the Faustian person always seeking, (3) the idealist always pursuing a non-ending quest, (4) the tragic lover whose romantic love only can be consummated by death
22.**Empathetic love that leads to joy and nurturing of another** but also may lead to blocking the awareness of the need for "self-transformation" or the nurtured person's need for "self-transformation"	22. Commitment to autonomy which prevents receiving or giving empathetic love.
23.Leadership style that empowers others and/or facilitates others; the "followers" are encouraged to feel good about themselves and only secondarily to admire the leader	23. Leadership style that seeks to dominate, command, and employ strategies for charismatic inspiration of others to follow the leader's ideas or programs

24. Employ strategies for <u>transcending conflicts</u>	24. Employ strategies for <u>resolving conflicts</u>

1. Assert one's position and then seek ways of achieving harmony that transcends the conflict; one's position may be modified but it is not compromised or totally abandoned.
2. Seek inner harmony (peace) in the context of external chaos or external defeat.
	1. Attack-defend
	   a. Aggressive: Confront-attack
	   b. Confront-withdraw (to avoid defeat)
	2. Seek Win or Stalemate
	   a. Manipulate/Stalemate: Confront-depend
	   b. Limited Win/Stalemate: Reduce-depend
	3. Seek to Accommodate: Reduce-withdraw
	4. Avoid Conflict: Reduce-attack

25. Living in and content with each moment of one's life	25. Ambitious, always living in the future, which holds promise of accomplishing some goal
26. Seek or promote situations where everyone wins; that is, prefer Win/Win or no deal interpersonal interactions	26. Competitive; prefer Win/Loose = Zero sum game in interpersonal interactions
27. Egalitarian, communal where status is not important	27. Hierarchal; seek status within a hierarchy
28. **Love others and want to be loved**	28. Have Power over others and want to be respected

Even though Sufism appears in historical times mainly within the pale of Islam, it should not be considered to be merely the mystical aspect of Islam. Rather, it has been one stream of direct, evolutionary experience of humanity that has influenced all the great schools of mysticism.[11] However, though orthodox, legal interpretations of the Quran sometimes opposed

---

[11] Idries Shah. *The Sufis* (New York: Anchor Books, 1971), p.60.

Sufi mysticism, Islam provided protection to Sufism and facilitated its spread in the world. This is because all that a person need do to be an Islam believer is to assert that he/she subscribes to the formula: "There is no God but Allah and Mohammed is His Prophet." There are no dogmas describing the nature of Allah and the relationship of Allah with the Prophet was fixed; so a Sufi would have no problem with subscribing to Islam by means of this formula.

Muhammad ibn abd Allah was born 570 AD in Mecca, Arabia (near the Red Sea) and in 622 migrated to Medina (Yathrib). This year marks the shift from a persecuted minority religion to an organized autonomous Islamic community. At the time of Muhammad's death in 632 Muslims by a mixture of preaching, persuasion and force controlled all of the main tribes and towns of Arabia. By 670 Islam had captured Jerusalem, Damascus, Syria, Persia, Cairo, Egypt, and Tunisia and in 708 reached the Atlantic Ocean thus occupying Morocco just south of Spain. In 711 Muslims conquered Spain that was under the control of the Visigoths. By 713 they occupied all of Spain except Asturias in the far north. In 718 Muslims crossed the Pyrenees mountains and invaded southern France, but in 732 they were turned back outside of Poitiers several miles north of the Pyrenees. The first and most powerful, classical Sufi school in Europe was founded in Spain in 711 AD. Sufis accompanied the Arab armies in their conquest of Spain.[12] This school brought forth the theory of *theosophy*, which accounted for different religious manifestations among different communities. Fariduddin Attar (1,110-1220 AD) described the stages of development of a Sufi in the book, *Parliament of Birds*. In this work, birds, who represent humanity, are called together by an enlightened Sufi who tells them that in their quest for enlightenment they have to traverse seven stages called valleys. The second stage is the Valley of Love, which is the limitless area in which the Seeker is completely consumed by a thirst for the Beloved.[13] This "thirst for the Beloved" is *romantic love* that is expressed in diverse ways producing diverse negative and positive consequences. Thus, Attar expressed the direct Sufi stream of romance teaching that antedated its appearance in southern France in the lyric poetry of troubadours. The theory of archetypes, first stated by Ibn El-Arabi,[14] provides a framework for conceptualizing the dynamics of romantic love. Attar's *Parliament of Birds* implies this archetypal theory of romantic love.

Romantic Love

Based on my own experiences of mysticism and romantic love, I propose an expanded understanding of the Jungian theory of this kind of love in the context of my vision of mystical, heroic creativity. Having worked out the details while in bed in the early morning hours, I was pleased to read a similar interpretation in a description of Attar's work.[15] The term, archetype, must be understood metaphorically as involving intuition rather than rationality that uses logical,

---

[12] Idries Shah. *The Sufis* (New York: Anchor Books, 1971), p.51.

[13] Idries Shah. *The Sufis* (New York: Anchor Books, 1971), p.121.

[14] Idries Shah. *The Sufis* (New York: Anchor Books, 1971), pp.58-59.

[15] Idries Shah. *The Sufis* (New York: Anchor Books, 1971), pp.119-123.

conceptual thinking. An archetype is a pattern stored in the collective, non-conscious, which is one of the three aspects of soul described in chapter 7. This pattern guides a particular aspect of the developmental evolution of individual consciousness, but the details of this guidance depends on the culture and historical context in which it occurs. The embodied soul as a feeling center points to the individuality of a personality living out his/her life story, but the soul as the collective non-conscious that contains archetypes points to membership in the archetypal world that contains all humans past, present, and future. In this perspective the individual soul takes on the aspect of a universal soul that manifests processes that all humans can carryout but in diverse ways. When a person is drawn into individuation guided by archetypes, his/her everyday awareness enters the archetypal world. The experience of this is overwhelming and mentally destabilizing. This is because a different aspect of human consciousness lives in the archetypal world that displaces the mind self-consciousness. As a result, the mind self sees that it is not in control and doesn't know what is happening.

The pair of patterns, anima and animus, are the major archetypes that guide the transformation to a persona self. In most males, the socially defined masculine traits are associated with the *animus* that defines the masculine persona; the socially defined feminine traits are associated with the *anima* that is repressed into the individual, collective unconscious. The converse of this development occurs in most females. The anima defines the feminine persona and the animus is repressed into the individual, collective unconscious, see chapter 7. When a male is overtaken by *romantic love*, he projects his unconscious anima on to a particular concrete female who becomes the object of his overwhelming yearning. Even though he thinks he loves and yearns for a particular woman, this is an illusion. He loves and yearns for the anima "stored" in his soul that is unavailable to him except as a projection onto a particular woman. The same dynamics of romantic love apply to a female. Her illusion is that she loves and yearns for the animus "stored" in her soul that is unavailable to her except as a projection onto a particular man. The overall process is non-rational and therefore disconcerting, especially to humans who have individuated to rational persons who are use to making decisions based on rational analysis. Such a person is temporarily insane and will tend to be irrational and self-destructive. However, at first, "falling in love" amounts to the ever present but hidden archetypal world and an aspect of one's soul invading the mind self's ordinary consciousness. The experience is exciting, fascinating and tantalizing in that the person glimpses his/her true self, the soul, but yet, he/she cannot unite with it. One looses his wits in rapture and emotionally over laden, spiritual ecstasy of seeking to be completed by union with soul that forever recedes from one's grasp. Thus, the love experience is "sweet sorrow."

The experience of romantic love may produce negative behaviors. The "lover" seeking union with the "beloved" may attempt to do this by dominating and controlling many aspects of the person who is the object of one's affection. If the "beloved" seeks to break off the relationship, the lover may become so distraught that he/she resorts to some form of self-destruction, even suicide, or violence to the beloved ranging from stalking to molesting or even murder. As the lover engages in concrete interactions with the beloved, he/she gradually will see that the real person is very different from the anima or animus projection. The disillusioned lover may remain in despair or may be drawn to "loving" the process of falling in love, which now begins to happen over and over again with new people each time. Alternatively, as exemplified by the troubadours in southern France during the 11th through 13th centuries, the romantics are in love with the ascetic yearning; so they choose situations where the lovers can never be united except, presumably, after death. The same psychological, archetypal dynamics applies to what may be called *erotic mysticism*. The lover yearns for the divine person in heaven or as the source of one's inner self, but again the lover really is yearning for an aspect of one's individual soul that is projected onto the imagined divine being. The person thinks he/she is yearning for Christ or Allah but in fact is seeking an aspect of one's individual soul. Not only does this illusion not produce mystical union with the divine, it is a road block to ever achieving it.

The experience of romantic love or erotic mysticism does awaken one to the interior life. The person becomes aware of a gap between the external, objective mind self perceiving the material world and the spiritual soul as an aspect of the inner self. The illusion associated with these kinds of ecstatic yearnings may lead one to existential despair. One thinks that the gap – what Attar calls a valley – can never be traversed. So one's spiritual yearning is forever frustrated. One may come to believe either there is no divine Being; or if there is one, it has left all humans engaged in a hopeless quest for meaning or completeness. Alternatively, the awakening to interiority may lead one, who is embedded in something like the Augustian, authoritarian Christian religion, to an intellectual, religious nihilism. In this view, God created (creates) each spiritual soul that, in turn, generates rational thinking, but this God cannot be reached by Neo-Platonic, intellectual contemplation. At the same time, the institution that generates this vision, such as the Catholic Church, may impose a patriarchal way to salvation that denies or prohibits the interiority one directly experiences in romantic love or in erotic mysticism. Therefore, the necessary renewal of that institution, such as, the Catholic Church, would have one discover one's true self by developing rationality. At the same time, one must deny as much as is consistent with one's survival, all desires of the body ego, including all sexual interactions, because these desires oppose the spiritual expression of the soul. In effect, one reinforces an awareness of soul, one's true self, by understanding the mind generated by the soul as spiritual and good in a continuous struggle against the body viewed as non-spiritual and evil. The extroverted, ascetic control of the body continuously focuses the mind on this core vision. This is one way of describing what the Cathars in southern France developed during the 12th and 13th centuries. It is the heresy of attempting to transform traditional medieval Christianity into a form of Manichaeanism described in chapter 13.

Romantic love or erotic mysticism also can lead to a positive spiritual transformation. As the lover sees the illusion of archetype projection, he/she may make two decisions that reinforce one another: 1) one rejects the ecstatic experience of romantic yearnings and 2) one incorporates the projected archetypes into one's ordinary self-consciousness – the masculine persona incorporates the *anima* perspective. As a result, the transformed mind self has greatly expanded its options and power to interact with new, challenging situations. For example, the masculine mind self now can be more empathetic thereby leading to an ethic of care and be more intuitive. Likewise the feminine mind self that incorporates the *animus* archetype can be more assertive and in some situations even aggressive and can be more analytical in confronting some problems. Now the transformed mind self is consciously connected to the Inner Self that provides energy for creativity. Thus, the mind self having passed through the "Valley of Love," moves to the "Valley of Intuitive Knowledge" in which the heart receives directly the illumination of Truth and an experience of God.[16] With respect to mystical, heroic creativity, see chapter 13, the newly empowered mind self can look to the Inner Self thereby bracketing the mind self in order to receive "intuitive knowledge," which Augustine called divine illumination.

I propose that the medieval awakening is analogous to the avant garde modernism that began in France around 1885, see chapter 21. The Sufi mystics from Spain were like avant garde artists and thinkers, such as Nietzsche. Both groups started a countercultural movement in small, elite communities – in the 13th century they were the monks and troubadours; in the 19th and 20th centuries they were the modern radical artists and thinkers. The countercultural movement led to the more extensive social awakening to subjectivity and interiority. In the 13th century this led to the medieval enlightenment as represented by the philosophy-theology of Thomas Aquinas. The countercultural revolution begun in the late 1960s led to the more extensive social awakening manifesting as degrees of radical, expressive individualism. The 1960s revolution produced postmodernism in the 1980s that generated social-cultural degeneration culminating in the financial crisis that emerged in September, 2008. American society now is evolving toward differentiation of the fourth enlightenment.

Chapter 15

CORE IDEAS OF THE MEDIEVAL ENLIGHTENMENET

EVOLVING RECEPTIVITY OF MEDIEVAL CULTURE

The medieval culture beginning with the last years of Augustine evolved toward a "mentality" structured to be receptive to some ideas and to be non-receptive to others. This evolution occurred in the secrecy of the dark ages, which we now can discern after the fact by noting what ideas were accepted and what ideas were rejected. Neo-Platonism was a theme that has a rejected aspect and a modified accepted aspect. On the one hand, the introverted intellectual reflection on the inner self as the divine Logos was rejected, but, on the other hand, extroverted intellectual reflection on the inner self that leads to a new understanding of nature was accepted. That is, the Plotinus version of Neo-Platonism that deemphasizes subjectivity and concrete

---

[16] Idries Shah. *The Sufis* (New York: Anchor Books, 1971), p.121.

experiences was rejected, but the Neo-Platonism that includes aspects of Aristotelian philosophy and concrete experiences was accepted. Correspondingly, Augustine's idea of divine illumination that leads to mystical union with God manifesting the Inner Self was rejected – except for a few mystics – but divine illumination that leads to a new understanding of nature was accepted. The Aristotelian idea of God as "the uncaused cause that created the world but then remained separate from it" was rejected, but God as continually manifesting the world and then being immanent in it even as IT transcends the world was accepted. God manifesting reality as a cyclical process or as a linear, evolutionary process was rejected, but God manifesting Order represented as eternal natures whose interactions define *natural law* was accepted. Correspondingly, each human viewed as a developmental, evolutionary life story was rejected, but each human having a *nature* that determines behavior was accepted. Human individualism emphasizing free choices that determine one's life story was rejected in favor of admonishing each human to submit to the dictates of the Church and passively receive Grace by participating in Church ceremonies and sacraments. Correspondingly, Jesus Christ's life, death, and resurrection metaphorically understood as an example of how one is to achieve salvation was rejected in favor of the death of Jesus Christ understood as ransom for the sins of humanity and his resurrection understood as the source of Grace automatically given to those who submit to the mystical body of Christ, which is the Church.

The rejection and acceptance of these ideas constituted the medieval Enlightenment as expressed primarily in the philosophy-theology of Thomas Aquinas. Islam, especially Sufism, transferred many if not most of these rejected and accepted ideas through Albertus Magnus (1193-1280) to his student, Thomas Aquinas (1225-1274). Albertus was well versed in Spanish Sufi literature and philosophy. He even dressed as an Arab as he presented in Paris the ideas of Aristotle from the works of al-Farabi, Avicenna (Ibn Sina), and Ghazale.[17] Overall the rejected and accepted ideas of the medieval Enlightenment came from five major sources: 1) Augustine, 2) Mu'tazilis, 3) Spanish illuminism, 4) Spanish theosophy, and 5) Sufism in general. Aspects of the legacy of Augustine, his idea of Neo-Platonic, intellectual contemplation, and his voluntarism as described in chapters 13 and 14, was rejected, but a modified version of his patriarchal way to salvation was accepted.

The *Mu'tazilis*[18] is the rationalistic school of Muslim theologians. They believed that the Quran was created in time and therefore could be interpreted with considerable allowances made for time-bound changes due to historical and social conditions. This is in opposition to the orthodoxy of Sunni Muslims (as put forth by Ahmad ibn Hanbal) that proclaimed the Quran pre-existed in Allah and therefore must be taken as literally true. With respect to Christian scriptures, medieval culture adopted views similar to Mu'tazilis and rejected views similar to the Sunni orthodoxy. Mu'tazilis defended the idea of free choice based on rational knowledge of

---

[17] Idries Shah. *The Sufis* (New York: Anchor Books, 1971), p.256.
[18] Malise Ruthven. *Islam in the World* (New York: Oxford Uni. Pres., 2000, 2nd ed.), pp.87, 97, 99, 106, 149, 196.

nature. Thus, God is obliged to reward the good. This is in contrast to the Sunni divine command theory of ethics in which an act is right because God commands it. The Mu'tazilis attitude was developed by Aquinas and others into the doctrine of natural law. Thomas' assertion was that it is impossible for the truth of faith to be contrary to the principles known by natural reason. The rationalistic tradition was further developed by the two great Muslim philosophers, Avicenna (Ibn Sina, 980-1037) and Averroes (Ibn Rushd, 1126-1198). Ideas in Avicenna's works were taken from Aristotle, but it primarily was Averroes who introduced Aristotelian philosophy to the West through Albertus Magus and Thomas Aquinas. The Mu'tazilis interpretation of the Genesis story was that Adam's act of disobedience was understood from a human evolutionary point of view not as the cause of original sin but as the unavoidable consequence of his humanity. That is, Adam's first act of disobedience also was his first act of free choice resulting from his emergence from a primitive state of instinctive appetites to the conscious possession of a free self capable of doubt and disobedience.

The essence of Spanish, Sufi illuminism is the formulation of schools of study and practice in which the teacher, what is taught, and the learner (the student) form a unity.[19] This unity contrasts with one extreme where the teacher transmits information to the student who temporarily stores it without digesting it, that is, without internalizing it in any way. An intermediate version of this teaching-learning unity involves the student to some degree conceptualizing the received information that then is stored as theoretical knowledge or as some practice, such as a technique or ability to solve a particular type of problem. The other extreme of this unity is *the Secret Doctrine* wherein the teacher's deep, subjective understanding based on personal experiences *elicits* an engagement of the student with the teacher and the teaching process. This engagement, in turn, leads the student to construct an individualized understanding of what is being taught. This individualized understanding, in turn, evolves as it is influenced by one or several of the following factors: 1) further interactions with the teacher; 2) interactions with other teachers represented by persons and books, and 3) a student's reflection on his or her personal experiences. Thus, the "secret doctrine" is the teacher acquiring wisdom from intuitive reflection on personal experiences who, in turn, elicits in his students an analogous experience of acquiring wisdom. By and large medieval culture rejected this doctrine that became labeled as occultism, illuminism, or Sufi mysticism.

As the West was closing the door on the influence of Islam, especially the influence of the Spanish illuminists, there were a number of "Sufi-like" thinkers who passed on the tradition of the secret doctrine. These include: 1) St. Brother Anselm (1033-1109), a Majorcan mystic also known as the sanctified Sufi Abdullah el Tarjuman, 2) St. Raymond Lully (1235-1315), 3) Roger Bacon (1214-1294), 4) St. Albertus Magnus (1193-1280), 5) Ibn Masarrah of Cordoba (800s AD), and the three intellectual giants of Muslim philosophy: 6) al Ghazali (1058-1111), 7)

---

[19] Idries Shah. *The Sufis* (New York: Anchor Books, 1971), p.265.

Suhrawardi (1154-1191) who provided Dante with his ideas, and 8) Ibn al-Arabi (1165-1240).[20] The idea of direct, experiential illumination is the core of Augustine's doctrine of divine illumination, which I described as Christian, pre-Socratic arete at the end of chapter 13. This and Sufi mysticism is an expression of mystical, heroic creativity also described in chapter 13. Though medieval culture seemed to have rejected this inspirational illuminism, this way of knowing generated in Thomas Aquinas the core of his thought, which is the analogy of being to be described later in this chapter.

The analogy of being initially cannot be rationally understood, but rather it first must be experienced and then conceptualized. Jacques Maritain described his experience of being" when he was twelve years old. This resonated with me because I had a similar experience when I was twelve. I was laying on my parents bed when it occurred to me that thirteen years earlier I did not exist but now I do. This was an overwhelming insight that as an adult some years later I could verbalize in the fashion of St. Augustine's introspective affirmation of God's existence. I am a finite being continually generated by the source of being, which is pure Being. As a result of my experience, when I as a 17 year old took "Introduction to Metaphysics" at the University of St. Louis, my mind immediately resonated with the Thomistic idea of the analogy of being. But this analogy of being presented at the leading center of Neo-Thomism in the US in the 1950s was taught as a rational idea. One supposedly intellectually grasped it by means of a higher level of abstraction meaning a higher level of rational conceptualization. As described in chapter 5, this is correct as far as it goes in that metaphorical, conceptual thinking involves conceptualization. Also, logical, conceptual thinking is grounded on this ordinary language kind of knowing. However, rationality was emphasized in lieu of mentioning the intuition of subjective, metaphorical thinking. The result was and is that many teachers of Thomism, most priests exposed to it and most students never really get the experience of Being. This is because even though Thomism absolutely depends on the collaboration between intellect and inspiration, the student usually is unable to grasp this collaboration. Why? Because since the Middle Ages, the secret doctrine has been shut out from both religious and secular university education. Father "Wild Bill" Wade, S.J., Chairman of the philosophy department at St. Louis U. in the 1950s was an exception to this generalization.

Theosophy may be described as the core mystical insight of any religion or spiritual tradition. Ibn al-Arabi (1165-1240) is the major Spanish theosophists who had a profound influence on Thomas Aquinas (1225-1274). Arabi[21] proclaimed that God is pure Being who is the creator of levels of finite being as a result of an on-going manifestation of them. Thus, using modern terminology, God is immanent in all finite beings by continually manifesting them and yet He also transcends all of them. This dual idea of manifesting and transcending all finite beings enables Arabi to avoid pantheism. The Thomistic idea of the analogy of Being reinforces

---

[20] Idries Shah. *The Sufis* (New York: Anchor Books, 1971), pp.272-276.
[21] Malise Ruthven. *Islam in the World* (New York: Oxford Uni. Pres., 2000, 2nd ed.), pp. 236-238.

and perhaps was influenced by Arabi's vision. Arabi viewed human consciousness as existing along a similar continuum of consciousness from the absolute One as pure consciousness through various levels of consciousness down to lowest animate or inanimate creatures. Arabi's vision is in line with that of Advaita Vedanta and to some extent with the Neo-Platonism of Plotinus. Aquinas retained the idea of this continuum but replaced consciousness with the Aristotelian idea of *intelligibility*. This replacement produces a radically different outlook. For Aquinas there is a separation of the intelligible known from the knower. This separation leads to the Aristotelian, extroverted control humanism in which humans can rationally know nature and thereby control it. Of course, this view fit in well with the already well differentiated patriarchal ideology of the Church. In contrast, Arabi considered such a disjunction as an illusion. Any finite being is not autonomous and intelligible. Any creature only is intelligible in the context of being known by a human knower. Therefore, the knowledge generated by the intelligible being is finite and like any finite being is a manifestation of God. This means that just as God continually manifests all finite beings, He also continuously manifests all finite knower-known collaborations. These finite collaborations may involve constructing logical, models to represent them. But there always must be an original "seeing" that enables one to construct a rational model. This "seeing" is divine illumination that, as stated by Augustine, occurs to believers, agnostics, and atheists alike. One way of stating this universal phenomenon of inspiration is that a new idea must be "felt," that is, seen by intuition, before it can be grasped intellectually. Such a statement is consistent with Arabi's vision.

Illuminism and theosophy are aspects of Sufism, but I felt it important to single them out as sources of ideas rejected and accepted by Medieval culture. Chapter 14 and Table 14.1 described Sufism as a whole as emphasizing participatory subjectivity even while also developing control objectivity. Much to the puzzlement of many scholars of Islam, El-Arabi was a conformist in religion while remaining an esotericist in his inner life.[22] Thus, he exemplifies both ways of knowing and being. Medieval culture further differentiated the patriarchal ideology that began to be associated with an emerging lessaiz-faire capitalistic vision. Understandably this produced a turning away from Sufism as a whole. However, one aspect of this Sufi vision was incorporated into Thomistic scholasticism. Thomas Aquinas influenced subsequent thinkers to seek God in their understanding of nature. This meant that any pursuer of truth, who gained insights about nature could be a guide to one seeking union with God. This is in accord with the Sufi vision that God is immanent in all of nature. Therefore, direct experiences of nature followed by appropriate reflection leads one to some kind of union with God. The sequence of "felt then intellectually known ideas" now must be understood as "experienced and then rationally known ideas."

---

[22] Idries Shah. *The Sufis* (New York: Anchor Books, 1971), p.161.

CORE OF THE THOMISTIC VISION

In radically modifying Augustine's vision, Aquinas prepared the way for the Scientific Enlightenment. Aquinas, so to speak, brought the One back into the Many. The core idea here is a radically transformed version of Aristotle's formulation of the *analogy of Being*. Each thing-event is in the same non-conceptual category according to the following analogy. Each being has an *essence* that specifies the way it interacts with any other being, and the *existence* of each being is the Energizing Source of all the changes and interactions of that being. While the essence of any being is distinct from its existence, there is a mutuality between essence and existence analogous to the mutuality between thing and event. Any finite being, according to Aquinas, is an *essence-existence mutuality*; so in this way all finite beings are similar, but no two "essence-existence mutualities" are exactly alike. No two beings fit into any conceptual category. In order for any two beings to be classified, that is, fit into some conceptual category, they must be identical in some respect. Since, as stated earlier, no two beings are identical, no classification scheme is totally adequate for describing interacting beings in the world. However, reason by itself begins and ends with conceptual categories. Thus, the only way to *truly* understand Nature is to comprehend it by a meta-rational analysis grounded on the analogy of being. This is what both Aristotle and Aquinas meant by meta-physical knowledge.

However, Aquinas went way beyond Aristotle in saying that God also is most fundamentally *BEING* and thus He is in the same analogical category as all finite beings. According to the Old Testament, when Moses asked God by what name He should be called, God answered: "I am who am." Aquinas interprets this to mean that in God, ESSENCE and EXISTENCE are non-different, just as in Plato's self-generating circle, descent and ascent are non-different though to finite humans each has a different meaning. Thus, according to Aquinas's insight, from an analogical point of view, each finite being is God and, what is more, God in His timeless Creative act is *descending into* and *sustaining in existence* each of His creatures.

Thus, Aquinas's vision reinstates harmony among the Many that transcends all disharmonies that happen to occur because God is in all finite beings. That is, *this world* is fundamentally harmonious and Good because each being analogically speaking is God and each being at every moment is being sustained in existence by God. If God is ultimate Harmony and ultimate Good, then as a result of *the analogy of being* and of God sustaining each being in existence, each finite being not only is Being but also is Harmony and Good. This seems like a strange way of talking about finite beings because in the human phenomenon of *objective knowing*, humans think of finite beings as autonomous objects "out there" existing independent of the knowing subject. But this way of thinking is "seen" to be an *illusion* when one applies Aquinas's metaphysical-religious perspective to the nature of objective knowing. Just as *running* is an aspect of a person who is running, that is, running exists only as an attribute of the person, so also objective knowing is an aspect of the person who is knowing some object "out there." According to Aquinas, the knowing and running are real because they are energized by the *existence* of the person who knows and runs. But in each moment in time with respect to the person and in each timeless moment with respect to God, this existence of the person is real, because it is analogically the same as and sustained by EXISTENCE descending into the person. Thus, the objective knowing and running by virtue of being sustained by existence of the person are indirectly sustained by EXISTENCE descending into the person. Any object of objective knowing, whatever else may be said about it, is an aspect of the objective knowing. Therefore, this known object also is sustained by EXISTENCE, and in fact, so also is any object supposedly out there that comes into a person's awareness as a result of his/her objective knowing.

Objective knowing has two defining characteristics. Firstly, objective knowing consists of the interaction between a knowing subject and the object being known. The subject and object imply one another; neither has existence independent of the other. Secondly, objective knowing is the foundation of all human language, in that it comes into being as a result of a consensus between two or more humans who agree on a commonality present in each of their individual subject-object knowing. That is, the consensual meaning present in all the individual subject-object knowing converts all of these subjective knowings into objective knowing. After the extensive differentiation of rational thinking such as occurred to a few Greeks, such as Aristotle, in classical Greece, humans having the shared experience of many instances of objective knowing can and did create the myth: each of us humans, as well as all the objects that we have become aware of, exist now and have been existing even before we started to have objective knowledge of them. This is a very useful myth because it facilitates discourse among humans that leads to an expansion of objective knowing. This, in turn, enables humans to survive more effectively in a society that, in turn, can more effectively continually adapt to its environment.

It is a useful myth to think that individual humans and each of the objects of their objective knowledge are autonomous beings out there in Nature. However, when humans commit themselves absolutely to the autonomous existence of these things in Nature, they are forgetting that their commitment is based on a story that humans made up at one time, perhaps without fully realizing that they were creating such a myth. The commitment to the autonomous existence of things thus becomes an *illusion*. Without realizing it, humans pretend that autonomous things exist out there in Nature and then humans construct new myths that explain how any human can come to know these autonomous things. Both Aristotle and Aquinas were committed to this illusion of *objective reality*. I believe that Aquinas had the metaphysical-religious insight to avoid this illusion, but up until the last few months of his life, he was a victim -- as all of us are in one way or another -- of the consensus of the age in which he lived.

If we ignore Aquinas's theory of knowing and focus only on his metaphysical-religious insight, we see that he has the non-dual perspective like that of Plato/Plotinus, Vedanta, Mahayana Buddhism, Tantra, and so on. According to Wilber, Vedanta includes three core ideas. 1. The One is the Good toward which all things via evolution ascend. 2. The One is the Goodness that by de-evolution descends to the Many. 3. A transcendental understanding of the One is that It is the absolute Non-dual Ground both of the One and the Many.[23]

In one sense, Aquinas has the same view of reality as the non-dual perspective, such as, the Vedanta. In Vedanta, a finite thing is neither Real nor non-real; rather it is *contingently real*. That is, the being is not Real in that it is not the non-manifested Brahman, but it is not non-real in that it *is* a manifestation of Brahman and the manifestation is such that any non-dualist would say that each being *is* manifested Brahman. In like manner, Aquinas would say that a finite being is neither EXISTENCE nor NON-EXISTENCE; rather, it is *contingent existence*. That is, the finite being is being only as a result of the analogy of being, but its essence is not identical (is not non-different from) to its existence. But the finite being is *existence* (determined and therefore limited in a certain way by its essence) by virtue of EXISTENCE at all times descending into it and sustaining it. The descending EXISTENCE is BEING the finite thing in ITS infinite act of creating the creature, so the creature is existence. Or to say the unsayable in still another way, though from the creature's point of view, the finite being is contingent existence, but from God's point of view, the finite being is God. Also, in Vedanta, humans delude themselves when they think that things are the way or approximately the way humans objectively know them as if things are simply there for the knowing subject to grasp. According to the Vedanta, it is an illusion to believe that the world out there is made up of interrelating, autonomous things. Because of his theory of knowledge, whose core ideas he took from Aristotelian epistemology (theory of knowing), Aquinas bequeathed this kind of illusion to post-Medieval culture. However, I believe, had Aquinas not subscribed to this Greek rationalism, he would have come to a similar view of human knowing as described by the Vedanta.

If the story told by Maritain is true then I believe that Aquinas saw through his illusion a few months before he died.

> One day, the 6th December, 1273, as he [Aquinas] was celebrating
> Mass in the chapel of St. Nicholas, a great change came over him.
> From that moment onward he ceased to write or dictate. Was the
> *Summa* [his life's work] then, with its thirty-eight treatises, its three
> thousand articles and ten thousand objections, to remain
> unfinished? Reginald [a fellow Dominican monk assigned to take
> care of Aquinas] ventured to complain: "I can do no more," said
> his master. Reginald insisted: "Reginald, I can do no more: such
> things have been revealed to me that everything I have written
> seems to me rubbish. Now, after the end of my work, I must await

---

[23] Ken Wilber. *Sex, Ecology, Spirituality* (Boston: Shambhala Press, Inc., 1995), pp.346-347.

the end of my life...." He died... on the 7th March, 1274, in the forty-ninth year of his age.[24]

Thus, on the one hand, Aquinas revived by implication a universal ecology similar to the one implicated by the vision of Plato and Plotinus. However, as indicated above, Aquinas learned from and expanded Aristotle's theory of knowledge of the world, called the science of nature. This *natural philosophy* was interwoven with and seemingly indispensable from Aquinas's metaphysical-religious insight. Soon after Aquinas died, the Church proclaimed him a Saint and designated him the Angelic Doctor of theology. This, of course, sanctioned Aquinas's illusion - his natural philosophy - along with his philosophical-religious insights. According to Aquinas's religious insights, God is present in Nature so that by studying Nature with this in mind, one can ascend to God. However, as soon as God becomes an object of objective knowing, He, as Heidegger would say, retreats from being comprehended in that way. Objective knowledge taken to be absolutely True or even approximately True, is a barrier to ascent to God. *Absolutely True objective knowledge is a human illusion!* Aquinas carefully pointed out that God is non-different from EXISTENCE, TRUE, GOOD, ONENESS which I interpret to mean HARMONY that produces BEAUTY. But for a community of humans to say that they can create via objective knowing TRUTH is blasphemy. God is TRUTH and IT creates humans who, via objective knowing, produce language that contains meaning. But meaning, like the humans who produced it, is *contingent meaning.* Just as no human at an intermediate level of consciousness in the descent-ascent circle between the One and the Many is not God, so also objective meaning produced by that intermediate consciousness is not TRUTH.

With this Vedantic attitude toward reality, that is, right = valid and wrong = non-valid, one may appreciate how Augustine and Aquinas each in different ways were right and wrong. In the Augustinian perspective, one may start with reason, but then one goes inward to the ground of reason which is its immediacy, the *internal witness,* that takes the reflective person beyond reason to the Ground of reason. At the same time, Augustine was quite correct in believing that God is not comprehended by the objective knowledge of pagans, but Augustine was wrong in thinking that God can be comprehended by the objective knowledge of Christian dogma. I believe, as described at the end of chapter 13, that Augustine's voluntarism corrected this error. According to this view, the soul is not mind that can objectively know God. Rather, the soul is pure will that can "see" God by means of supernatural Grace of Faith, Hope, and Love. When either or both pagan and Christian objective knowledge is seen to have contingent truth that evokes one to come into the immediacy of God present in the world and in the depths of one's self, which can be experienced but never comprehended, then and only then can objective knowledge of the world put one on the path back to God. (I believe that Heidegger had this insight when he spoke of letting-oneself-into-nearness of being in his book, *Discourse on*

---

[24]Jacques Maritain, St. *Thomas Aquinas,* trans. J. F. Scanlan (London: Sheed & Ward, 1946), pp. 26-27.

Thinking). Likewise, Aquinas was right in understanding that his *analogy of being* may lead one to know this world in a way that evokes one to come into the immediacy of God present in the world. Thus, the descended God that one is evoked to find in this world is the same God that one is evoked to find via an interior reflection that carries one up to the ascended God. The descending and the ascending are found together non-different from one another, both in extroversion to this world and in introversion to the Inner Self. Aquinas was wrong in thinking that objective knowledge expanded to include pagan as well as Christian knowledge of itself could lead one to God. As indicated earlier, Aquinas probably rejected this illusion a few months before he died. However, the Church made Aquinas's illusion a corner-stone of its dogma. In teaching this dogma of illusion to Western Europe, the Church unwittingly produced the diversity problem that paved the way for the Scientific Enlightenment.

AQUINAS'S LAST INSIGHT: NON-DUALITY OF GOD

The core idea, the central pillar upon which all else depends, of Thomistic philosophy is the analogy of being. This is a high level form of metaphorical, conceptual knowing. The analogy of being is a conceptualization of a pattern, the mutuality of existence and essence, in all beings except God who is designated as Pure Being. God is not a duality of existence and essence. It is a "cop-out" to say that God is this duality with the proviso that His existence is identical to His essence. My interpretation of this aspect of Thomism is that *God is not a being*. Rather, God is something like the Buddhist idea of absolute nothingness or Void that generates both being and non-being. Besides this fundamental discontinuity in Thomistic philosophy, there is a more fundamental flaw. The analogy of being, like any metaphorical concept, has a subjective as well as an objective aspect, as described in chapter 5. This means that one only can understand this idea in relation to one's subjective, existential, non-conceptual perceptions of beings. As a result, all metaphysical principles that use ideas derived from the analogy of being have a subjective aspect. These principles cannot be objectively true; there always will be a subjective interpretation aspect.

For example, the traditional interpretation of the analogy of being is that God is a being. But another, to me, more valid interpretation is that God is not a being; that is, It is not a creator being separate and fundamentally different from finite beings. Rather, this interpretation replaces the Judao-Christian duality of creator God versus creatures with the Eastern insight of non-duality of ultimate Reality. Beginning with Thomas Aquinas, Christian thinkers have partly acknowledged the importance of subjective interpretation. Aquinas's five ways for acknowledging the existence of God are not rational proofs. They are guidelines for any thinking individual to interpret his experiences of reality in relation to the idea of a creator God. Cardinal Henry Newman affirmed in his masterpiece, *Grammar of Ascent*, the importance of subjective commitment and understanding. In essence Newman proclaimed that his or anyone else's commitment to the existence of a creator God is grounded on the subjective interpretation that not only "makes sense," but also it gives meaning and value to all of one's subjective experiences of life. Thus, one knows the existence of God by an act of Faith.

Aquinas made a first step toward spiritual constructivism. He transformed Greek philosophy into narrative, rational thinking. The pre-Socratic philosophers created metaphorical, conceptual narratives of experienced reality. Socrates, Plato, and Aristotle converted these narratives into logical, conceptual models while rejecting the narrative type of knowing. Based on mystical, heroic creativity guided by the Christ-event, Aquinas proclaimed the fundamental importance of narrative knowing, which, I propose, is similar to the fundamental importance of this kind of knowing in narrative, scientific constructivism, described in chapter 1. A narrative understanding of ideas in science always can be made useful for accomplishing tasks by reducing metaphorical conceptual models to logical, conceptual models. In like manner, Aquinas's analogical ideas can be reduced to logical, conceptual theories that define morality and guide the creation of social institutions. However, Aquinas did not appreciate that metaphorical, conceptual ideas are subjective interpretations of reality. As such, they are assumptions about reality. The assumptions cannot be known to be true or false. The logical theories derived from these assumptions are not necessarily true because the assumptions are not known to be true. If these rational theories turn out to be useful, then they may be said to be valid and the assumptions from which they are derived also are valid.

The medieval commitment to the patriarchal ideology influenced Aquinas to formulate his vision into a logical, conceptual philosophy analogous to that of Aristotle. Like Aristotle, in committing to this philosophy one starts with "immediately evident truths" and then deduces various propositions of the philosophy. Aquinas's foundation truths included the dogmatic interpretations of the scriptures developed by the Church, especially since the 4th century. Thus, so long as one adheres to the laws of logic, the Catholic philosopher could achieve absolutely true knowledge of natural order, Nature, and the supernatural order described by the Old and New Testaments. However, the first principles of this philosophy were formulated in terms of ideas related to the analogy of being. As stated earlier, these analogical statements involve subjective, metaphorical intuition and therefore are open to diverse interpretations. Moreover, the accepted position by the Church was that scriptures contain "true" metaphorical stories open to rational interpretation rather than literally true stories. Therefore, the Catholic philosopher only can achieve true knowledge in the context of the Church's prescribing the "correct" orthodox understanding of the first principles. Yet, if Maritain's report of Aquinas's last days is valid, then he had an experience analogous to what Augustine experienced toward the end of his life. That is, similar to what Augustine experienced, Aquinas realized that even if one had a true, rational understanding of God (as Aquinas's theology claimed to present), he/she would not experience union with god. In fact, such knowledge would be a hindrance to such a union. Aquinas, like Job, "saw" Truth "face-to-face." When one has such an experience, then one's supposedly, absolutely true knowledge of reality is "rubbish."

Aquinas's philosophy, like that of Aristotle's, is called *realism* that claims to achieve absolute certitude, but his mystical experience shattered that certitude. Perhaps like what happened to al Ghazali, he could have transcended scholasticism and develop a mystical vision that included the idea of constructivism not only applied to forming opinions about nature as Plotinus proposed but also applied to all speculations including philosophical-theological ones. We can say, after the facts of his situation, that Aquinas felt he could not drastically reformulate his philosophical realism. Nevertheless, just as al Ghazali (1058-1111) abandoned his career of teaching Neo-Platonism-Aristotelian scholasticism to become, after 10 years of wandering, a great Sufi mystic,[25] Aquinas could no longer pursue his meta-rational quest to know God. In the grand scheme of events this is perhaps fortuitous. Aquinas's legacy included emphasizing experiential knowing of nature as a spiritual quest to discover God immanent in nature. It also generated the intellectual diversity problem. Both of these legacies provided the context for the emergence of the scientific enlightenment beginning in the late 1400s.

---

[25] Idries Shah. *The Sufis* (New York: Anchor Books, 1971), pp.166-171.

UNIT VI MANICHAEAN THIRD ENLIGHTGENMENT: MODERN SCIENCE
Chapter 16
PATRIARCHAL IDEOLOGY BECOMES SCIENTIFIC CONSTRUCTIVISM

DECLINE OF MEDIEVAL PARTICIPATORY SUBJECTIVITY

Turning from Ascent to Descent

While Thomas Aquinas's philosophy (1225-1274) legitimized an extroverted interest in and knowledge of the world as a way of glorifying God and ascending to Him, it pushed Western Christian culture toward a *new kind of disharmony* and despair. Aquinas's vision pushed Medieval intellectuals toward objective knowing and away from the participatory, subjective knowing that is the core of mystical contemplation. This is how Aquinas's ideas led to this turning away from ascent toward God to a descent toward the world viewed as independent of God (the One).

Aquinas retained the ideas of hierarchy of being and the distinction between natural and supernatural orders in reality. However, Aquinas anticipates one way of understanding an aspect of the modern theory of evolution. In evolution, a lower level pattern of organization is radically modified so as to be incorporated into a higher level pattern of organization. For example, the lower level pattern in bacterial cells, including the molecular language (DNA → RNA → protein synthesis) was modified and incorporated into the higher level pattern in plant and animal cells; likewise, the central nervous system (CNS) pattern in apes was modified and incorporated into the CNS of hominids, which in turn was modified and incorporated into the CNS of the first humans. In like manner, according to Aquinas, "God's Grace" does not replace nature but rather transforms it so that it becomes incorporated into the supernatural order. With this "evolutionary understanding" of the transforming power of Grace, Aquinas distinguishes two legitimate kinds of knowing: secular knowing, which understands the natural order and sacred knowing which understands the supernatural order. Sacred knowing builds upon and therefore depends on secular knowing, which a pagan such as Aristotle may possess to a far greater degree than most Christians. Thus, Aquinas radically separates secular and sacred knowing and then integrates them.

Aquinas' radical distinction helped initiate and then unify medieval culture. Now Christian intellectuals were encouraged to participate in Nature more fully than they were in the Augustine, Hellenic phase of Christianity. However, this participation was internally conflicted but maintained up to the Renaissance by dogma, the Inquisition, and the Hermetic tradition. As indicated in chapter 15, p.170, the ideas related to the analogy of being involve subjective interpretations of reality. They cannot be known to be true or false. Therefore, the logical theories derived from them likewise are not necessarily true. But the Church can and did proclaim that particular interpretations are "correct," so that all that logically follows from them also is correct. One vital interpretation is that God, as pure Being, is immanent in all reality in such a way that humans who study and correctly reflect on this will come into some degree of intellectual union with God. The Church maintained an ideology that provided a participatory, subjective engagement with nature for the educated elite. Most people in the Middle Ages were not educated, and most of those who were probably did not have an "experience of being," which is a feeling intuition of being necessary for a full intellectual understanding of the analogy of being. Thus, an intellectual, participatory, subjective engagement with nature was not available to most people up to the Renaissance. At the same time, the Thomistic-Aristotelian realism that produced the idea of natural law puts an absolute barrier to participatory union of the knower and the known object. This rejection of Sufi illuminism and the secret doctrine (see chapter 15, p.162, where the teacher elicits individualized understanding in the student) was incorporated into medieval universities and continues in US universities to this day. This style of teaching not only tends to prevent one having the "experience of being," it is a denial of participatory subjectivity in all of one's activities including learning new knowledge.

The *Hermetic tradition* comes from the occult sciences that include alchemy, magic, astrology, animism, and metaphorical, conceptual, mythical knowing that, as described in chapter 10, became pre-Socratic arete. Though the insights contained in this tradition, especially alchemy and magic, officially was banned by the medieval Church, they permeated medieval consciousness to a significant degree.[1] As a person raised as a Catholic in the 1940s and 1950s, I can see how this statement could be true. Participation in Church rituals often generated a kind of numinous experience. This was especially true of the "High Mass celebration" in which there were a lot of candles burning, the priest wore special vestments and there often was more than one priest celebrating the mass. The priest faced the alter rather than the congregation; he sung in Latin, a language I did not at all understand, and the consecration of the host as literally turning into the body and blood of Christ was a dramatic high point of the mass. Bells rang, some priests made very dramatic gestures, and we in the congregation bowed our heads. We were not to look upon the elevated host as it was raised above the alter. Then, when we received communion, we believed we literally were eating Christ, God, who in some manner was present in each consecrated host. Thus, we were totally engaged in what appeared to be a magical ritual. Even several years after graduating from college and after receiving a Ph.D. in cell biology, first minor in mathematics and second minor in physics, and two years of doing research for the Ph.D. at a biophysical research institute, I totally believed in the myth concerning Original Sin. So much was this the case that if I had committed a sin, such as looking at a woman with "impure thoughts," I had to go to confession before taking a trip on an airplane. After all, if the plane went down, I would go directly to Hell. Thus, magical and mythical consciousness that I experienced probably was even more intense and pervasive in the Middle Ages. Therefore, people at this time were steeped in what Berman quoting Owen Barfield, calls "original participation."[2]

Religious Diversity

The medieval Church enforced acceptance of dogma that protected believers from error, which would lead them away from salvation. This "true knowledge" would lead the faithful to fully participate in Church ceremonies, especially the mass, and thereby receive the saving Grace

---

[1] Morris Berman. *The Reenchantment of the World* (Ithaca, New York: Cornell Uni. Press, 1981), pp.73-82.
[2] Morris Berman. *The Reenchantment of the World* (Ithaca, New York: Cornell Uni. Press, 1981), p.71.

of the Holly Spirit. The full participation was guaranteed by the Hermetic tradition maintained by the magic and mythology intrinsic to the Church ceremonies and teachings. By the beginning of the 1500s the Catholic Church had become like the bureaucratic hierarchal structures of the Holy Roman Empire. It had become very powerful, but internally corrupt, and in the context of the Renaissance, its teachings and ceremonies had become sterile and irrelevant to the times. Church dogma, bureaucracy, and power brought control objectivity to totally dominate participatory subjectivity. The teachings of the Renaissance humanists inspired others to create for themselves their own understanding, especially of the scriptures, rather than merely accept the standard dogmatic interpretations. In other words, many were drawn to this religious individualism in which one directly experiences Christ as manifested in the scriptures.

What began to emerge was a choice between a subjective, experiential faith in one's interpretation of the Scriptures and an objective, externally imposed verbal faith in dogmas of an internally corrupt Church. Each religious individualist applied his/her interpretation of scripture to life situations and thereby corrected, validated and expanded it over time. In contrast, the Catholic believer was required to accept the truth of the dogmatic interpretation based on the authority of a corrupt priesthood hierarchy. As I was growing up as a Catholic, we all were told not to read the scriptures in that we may make a false interpretation. At a deeper level it was a choice between experiencing ultimate Reality, Christ, directly versus. accepting the promise that one will be "saved" by absolute obedience to a corrupt Church. In 1517 Martin Luther (1483 – 1546) nailed a document to the wooden door of Castle Church in Wittenberg. The document listed 95 statements that argued that the Church's sale of indulgences all over Germany was contrary to the Bible. The indulgences were worthless pieces of paper that in exchange for money promised that one would be released from so many years of punishment in purgatory for the sins he/she commits.

This event started the Reformation that generated Protesters, that is, Protestants. Luther wrote the 95 statements in Latin, rather than German, in hopes of beginning a debate with professors at the University of Wittenberg concerning his core thesis, "justification by faith." Luther believed that he was fighting for the Gospel, which he believed clearly implied his major thesis. He stated that he would happily yield on every point in his dispute represented in the 95 theses if only the Pope would affirm justification by faith. Luther's 95 statements and his later teachings explaining justification by faith were translated into German and printed throughout Europe. Justification by faith is a special case of participatory subjectivity. When associated with control objectivity, it produces diverse creative expressions. Luther's protests produced a diversity of other forms of protests each, like Luther, claiming to explain the absolutely true way to individual, experienced salvation. These diverse forms of Protestantism caused the Catholic Church to generate a Counter Reformation. Doctrinal conflicts among diverse sects of Protestantism – Lutherans, Calvinists, Separatists that further divided into Pilgrims and Puritans, Quakers, and others – and the Catholic Counter Reformation produced the religious diversity problem.

This religious diversity problem produced a north-south split in Europe. In general, the northern countries became Protestant while the southern countries remained Catholic. The Protestantism of the north grew out of and was partially synergistic with Renaissance humanism. As such the Protestants: (1) initially affirmed participatory subjectivity, that is, "justification by experiential faith, but later rejected participatory subjectivity in its attack on magical aspects of humanist thought; (2) rejected mythological thinking in the form of rejecting Catholic rituals and ceremonies that had become sterile, and correspondingly this led to a rejection of visual arts; (3)

emphasized verbal, conceptual understanding that is complementary to Socratic, Greek humanism and Neo-classical Greek art rediscovered by Renaissance humanists; and (4) validated subjective interpretations in terms of the pragmatics of living one's life. The Catholicism of the south: (1) affirmed control objectivity expressed as a commitment to faith based on authority of the Catholic Church as the "mystical body of Christ" – one believes because only the Church can dispense Supernatural Grace through its teachings and ceremonies (sacraments) that brings salvation, (2) celebrated mythical thinking as expressed in traditional Church ceremonies, (3) emphasized a literal metaphorical and "sacramental" understanding, for example, in Communion celebrated in the Mass, the bread and wine are literally transformed into the Body and Blood of Christ, which upon eating it gives one Supernatural Grace that moves one toward salvation, (4) rejected participatory subjectivity in order to submit to control objectivity of the Church. The Protestantism of the north was receptive to the scientific revolution, in contrast to the Catholicism of the south, which from the very beginning opposed the new science and only later begrudgingly adapted to it.

Renaissance Humanism

The philosophy-theology of Thomas Aquinas was the beginning of Western thinkers rediscovering the writings of classical Greek thought, especially Greek mathematics and the writings of Plato and Aristotle. The metaphysics and the philosophy of nature of Aquinas encouraged thinkers of the 14th and 15th centuries to believe that Nature has a definitive structure, that humans can know that structure and represent it by conceptual language, and that pursuing knowledge of Nature will facilitate humans to "experience the transcendent God," who also is present in Nature. The combination of a corrupt Church coupled with the Protestant revolt, the rejection of the Thomistic-Aristotelian Scholasticism, and the Hermetic tradition pervasive in medieval consciousness led to Italian thinkers in particular to embrace a magical worldview coupled with the occult/craft tradition of trial-and-error experimentation. These "magicians" exploited the "powers" within nature as well as the power of humans to manipulate nature to accomplish their goals. This approach eventually led to the experimental method and utilitarian perspective of modern science.[3] According to Berman, these Italian magicians/scientists were analogous to the French avant-garde (1885-20th century) of the 1960s countercultural revolution and like them provided a transition from both the Greek and medieval intellectual attitudes toward a magical participation in nature as an aid to knowing her. These Renaissance avant gardists included Giordano Bruno, Pico Della Mirandola, Marsilio Ficino, Francesco Giorgi, and the English alchemist Robert Fludd. These thinkers along with others such as Tommasco Campanella, Johannes Kepler, and *the early Isaac Newton* consciously or unconsciously attempted to make humanist magic the new way of knowing that was to become modern science. In this regard, Bruno (1548 – burned at the stake 1600) believed in the Copernican, mathematical heliocentric theory (see below) because it symbolized the divine light present in every human that was accessible by means of magical practices. Also, the Copernican sun was thought to be the ancient Egyptian light that would dispel the present darkness of the world. Thus, heliocentricity was thought to be the astronomical confirmation of the Egyptian/Hermetic core of the original Christian religion.[4]

---

[3] Morris Berman. *Coming to Our Senses* (New York: Bantam Books, 1990), pp.222-224.
[4] Morris Berman. *Coming to Our Senses* (New York: Bantam Books, 1990), pp.225-226.

In a similar vein Pico Della Mirandola (1463-1494) wrote that natural magic is the practical part of the new science that provides the link between heaven and earth that enables one's soul to leave the body and ascend to union with God. The dignity of man for Pico and Marsilio Ficino was that by natural magic he operates on nature and dominates it. Tommasco Campanella practiced magic, saw heliocentricity as a return to ancient wisdom, and viewed Christ as a very great Magnus and thus sought a magical reform of Christianity. All of these Renaissance humanists were like the pre-Socratic philosophers who created ways of understanding nature expressed in metaphorical, conceptual language. As described in chapter 10, this Greek arete led to the dissolution of classical Greek culture, but during the chaos of those days there arose Socratic rationalism expressed in the writings of Plato and Aristotle. The Renaissance humanists already had rejected Aristotelianism as a way of purging ambiguity from their conceptual models. Neo-Platonism's perspective of mathematics possibly provided a solution to the problem of ambiguity, but this way out also generated the intellectual diversity problem.

Intellectual Diversity Problem

Conflicts arising within objective knowledge result from the diversity problem which may be summarized as follows. Greek rationalism assumes a subject-object duality in which reality has an unchanging structure (form or order) that is to some extent potentially intelligible to a knowing human subject. The human subject has modes of knowing, which provide knowledge models that represent the structure of reality. The model may be an absolutely true or an approximately true representation of some aspect of reality. The diversity problem emerges when two objectively "true" models contradict one another. Within modern science we have this diversity of rational models representing the world. For example, Newtonian mechanics was thought to be an approximately true representation of the world involving a universal force of gravity between any two mass objects in the world. In Einstein's general theory of relativity there is no universal force of gravity between any two mass objects in space. Rather, mass objects appear to be attracted to one another as a result of their motion being determined by the properties of space. Any two models that claim to give an absolutely true or the best possible approximately true representation of the same aspect of reality must be equivalent to one another. If the models are not equivalent, they conflict with one another.

The diversity problem results from having to choose between two conflicting models each thought to be absolutely or approximately true based on a social consensus of appropriate criteria of truth. That is, each model is said to be "objectively true," meaning that the truth of the model is independent of any one person's subjective insights. However, "truth criteria" are expressions of a particular rational point of view whose "truth" is guaranteed by a more fundamental set of truth criteria whose "truth" depends on a still more fundamental rational model whose "truth" depends on …and so on to infinity. Another way of saying this is that knowing and truth criteria for knowing, that is, ontology or science and epistemology, imply one another; neither can stand on its own so that it can prop up the other. Ontology by itself cannot give you a "true epistemology," and epistemology by itself cannot give you a "true ontology or science." Aquinas, following Aristotle, assumed that the universe has an unchanging fundamental structure, the human mind is capable of understanding at least some aspects of that structure, and human understanding generates an absolutely (or even an approximately) true language representation of structure in nature.

But how do we know that these assumptions, which may lead to "so-called absolutely true" language representations of the universe, are true? What I saw as a seventeen year-old college freshman (and I believe what Thomas Aquinas also realized) is that we don't know the assumptions are true. We may commit to them absolutely – as I did at one time and as I believe Thomas Aquinas did – by a leap of faith. Participatory subjectivity leads to such a leap of faith in one's personal understanding of some rational model. So long as the rational model "makes sense" and solves at least some of life's problems, then one will continue to make this leap of faith. Jacques Monod, quoted in chapter 18, had the courage to realize and to state that scientists make an analogous leap of faith in their commitment to mechanistic theories. Thus, Monod's realization is a special case of what Nietzsche boldly proclaimed, that absolutely true knowledge results from a human act of will. Either true knowledge results only from a recognized or unrecognized leap of faith; or it results from a Will-to-meaning. If one is unable or unwilling to make a leap of faith, then there can be no True knowledge, and correspondingly, the universe has no meaning; nor does one's life in the universe have meaning. If one demands to have meaning, then he/she must *will* to have meaning, even though the person has no rational basis for asserting that there is meaning. Nietzsche, who believed himself to be the exemplar of his idea of the *superman*, did this *Will-to-meaning*, but then Nietzsche went mad.

During the Renaissance, the most dramatic example of this had to do with cosmology. The medieval Catholic Church held that Ptolemaic theory, created by Ptolemy (?100 – 168 A.D.), was absolutely true with respect to the idea that the earth is the center of the universe and the sun revolved around the earth. This fit very well with everyday observations, in that the sun always arose in the east and set in the west. However, over the centuries accumulation of astronomical observations forced thinkers to revise the Ptolemaic theory to accommodate these new astronomical observations. As a result, the theory became hopelessly complex and unable to explain many new observations. Nicolas Copernicus (1473 – 1543), born in Torun, Poland wrote in about 1513 an account of cosmology that was much simpler and more elegant than the Ptolemaic theory. Copernicus modified his theory over a thirty year period and was reluctant to publish it, but relented as a result of the insistence of a German mathematics professor, George Rheticus; this heliocentric theory was published the year Copernicus died, 1543. This theory not only explained more of the new astronomical observations, but it presented the possibility of being much more useful for making navigation calculations. This was a time when European sailors were venturing out into the unknown seas to explore new continents, for example, Columbus landed in America in Cuba, 1492.

In Copernicus' cosmology the sun was the center of the universe and the earth revolved around the sun. Such a view contradicted one of the dogmas of the Catholic Church, and Copernicus was a devout Catholic. His approach was that his theory was not really true; rather it distorted the true Ptolemaic theory in a way to make it useful. Thus, one can live with such a discrepancy by distinguishing between true knowledge and utilitarian knowledge. (Some Jesuits at St. Louis University in the 1950s tried to convince me of the same distinction with regard to true Thomistic philosophy of nature versus merely valid natural sciences.) Other thinkers after Copernicus, such as Galileo (1564 – 1642), became caught up in the spirit of Greek humanism that celebrates logical consistency, which also was a hallmark of the Aristotelian-Thomistic synthesis. According to this view, practical sciences, that is, utilitarian sciences, are derived from theoretical sciences and therefore a true theoretical science will lead to a true practical science that achieves one's utilitarian goals. Also, if two theories contradict one another, one must be true and the other false. However, the distinction between true knowledge and valid knowledge provided a hint for a solution to the diversity problem. This hint was taken up and radically modified by the scientific revolution, culminating in Newton's theory of motion published in 1686.

RENAISSANCE LEAP OF FAITH: CONSTRUCTIVISM

Constructivism

Constructivism is a perspective about how humans know reality. There are two types of constructivism: intellectual, participatory constructivism, which is realism, and non-participatory constructivism, which includes scientific constructivism.

Realism: intellectual, participatory constructivism. Realism is the Thomistic-Aristotelian philosophy of nature that claims that nature consists of independent objects that are interrelated by cause and effect interactions. Each of these objects contains forms that together make up its nature or essence. For example, a circular object "contains" the form "circleness;" a square object "contains" the form "squareness." Each object is *intelligible*, that is, *knowable* as a result of humans potentially being able to abstract a form manifested in concrete experiences of an object into a *mental concept* that exists in the mind. This kind of knowing begins in a person who has individuated to the persona self, see chapter 6, p.57. Conceptual knowing plus concrete experiences of nature lead one to construct by abstraction the concept of causality and then construct cause and effect interactions among objects. When an individual knows by concrete intuition or assumes that objects exist, then any mentally constructed pattern of causal interactions represents subjective knowing of nature. Language consists of symbols that represent mental concepts or that designate the existence or non-existence of objects. As a result of language, two or more people can communicate to one another their subjective knowing of nature. Shared knowing by language communication thus is the process of converting subjective knowing into objective knowing. When objective knowing is purified by the logical principles of induction and deduction, then it is said to be absolutely or approximately true. Thus, the realism perspective allows several possible types of subjective knowing that are mental constructs.

Subjective language knowing, that is, conceptual knowing, may be objectively true or approximately true or objectively false. Subjective non-language knowing, that is, non-conceptual knowing associated with purely intellectual insights or with feeling insights cannot be communicated to others and therefore can never be said to be objectively true or approximately true or false. One may speculate, however, that the subjective, non-conceptual knowing that guides a person to carryout objectively acknowledged great accomplishments, such as raising children to be responsible, moral, and psychologically healthy persons, is in some sense true.

Degrees of Constructivism related to skepticism refers to the idea that individual humans are *constructivists* who construct conceptual representations of reality. Thinkers committed to realism are constructivists, but thinkers who identify themselves as constructivists usually are committed to *constructivism associated with some degree varying from a weak to a strong form of radical skepticism.* For example, Jean Piaget was a modern biologist-scientist committed to Nominalism, but he probably would reject the subjectivism of postmodern thought. Piaget's major thesis is that knowledge construction follows preset stages of development corresponding to the stages of the emergence of rational, ego self-consciousness. The final stage is logical, conceptual knowing that may be the same for all humans who reach this stage and thus is equivalent to the realist's view of conceptual knowledge obtained by abstraction. Seymore Papert, who was a student of Piaget and who worked with him in the 1950s and early 1960s, proposed to the contrary (1990, see http://www.papert.org) that concrete knowing, what I call metaphorical, conceptual narrative knowing, is a more universal type of knowing that may evolve to an understanding that is higher than logical, conceptual knowing. Thinkers such as Paul Ernest and E. von Glaserfield, commit to a strong variation of radical skepticism. Some constructivists such as Vygotsky (early 1930s), Saussure's structuralism, and Peter Berger's & T. Luckman's 1967 book, emphasize the social aspect of constructing knowledge. Most constructivists embrace some epistemology and/or philosophy as the basis for their particular version of constructionism/constructivism. In contrast, I agree with the great Buddhist spiritual thinker, Nagarjuna (?150 – 250 A.D.) who rejected any epistemological or metaphysical basis as a starting point for his perspective. For him personal experiences were the basis for his perspective that coupled radical skepticism with the experience of ultimate Reality, which he called Emptiness. My perspective gradually emerged from experiences of failure, meditation, experiential trust in ultimate Reality, which for most of my life I called Christ, and reflection beginning in my freshman year of college. My reflections have been influenced by the Catholic version of the Judeo-Christian tradition, especially as embodied in the theological religious thought of Thomas Aquinas. My current interpretation of ultimate reality is similar to that of Nagarjuna. I also believe that ultimate Reality is Emptiness, but I chose to call it SOURCE that manifests individual soul that, in turn, manifests self-consciousness. The ideas of my perspective are similar to those of Nagarjuna, but these ideas emerged in the context of a Western cultural orientation and are expressed in terms of modern science and postmodern thought. My perspective emerged before I knew Nagarjuna's thought.

Pre-Socratic, arete constructivism. This kind of constructivism involves a metaphorical, conceptual participation in nature rather than a logical, conceptual knowing of her. Thus, it is not realism but does involve a kind of intuitive, intellectual participation in nature. According to Wikipedia: **http://en.wikipedia.org/wiki/Geocentric_model** , Anaximander (611-547 BC), a pre-Socratic philosopher, proposed that the earth is shaped like a cylinder and is at the center of everything. Pythagoreans, associates of Pythagoras (582-500 BC) believed that the earth was a sphere in motion around an unseen fire. Later these ideas were combined to be the view held by most educated Greeks in the 500s BC that the earth is a sphere at the center of the universe. According to Plato (427-348 BC), the earth was a stationary sphere at the center of the universe. Eudoxus, who worked with Plato, developed a mathematical theory of how planets rotate around the earth in uniform circular motion. Aristotle's expanded version of Eudoxus' theory also proposed that the spherical earth was at the center of the universe. The models proposed by Plato and Aristotle are not primarily deductions from their respective philosophies and therefore may be thought of as epistemological constructs that fall between Pre-Socratic constructivism on the one hand and the non-participatory constructivism of the Ptolemy geocentric model on the other.

Ptolemy's mathematical, geocentric model. Ptolemy's model is an expansion of constructivism described by Plotinus (204-270 AD), who may have been influenced by Ptolemy. Plotinus, see chapter 13, believed that the Higher Soul is diffracted by passing through matter into the lower soul, which is a human's inner self. The Higher Soul contemplates Logos that generated it and is Intelligence containing Plato's eternal Forms. Thus, by intellectual introspection one achieves true knowledge and progressively comes to union with Higher Soul, Logos, and eventually the One that eternally manifests Logos. One only can achieve various degrees of valid opinions of nature because material objects are continually changing in contrast to Logos containing eternal Forms. Matter is totally passive but eternally receptive to Forms. As a result, there are two types of matter: intelligible matter that has received a particular Form (corresponding to energy as Order that may have been generated by Eros-order, see chapters 1 & 2) and "fecund darkness" equal to the "depth of indeterminacy" that presents new possibilities of new collaborations to produce a new order, see chapter 2. According to Plotinus, if one constructs and then validates a mathematical model to represent the possible order in a material system, he attains the highest level of certainty in his opinions about nature. Of course one also may construct non-mathematical models or mathematical models of various degrees of validity and correspondingly have various degrees of certain opinions

During 127-141 AD, Ptolemy made astronomical observations from Alexandria in Egypt. "He created [constructed] a sophisticated mathematical model to fit [represent] observational data, which before Ptolemy's time was scarce, and the model he produced, although complicated, represents the motion of Planets [and other astronomical data] fairly well.[5] Both Plotinus' theory of opinions about nature and Ptolemy's model represent non-participatory constructivism that became the "new way of knowing nature of the third enlightenment," called in chapter 23 a limited version of *scientific constructivism*. In this regard, Augustine's rejection of Plotinus' intellectual introspection or the Christian version of it opened up psychic space for the Thomistic-Aristotelian extroverted attitude toward knowing nature. The disenchantment with medieval scholasticism opened up for Catholic, Renaissance humanists still influenced by the Hermetic tradition a magical or alchemist, participatory way to understand nature. When later Renaissance humanists and Protestants wanted to eliminate the ambiguity of this magical, participatory, subjective knowing, they turned to the Copernican model exemplifying Hellenic constructivism. This approach combined with experimentation pioneered by the Renaissance magician/scientists and alchemists to produce scientific constructivism.

SCIENTIFIC CONSTRUCTIVISM

Measurement: the Core of Scientific Knowing

Measurement is metaphorical. The essence of metaphorical understanding is that we compare and "see" a likeness between something we do not understand and something else we do understand as a result of picturing it in our imagination, see chapter 6. Measurement is metaphorical in that it always compares two quantities: an unknown quantity is compared to a known quantity. The two quantities, $\underline{a}$ and $\underline{b}$, are imagined to be similar in some respect and are judged in one of the following three ways: $a < b$, $a = b$, or $a > b$. For example, the distance between the two ends of a table is imagined to be different than the distance between two pencil marks on a basket ball but is similar to the distance between two points on a ruler. As a result of dissimilarity, we cannot understand table distance in terms of distance on the surface of a ball, but as a result of similarity, we can understand table distance in terms of ruler distance.

A quantitative, metaphorical comparison between a table and a ruler would lead to one of three statements: the table distance is less than, equal to, or greater than the ruler distance. However, such a comparison still is not necessarily a measurement. If the table and the ruler are about the same size and two people carry out this comparison from different vantage points, they

---

[5] http://www.newuniverse.co.uk/ptolemy_print.html

may come to very different conclusions; one may say the table is larger than the ruler and the other, the ruler is larger than the table. Measurement must be defined in such a way that a subjective comparison is transformed into an objective piece of information.

Objectivity of measurement. A particular type of measurement is objective when any two people who perform the measurement can reach a consensus on assigning a measure to some quantitative, subjective experience. For example, length is a measure assigned to the subjective experience of distance; temperature is a measure of the "hotness" (or "coldness") of an object; a positive whole number is a measure of the "many" of a collection of objects imagined to be similar in some way. The assigned measure must be operationally defined by: 1) a physical procedure for assigning a measure and 2) a number called the value of the measure, which may be determined by mathematical operations.

Any particular type of measure, such as length, temperature, area, counting number, must be a quantity which exhibits three characteristics. Firstly, it consists of autonomous parts, that is, each part can be imagined to have meaning independent of any other part. Any positive whole number is autonomous in contrast to human mother and offspring which are non-autonomous "parts" of a mother-offspring relationship. The idea of mother has no meaning independent of the idea of offspring and vice versa. Secondly, autonomous parts flow into one another meaning that there always is some unique antecedent which implies a particular part which in turn implies a unique subsequent part. For example, the positive whole number, 1, has the unique antecedent of no number and the unique subsequent of 2; likewise 2 has the unique antecedent of 1 and the unique subsequent of 3, and so forth for any positive whole number. Thirdly, the measure is equal to the aggregate of all its parts.

A measure only has meaning in relation to an imagined or real quantitative, subjective experience, but conversely, not all such experiences can be measured. For example, suppose the only measuring tools available to us are rulers divided into inches and the only mathematics we know about is counting by using positive whole numbers, that is, we do not understand fractions or the concept of proportion. The only distances that are measurable are those that can be expressed as some whole number times an inch. A distance that is greater than three inches and less than four inches is a non-measurable quantity relative to our technical and mathematical skills. This illustration shows how the evolution of science depends on innovations in technology and mathematics. Also, one can describe the growth of an embryo/fetus in terms of measuring its progressive increase in length or in terms of the progressive decrease in entropy, see chapter 1. But these measurements do not capture the more fundamental developmental changes, which are the progressive increase in Order that only can be understood by qualitative descriptions.

Beginning in the 10th century AD Greek science-philosophy and mathematics preserved and modified by the Arabs for several centuries gradually were transmitted to the West. This tradition consisted primarily of the qualitative cosmology of Aristotle and the mathematical astronomy of Ptolemy (100-168 AD), which built on the mathematics of the classical period, that is, Eudoxus (408-355 BC) and later the Hellenistic period, for example, Euclid (330-275 BC), Archimedes (287-212 BC), and Apollonius (262-200 BC). Because of the Thomistic grand synthesis of Aristotelian philosophy with Christian theology, qualitative cosmologies overshadowed mathematical descriptions. Ptolemy's astronomy was accepted because to some extent it explained astronomical observations and it fit well with Aristotle's physics.

Both the Greeks and medieval thinkers described spacial relationships in terms of measurement, but they understood motion partly in terms of measurable quantities and partly in terms of causality, which is not measurable. Motion referred to an object's change in location in space. Both the distance it traveled and the duration of travel were measurable, but some efficient cause was thought to have brought the object to its new location understood as the final cause of the motion. Neither efficient nor final causality were measurable. In the 16th century some thinkers attempted to describe motion totally in terms of measurement in spite of Zeno's paradoxes demonstrating that our subjective experience of motion is not measurable. This drive to measure motion culminated in Newton's new mathematical theory of measurement, called the calculus, applied to motion. Newton's calculus was an elaboration of the Greek theory of measurement.

Newton's Constructivism Theory of Motion

Concept of momentum as quantity of motion. Newton generalized Galileo's analysis of a projectile such as a cannon ball and proposed that any motion in a plane can be represented as the resultant of two functions: $D_x(t)$ = distance moved in the + or − direction along the x axis as a function of time and $D_y(t)$ = distance moved in the up (+) or down (-) direction along the y axis as a function of time. From these functions one can derive two new functions that represent the magnitude and direction of the *velocity* as a function of time. Velocity is something like speed. By means of the calculus one can imagine speed at a point in space. Without the calculus this would be an irresolvable paradox. Since speed is distance divided by time, at a point both distance and time are zero so that speed would be zero divided by zero, which has no meaning in any theory of numbers. But velocity is a particular ratio that emerges as the distance interval is made ever closer to zero. Likewise from these velocity functions one can derive two functions that represent the magnitude and direction of acceleration as a function of time.

Having imagined describing motion in this way, Newton may well have asked how much motion does an object have at a particular point in space. The quantity of motion must have a goal, which in this case would be given by direction. The quantity also must describe to what extent the object can change the motion of some other object. In the imaginary world of mathematics velocity quantifies the motion of an object at a point in space at a particular moment in time, but in the empirical world of experience some new factor comes into play. For example, consider two balls of the same diameter, each traveling at 80 miles per hour, in a straight line towards me. If the one object is a tennis ball, when it hits me, this would smart a bit but the ball would not knock me over. If the second ball was made of marble, it would knock me over upon impact. Thus, though both balls have the same velocity, the marble ball has a greater quantity of motion by virtue of being able to change my motion to a greater extent than the tennis ball is able to do.

Of course we know that the marble ball weighs more than the tennis ball. Newton assumed that weight is a measure of "quantity of matter," which he called *mass*. Mass is something like distance in that both are assumed to be continuous and can be measured. That is, just as we define a standard unit of length and then compare any other distance to this standard length, so also we can define a standard unit of mass and compare any other mass to this standard mass. Having created the physical concept of mass, Newton could choose between the two simplest ways of defining quantity of motion: either it is mass + velocity or it is mass x velocity. In either case motion always is an object's change of position in space. But if quantity of motion is defined as mass + velocity, then when velocity = 0, the object would still be considered to have a quantity of motion = mass + 0 = mass even though the object is not changing its position in space. This is a contradiction; therefore, instead Newton defined quantity of motion called *momentum* = p as p = mass x velocity = Mv.

At this stage in Newton's theory causality easily can be visualized as one object transferring some or all of its momentum to some other object. Thus, during any causal interaction, momentum is conserved. Actually it was Rene Descartes who first proposed this idea and later Christian Huygens generalized it to the law of conservation of momentum. Transfer of momentum describes in a general way what happens during efficient causality of motion. Newton was in possession of the calculus which enabled him to propose a measure of the quantity of "efficient causality of motion" that is occurring at a point in space. At that time it was reasonable to assume that mass = quantity of matter does not change during any change in motion. Thus, when one object changes the momentum = p of another object, the mass remains constant and only the velocity changes. Acceleration defines the change of velocity with respect to time at a moment in time and at a point in space; so the quantity of efficient causality of motion, which Newton called *force*, is equal to mass times the change of velocity with respect to time, which is acceleration, that is, it equals to mass x acceleration, **f = ma**.

Newton referred to efficient causality of motion as *action*. Force, then, is a measure of action. When one object increases its momentum in a positive direction, it has expressed action, which is represented by the measure called force. That is, just as area is a measure associated with a space in a plane surrounded by straight and/or curved lines, so force is a measure applied to action. An object which loses momentum or even reverses its momentum during a causal interaction expresses a *reaction*. According to the law of conservation of momentum, the momentum gained by one object must always equal to the momentum lost by the other. Thus, the reaction must be equal but in the opposite direction to the action. Newton expressed this as "for every action there is an equal and opposite reaction." For every force we can measure, there is another force we could measure which is of the same magnitude but in the opposite direction.

It is important to note that with respect to this aspect of Newton's theory, force is *not* a cause of motion, but it is a measure of action (or reaction) which is the process of an object being caused to change its motion. As far as Newton was concerned, when an object is moving with a constant velocity, it is not causally interacting with any other object. If and when such an interaction does occur, there will be an action (change in momentum) that is measured by force. This was in direct conflict with Aristotle and the prevalent common sense notion that if a body is moving, something is causing it to move. Newton expressed this radical departure from Aristotelian physics as his first law of motion, the so-called law of inertia: a body at rest will tend to remain at rest, and a body in constant velocity (with respect to magnitude and direction) will tend to remain in this state of motion.

It will be helpful to summarize the key features of Newton's theory so far:

1. Material objects exist in an ongoing duration which is the same for all of them and in absolute space which extends indefinitely in three dimensions.

2. The measures length, area, volume, and time may be rigorously defined and assigned to the corresponding aspects of a space and duration.

3. Most fundamentally, motion is the change of position of an object in space during some interval of time. Some motions are continuous and may be represented by a family of continuous mathematical relationships of distance as a function of time and its derived functions of velocity and acceleration.

4. The quantity of motion an object has at a particular point in space and moment in time is momentum defined as p = mv.

5. When one object causes another to change its motion, there may or may not be a transfer of momentum from one object to another, but in either case, the law of conservation of momentum always is followed.

6. The process of one object causing another object to change its motion (change its momentum) is called action, and force, defined as $f = ma$, is a measure of action.

7. For every action there is an equal and opposite reaction; the force that is a measure of a reaction is equal in magnitude and opposite in direction to the force that is a measure of the action associated with this reaction.

8. Law of inertia: a body at rest will tend to remain at rest and a body in constant velocity (with respect to magnitude and direction) will tend to remain in this state of motion.

Systematic Experimentation

In the 16th and 17th centuries several thinkers culminating in Isaac Newton (1642 - 1727), created the mechanistic theory of motion. In so doing they participated in the creation of a modified version of objective knowing that enabled humans to obtain public consensus on the validity of the proposed solutions to certain types of problems. The ideas of operational definitions, empirical evidence, induction, analytic synthetic thinking involving rational intuition and deduction, and public consensus of objective truth were modified to fit in with a radically new approach to studying nature called *systematic experimentation*. The integration of all these ideas with systematic experimentation became known as the scientific method for producing objectively true knowledge of the world.

I use the word "producing," rather than "discovering" to indicate that this method involves two new features in knowing: First, the knower interferes with nature, actually disrupts nature, in order to comprehend her; second, both the knowing process and the knowledge that is produced are mechanical. The distinction between theoretical and practical knowledge so cherished by Aristotle is blurred. Theoretical knowledge no longer is wisdom defined as rational contemplative participation in nature. Such contemplation may be ecstatic, producing the highest form of happiness, and has value in itself. Contemplative knowledge is to the mind what health is to the body. Scientific knowledge, whether pure or applied, is like a set of directions explaining how things or events interact. The directions contain pieces of information.

There is nothing new about experimentation. All humans and even many other mammals do it. Experimentation is a kind of manual curiosity, irresistible in children, about one's environment: if I do this, what will happen? When the experimental observation is repeated many times under similar circumstances, it may lead to a useful generalization; for example, gas at a constant volume will always increase in pressure when the temperature is raised. Primitive science contains many generalizations of this sort obtained by experimentation. The facts and experimental generalizations in primitive science give one only a very limited control over nature. The key difference between primitive science and mechanistic science is that the one is grounded on mere experimentation whereas the other is grounded on *systematic experimentation* defined as: the association of "scientific mythology" with observations obtained by experimentation. Scientific mythology is a story that is taken to be literally true because the metaphors in the story can be related to operationally defined terms. This type of definition allows the story to be converted into a rational model that can be validated by empirical observations (see later description of the scientific method). In contrast, "non-scientific mythology" depends on metaphors, which lead to diverse, subjective interpretations of the story.

"Scientific mythology" may also involve metaphors, for example, DNA is like a sentence on a printed page that contains information, but one way or another the metaphors are transformed into concepts that lead to a uniform understanding of the story.

For example, the modern scientific thinker may tell the story that even though we cannot see gremlins, millions of them are in the air in this room. Some of these gremlins are evil in that if they somehow get into the blood stream of an animal such as ourselves, the animal will die in a day or two. These gremlins, so the story goes, are able to rapidly reproduce themselves when in a liquid nutrient broth. The scientist then associates this story with observations from two experiments. In the first he inserts a wire loop exposed to the air into a nutrient broth, which is then sealed so that none of the gremlins can get out. After a day or so, the clear broth becomes cloudy, due to the many new gremlins, and then after several days becomes clear again as all the gremlins die and sink to the bottom of the test tube. In the second experiment, he injects some of the cloudy broth into the blood stream of a rat and observes the animal developing several symptoms of distress, culminating in death. Of course the enormous practical value of this scientific myth is that surgeons sterilize their instruments and do whatever else is necessary to prevent evil gremlins from entering the blood stream of patients during an operation.

William Harvey (1578-1657) observed hearts dissected from dead humans and noted that the bulk of the heart is muscle and that the volume of the heart chamber is about four and a half ounces. Harvey reasoned that when the muscle contracts corresponding to a heart beat, blood would be forced out of the vessels leading away from the heart. Assuming that the heart beats an average of 75 times per min. and empties 4¼ oz per beat, he easily calculated that the heart empties 315 quarts per hour or 7,875 quarts per day. There is no way the body could produce 315 quarts of blood in one hour. Therefore, Harvey created the very useful and fundamental scientific myth that the heart causes blood to circulate in the blood vessels in the body.

What is even more wonderful about scientific myths is that often individual stories fit together to form a grand myth with far-reaching applications. The experimental observation of a burning candle going out some time after being covered by a bell jar is associated with the story that fire is the result of invisible oxygen chemically combining with some substance, in this case wax. A mouse enclosed by a bell jar will eventually die; this, of course, is associated with the story that breathing involves taking oxygen into the body where it is used to burn up some substance in the body (Claude Bernard referred to this as "life is death," that is, the body stays alive by burning itself up). Other experimental observations and corresponding stories lead us to understand that hemoglobin in blood carries oxygen to all parts of the body, along with the glucose that is dissolved in blood. Most cells take in glucose and oxygen, which combine to produce a kind of stepwise burning that releases energy. The released energy is used to carry out life processes in each cell that keeps it alive. Thus, the ancient myths involving the highly charged metaphors of fire, inspiration of breathing, and blood are related to interconnected scientific myths that use these terms to create a very powerful mechanistic story that relates to many kinds of animals.

Characteristics of the Scientific Method.

The scientific method is to all aspects of modern Western culture what water is to all life forms in the ocean. This thought process has profoundly influenced all our modes of thought and has transformed specialized areas of inquiry even in sociology, psychology, and some sub-disciplines in the humanities into science-like disciplines. To some degree, all these disciplines use the scientific method, which is characterized by the use of operational terms, empirical observation and induction, the application of analytic-synthetic thinking, and the use of scientific procedures for establishing public consensus of objective truth.

Operational terms: No matter how abstract and technical a scientific theory is, all its concepts must be derived from operational terms. Operational terms are defined by a process which will lead any two or more people who correctly perform the process to have similar sense perceptions. This characteristic of using operational terms allows only certain types of problems to be pursued. Thus, problems in such areas as religion, philosophy, and ethics are off limits to scientific discourse. For example, the abstract concept of beauty cannot be defined using operational terms.

Empirical observation: Science is empirical in that it deals with things or events that can be observed many times. Furthermore, the scientist may observe many instances of a type of thing or process and make the induction that all these instances have some characteristic in common. This empirical aspect of science is another limitation to the kinds of questions with which science deals. For example, one cannot observe instances of value; rather one observes things or events that he judges as good or bad, beautiful or ugly. Thus, science is without values other than the ones associated with the characteristics of the scientific method. As a result, scientists tend to avoid explicit value discussions in relation to scientific issues. It is much more difficult to precisely define values and reach a consensus about them than it is to resolve scientific problems.

Analytic-synthetic thinking: The scientific approach to studying nature is to: (1) analyze a thing or process, that is, imagine it to be composed of subunits; (2) describe the properties of the subunits; (3) describe how the subunits interact; and (4) make a synthesis, i.e., describe the whole thing or process in terms of the interactions among its subunits.

Scientific public consensus of truth: Public consensus of truth refers to the prescribed procedures for reaching a consensus about the truth of objective knowledge. In science there are two major criteria for scientific validity: (1) experimental prediction and (2) falsification. Three other criteria may also be used to judge the usefulness of a theory: (3) The theory accounts for and represents all of the thousands of relevant facts known so far; (4) the theory shows interrelationships among heretofore seemingly unrelated sets of information; and (5) the theory provides the mental disposition or receptivity to discover new facts, or better still to see new unsuspected interrelationships among already known facts.

One type of systematic experimentation (described previously), which at the same time is a way of obtaining scientific public consensus of truth, is experimental prediction. This involves: (1) specifying a particular set of circumstances in terms of a scientific theory; (2) useing the theory to predict what will happen under these circumstances; (3) setting up the specified set of circumstances, which is called an *experiment*, usually performed in a laboratory); (4) observe what happens in the experiment; (5) repeat the experiment several times; and (6) in some systematic manner conclude whether the experimental observations confirm or deny one's theoretical prediction. Experimental prediction is a type of hypothesis testing, which may be exemplified by the process of rectifying the situation where a light bulb in a lamp no longer gives off light. The process may be divided into six steps:

1.   Story (hypothesis): the light bulb is burned out (or some other hypothesis such as the fuse is blown).

2. Prediction based on the story: If I replace the bulb with one that I know will give off light, then the problem will be solved.
3. Action (experiment): I replace the light bulb with one I know works.
4. Observe results of the action: the light bulb still does not give off light.
5. Compare observed results of action with the prediction: the results of the action (experiment) contradict the hypothesis.
6. Conclusion: the hypothesis is false.

Falsification is the process of observing an event that contradicts some aspect of a scientific theory. Lack of falsification is a basis for accepting the validity of a theory; however, it is important that the theory be formulated in such a way that it can be falsified. Aristotle's theory that change is the process of the potential of a being receiving a form (a potential being actualized) is a "good" philosophical theory but a "bad" scientific theory. No observation can be made that could falsify this theory. Darwin's biological theory of evolution can be falsified. According to this theory multicellular plants and animals evolved from bacteria. If we ever discovered fossils more ancient than fossil evidence of the first bacteria, about 3.5 billion years old, then Darwin's theory would be invalidated.

SPIRITUAL CONSEQUENCE OF SCIENTIFIC CONSTRUCTIVISM

A core idea of Protestantism was that each person, based on his/her interpretation of the Bible, must choose to believe in the Christian message. Theoretically this means that neither a Church bureaucracy nor a political system should force one to be a believing Christian. However, in practice what happened is that some minority group within a church community would disagree with the majority opinion on some fundamental issue. The resolution of such a conflict was either the majority would force its will on the minority or the minority would split off and form a new Christian cult. Sometimes a particular Protestant religion, i.e., Puritans in the Boston Bay Colony in the 1560s, became intertwined with a state power. Then those state Protestant religions operated in the same way as the Counter Reformation Catholic religion. Each of these dogmatic bureaucracies persecuted any individual or group of individuals that dissented from the state enforced norm. The resulting diversity of Christian religions produced religious wars and widespread discrimination. This is why the Puritans, for example, left England to resettle in the American colonies. There they could set up their own Puritan colony and, of course, persecute any individual or group who openly disagreed with the Puritan perspective. However, the Scientific Enlightenment replaced so-called absolutely true knowledge with validated, logical, conceptual theories. This *spiritual insight* combined with the laissez-faire capitalism in the American colonies and with the individualistic, pragmatic, frontier mentality to produce the Declaration of Independence leading to the Revolution that culminated in a Constitutional, representative Democracy.[6]

---

[6] Donald Pribor. *Spiritual Constructivism: Basis for Postmodern Democracy* (Dubuque, Iowa: Kendall/Hunt Pub. Co., 2005), pp.366-381.

Chapter 17
WILL TO MANICHAEAN MODERNISM

NEWTON'S METAPHORICAL MATHEMATICAL THEORY OF GRAVITY

Newton used Descartes' analytical geometry, Kepler's three laws, the calculus (in particular, a modified version of the fundamental theorem of calculus first stated and proved by Isaac Barrow [1630-1677] but now used to generate functions represented by indefinite integrals), and his laws of motion involving contact force, see chapter 16, to formulate a law to describe the motion of planets about the sun. This limited law of gravity, which only applied to "celestial bodies," implied that a gravitational force is "causing" a planet to accelerate toward the sun. The velocity of the planet that always is perpendicular to the line of force interaction between the planet and the sun is what keeps the planet circulating around the sun rather than actually moving toward it. Newton also used his "law of gravity" to describe the circulation of moon(s) around planets. In particular, he accurately described the motion of the earth's moon around the earth. Like the planets, the moon is continuously accelerating toward the earth. In effect, the moon is falling toward the earth, but because of a velocity perpendicular to this "falling" represented as a force interaction, the moon continues to circulate around the earth. This description of the moon circulating around the earth served as a metaphor and analogy for a more general idea. This *metaphorical conceptual knowing* led Newton to propose that any free-falling body toward the earth results from the gravitational force interaction between the earth and the falling body.

> The bold idea that the falling body is governed by the same law as the cellestial bodies had been conceived by Newton in 1666. At the age of twenty three he was in possession of the whole theory which enabled him to make the calculation. But with the numerical values then at his disposal his result differed by about 20 per cent from the value of *g*; it was disastrous.... We may imagine the suspense with which Newton, in 1682, was awaiting the results of degree measurements which Picard had begun in 1679. Legend has it that, when Newton learned of the corrected radius [of the earth], he was too excited to insert it into his formulas to see whether now his theory would be verified and that he asked a friend to do it for him. The result was thoroughly satisfactory, and only then did Newton publish his work containing his entire theory.[7]

The idea that any free-falling body is like the moon falling toward the earth led Newton to propose an even more general law. According to this *universal law of gravity*, any two mass objects anywhere in the universe continuously undergo a gravitational force interaction that is directly proportional to $1/r^2$, where *r* is the distance between the two mass objects. Thus, in general, let m_1 and m_2 be two masses anywhere in the universe, on earth or in the heavens that

---

[7] Otto Toeplitz, *The Calculus: A Genetic Approach* (Chicago: The Uni. of Chicago Press, 1963), pp. 164-165.

are separated by a distance, r, then there is an action at a distance measured by the force, F, as follows:

$$F = G \, m_1 m_2 / r^2 \text{ (G is a universal constant)}$$

> The motion of the freely falling body showed that terrestrial objects of any shape or form are attracted by the earth in accordance with the law of gravitation. But it was a mere hypothesis that any body, no matter of what size or shape, having its specific gravitational factor, imparts an acceleration to any other body, like the sun or the planets.... It was not until 1789 that Cavendish demonstrated with his torsion balance that two lead balls do exert a gravitational effect on each other.[8]

The formulation of the universal law of gravity was a culmination of a grand synthesis, usually recognized as one of the greatest achievements in the history of human thought.[9]

SCIENTIFIC POSITIVISM

However, the law of gravity makes the idea of force problematic. Force is a measure of action, which represents one object causing another object to change its motion. Newton's mechanistic theory of motion of this physical causality is that it must involve contact to allow a transfer of momentum from one object to another, and that it is finite in having a beginning and an end in space and in time. Newton's force of gravity denies all of these properties of physical causality: (1) There is action at a distance rather than contact action; (2) sometimes there is no net transfer of momentum such as when planets circulate around the sun; (3) there is no beginning (other than at the "creation" of the universe) or end to this action at a distance since all bodies in the universe always have and always will continue to attract one another; (4) this "action at a distance" is present everywhere in space; and (5) the force of gravity operates in all time intervals.

Newton believed that, as a secret Unitarian and one committed to a tradition of magic going back to ancient Babylonian culture, he would discover through experimentation and intense introspection a more fundamental law that would resolve the paradox of his non-contact, gravity force acting over a distance. He was convinced that he would discover some material agent to transmit the causality represented by force at a distance. However, he eventually abandoned the controversy about this mysterious gravitational force by saying: "hypotheses non fingo," which translates roughly as "I do not engage in idle speculation."[10] I take this to mean that he gave up trying to subjectively understand what this force is. Instead he gave himself over to the idea that we have a validated, objective description of this force independent of any

---

[8] Otto Toeplitz, *The Calculus: A Genetic Approach* (Chicago: The Uni. of Chicago Press, 1963), pp. 168-169.
[9] Edwin Arthur Burtt, *The Metaphysical Foundations of Modern Physical Science* (Garden City, N. Y.: Doubleday & Co., The Anchor Books ed., 1954), p. 207.
[10] Robert March, *Physics for Poets* (New York: McGraw-Hill, 2nd ed., 1978), pp. 52-53.

subjective comprehension of it. As a result, Newton's approach to constructing mathematical descriptions of motion congealed into a new approach to understanding nature.

Sixteenth century thinkers, including Galileo, were profoundly influenced by Descartes' (1596-1650) description of the sudden, intense, feeling inspiration he experienced on November 10, 1619. According to this vision, empirical patterns that were subjectively recognized as similar could be represented by the totally objective, logical, conceptual formalisms of mathematics.[11] [12] This kind of pursuit of knowledge of nature simultaneously was the pursuit of God in accordance with Thomistic philosophy-theology where God is immanent in nature and in accordance to a way of realizing the goal of ancient Manichaean thought adopted by the Cathars' movement beginning in the eleventh century. This movement sought a way of escaping the limitations of the material body. Perceptions based on the sensations of patterns in nature now could be reduced to mathematical patterns that are purified of the gross materialism of sensations.

The new way of knowing is mechanistic analysis associated with a positivistic, scientific method. By limiting this way of knowing to a very narrow range of problems it provided a way of reaching consensus of truth without any appeal to subjective insights or to secular or religious authority. But for many people, such as Einstein, this limitation is, paradoxically, coupled with a *subjective faith* in the all-encompassing applicability of scientific knowing:

> But what can be the attraction of getting to know such a tiny
> section of nature thoroughly, while one leaves everything subtler
> and more complex shyly and timidly alone? Does the product of
> such a modest effort deserve to be called the proud name of a
> theory of the universe?
> In my *belief* [italics is mine] the name is justified; for the general
> laws on which the structure of theoretical physics is based claim to
> be valid for any natural phenomenon whatsoever. With them, it
> ought to be possible to arrive at the description, that is to say, the
> theory, of every natural process, including life [and MIND], by
> means of pure deduction, if that process of deduction were not far
> beyond the capacity of the human intellect. The physicist's
> renunciation of completeness for his cosmos is therefore not a
> matter of fundamental principle.[13]

Mechanistic analysis is a totally objective way of understanding nature. Science after the Newtonian synthesis became identified with *scientific humanism* that is grounded on the belief that all "legitimate" human problems can be solved by the scientific method. The criterion for deciding whether a problem is legitimate or not is either *gross reductionism* or *scientific positivism*. Gross reductionism is mechanistic analysis implying that any object of study is a whole system that can be totally understood in terms of the interactions among its autonomous parts. *This is a defiant choice to completely suppress subjective, participatory consciousness,*

---

[11] Morris Berman. *Coming to Our Senses* (New York: Bantam Books, 1990), p.248.

[12] Edwin Arthur Burtt. *The Metaphysical Foundations of Modern Physical Sciences* (Garden City, N.Y.: Doubleday & Co., The Anchor Books ed.), pp.105-108.

[13] Albert Einstein. *Ideas and Opinions* (New York: Crown, 1954), pp. 225-226.

based on the belief that in principle everything in Nature is knowable by means of mechanistic analysis.[14] Thus, though scientific mechanism limits objective knowing to a very narrow range of problems, theoretically it is applicable to all of Nature.

Creative scientists, especially Einstein, would hasten to add that imagination and subjective insight are essential for producing new theories. However, subjective insights only present one with *possible* ways of understanding Nature; the insights themselves are neither true nor false. They only become true knowledge when they are shown to be objectively true knowledge by means of the scientific method. In fact, subjective participatory experience produces illusion, which always must be replaced by "true knowledge" gotten by the method of science. This certainly was true for Newtonian mechanics, and remains true today for those scientists who reduce all legitimate knowing to mechanistic analysis.

Scientific positivism may be defined-described by the following four statements: (1) All subjective insights must be doubted; (2) only those subjective insights that are converted into objective knowledge can possibly be valid or invalid, but objective knowledge never can be absolutely or approximately true; (3) only those forms of knowledge validated by means of the scientific method can be accepted as "legitimate" objective knowledge; (4) since scientifically tested objective knowledge involves specialized models, facts, and testing techniques, all of us must rely on the valid, specialized knowledge of experts.[15]

Thus, the emergence of the third enlightenment is an order, chaos, hierarchal new order process. The initial old order was a Thomistic analogical philosophy-theology that included an explicit mind-body duality and an underground neo-Platonism, hermetic magic, and Manichaean traditions that in different ways attempted to transcend the gap and opposition between mind and body. While medieval thought had degenerated to sterile, rigid scholasticism, the Thomistic vision that humans' direct experience of nature could lead to true knowledge of God in nature continued to exert its influence. This influence occurred in the context of the rebirth to Greek humanism of the Renaissance (14th to 17th centuries). These early humanistic thinkers constructed diverse ways of viewing nature that sometimes conflicted with one another as well as opposing the rigid scholasticism and were open to diverse subjective interpretations. All this led to the intellectual diversity problem collaborating with the Christian diversity problem started by Martin Luther and to radical destabilizing the magical perspective of humanist magicians/scientists to produce great chaos of the 15th and 16th centuries. Newton emerged from this chaos as one of the world's greatest mathematicians and scientists, but his thinking embodies all of these conflicting points of view. He was committed to Aristotelian causality and to neo-Platonism that sought to represent this causality by mathematical formalisms. At the same time he was steeped in the occult sciences, particularly alchemy. This led John Maynard Keynes, who bought masses of Newton's manuscripts at Sotheby's in 1936, to surmise that Newton was not the first of the age of reason but rather the last of the magicians.[16]

---

[14] Jacques Monod. *Chance and Necessity* (New York: Vintage Books, 1972)

[15] D. Pribor, *Spiritual Constructivism: Basis for Postmodern Democracy* (Dubuque, Iowa: Kendall/Hunt Pub. Co., 2005), pp. 22-24.

[16] Morris Berman, *The Reenchantment of the World* (Ithaca: Cornell Uni. Press, 1981), pp. 117-118 [Keynes quoted in B.J. T. Dobbs, *The Foundations of Newton's Alchemy* (Cambridge Uni. Press, 1975), pp. 13-14.]

MANICHAEAN MODERNISM

Newton's positivistic science perspective asserted that scientifically validated objective, mathematical constructivism should replace subjective, metaphorical knowing. There still is a place for subjective, metaphorical knowing, which is to provide metaphorical narratives that can be represented by mathematical formalisms and then empirically tested for validity. In effect, this Newtonian constructivism that ends with mathematical formalism accomplishes the goal of the Manichean heresy, which is to live the life of the spirit that now is life of the mind that is totally detached from interactions with the body and the material world.

By the mid 1800s Newton's implicit collaboration between subjective and objective knowing had evolved to objective knowing obscuring subjective knowing to the point where science textbooks and science teachers ignored or even were ignorant of the metaphorical basis for constructing a hypothesis to be tested for validity. In this same period scientific, objective thinking differentiated into two radically opposing perspectives. The holistic, hermetic aspect had degenerated into the radical skepticism of *logical positivism*. In this view, all of nature including humans is unknowable; all science can do is propose theories that give humans progressively greater control over nature thus generating *radical, utilitarian individualism*. The Manichaean aspect of science evolved to the objective, spiritual vision exemplified by Albert Einstein. Einstein, along with many other physicists such as Max Born and Boltzmann (who constructed a probability representation of the second law of thermodynamics), "… considered the second law [of thermodynamics] the result of approximations, the intrusion of subjective views into the exact world of physics."[17] These many scientists rejected what is implied by the second law that empirical events that occur over time are irreversible.

> Einstein appears as the incarnation of this drive toward a formulation of physics in which no reference to irreversibility would be made on the fundamental level.
>
> An historic scene took place at the Societe de Philosophie in Paris on April 6, 1922, described by H. Bergson (1972) when Henri Bergson attempted to defend the cause of the multiplicity of coexisting "lived" times against Einstein. Einstein's reply was absolute: he categorically rejected "philosophers' time" [subjective, metaphorical understanding of time]. Lived experience cannot save what has been denied by science ….
> <u>Duree</u>, Bergson's "lived time," refers to the basic dimensions of becoming, the irreversibility that Einstein was willing to admit only at the phenomenological level.[18]

In regard to any narrative, which always involves an irreversible becoming:

---

[17] Ilya Prigogine and Isabella Stengers, *Order out of Chaos: Man's New Dialogue with Nature* (New York: Bantam, 1984), p. 235.

[18] Ilya Prigogine and Isabella Stengers, *Order out of Chaos: Man's New Dialogue with Nature* (New York: Bantam, 1984), p. 294.

This discovery that the universe is expanding [as reported by
Edwin Hubble in 1929] was one of the great intellectual
revolutions of the twentieth century … Yet so strong was the
belief in a static universe [manifesting eternal forms described by
Plato] that it persisted [for some people who still rejected Darwin's
theory of evolution] into the early twentieth century. Even
Einstein … was so sure that the universe had to be static that he
modified his … [general theory of relativity] to make this possible
… [he] introduced a new "antigravity" force, which unlike other
forces, did not come from any particular source, but was built into
the very fabric of space-time.[19]

Thus, true to the vision of Manichaean heresy and the neo-Platonism exemplified by Einstein,
the mind-self must affirm the eternal truths, now expressed by scientific constructivism, and
reject the subjective feeling insights about becoming that come from the subjective Body Ego.

Chapter 18
IDEOLOGY OF SCIENTIFIC, OBJECTIVE KNOWING

OBJECTIVE RATIONAL KNOWLEDGE VERSUS SUBJECTIVE VALUE

In the last chapter of his book, *Chance and Necessity*, Jacques Monod makes a radical distinction
between objective rational knowledge and subjective values and then claims that all knowing
depends on a subjective choice of values. The basis for my agreement with Monod's distinction
is as follows. Suppose one wants to make an argument for the truth of a particular thesis. One
would refer to a particular rational model to give reasons why the thesis is true, but like all
rational models, one starts with one or many assumptions that one subjectively chooses to be
true. If one wants to avoid this "will to objective truth," he/she must refer to some other higher-
level model that justifies the truth of the assumptions of the lower level model. However, this
higher-level model also has assumptions whose "truth" depends on a still higher-level model that
also has assumptions. To avoid an infinite regress one must acknowledge that there is a set of
assumptions that one subjectively "sees" are true or one *chooses* to accept as true. A metaphor
for such a situation is a sequence of dominos wherein the falling down of one domino causes the
domino next in the sequence also to fall down. In such a sequence one can visualize that there
must be some first domino that either causes itself to fall down or while remaining standing it
causes the domino next to it to fall down. This, of course, is analogous to Aristotle's idea of God
as the uncaused cause of all creatures. With regard to one directly seeing the truth of a set of
assumptions, he/she appears to everyone else who do not directly see this truth to be making the
claim of truth for these assumptions as a subjective choice analogous to Nietzsche's will-to-
power.

The occurrence of the third enlightenment results from an evolutionary process that gave
"birth" to modern science. The perspective of the first and second enlightenments did not
acknowledge a distinction between objective knowledge and subjective choice of truth; that is,
thinkers in these two enlightenments were committed to the idea of immediately evident truths to

---

[19] Stephan Hawking, *A Brief History of Time* (New York: Bantam Books, 1988), pp. 39-40.

any properly trained rational person. This subjective assumption was the starting point for the evolution to the third enlightenment. Some of the immediately evident "truths" were that the universe has a definite, unchanging structure and at least some rationally educated humans can know this structure. This implies that the theories proposed by appropriately educated humans of "good will," that is, they did not have any "hidden agenda" for proposing their theories, would never contradict one another. However, Renaissance humanists of the 15th and 16th centuries proposed theories that did contradict one another. This intellectual diversity problem caused thinkers in this period to despair and even claim that the universe is absurd or that God played a dirty trick by causing humans to think they can know the structure of the universe when, in fact, they cannot know it. One solution to this absurd situation is that Christ's Church has received via revelations the absolute truth of certain propositions such as the assumption that the earth is the center of the universe in opposition to the claim thinkers, like Galileo, that the earth revolves around the sun, which is the center of the universe. In fact, this is what I was taught as a college student attending the University of St. Louis; that is, my teachers acknowledged that the earth revolves around the sun, but other statements are true because of revelations to the Church. But the Renaissance humanists would have none of this dogmatic restriction of human intellectual creativity. It appeared that the only way out of this absurd situation that also was consistent with the humanists' value of intellectual freedom was to propose a new set of assumptions, now designated as *scientific, radical skepticism*, that states the following propositions. <u>One</u>, either the universe does not have an eternal structure, which is one interpretation of Heraclitus' view that all is change; or if the universe does have a definite structure, humans cannot know it. In either case, the universe is not intelligible. <u>Two</u>, the universe does exhibit empirical patterns to humans who represent these patterns by metaphorical, conceptual narratives. <u>Three</u>, *empirical patterns can be adequately represented by the construction of logical conceptual models such as a a a mathematical model that can be tested for predictive validity.* Furthermore, the fundamental value guiding humans to choose scientific, radical skepticism is that *the will to objective control of nature* must dominate or in, the extreme, totally suppress the will to subjective, participatory engagement with reality.

Monod calls the subjective choice of scientific, radical skepticism, the *ethic of objective knowledge*; or often he refers to it simply as the *ethic of knowledge*. Firstly, the will to power of this ethic of knowledge implies the progressive cultural ascendance of *radical, objective, control individualism* that came to dominate Western culture by the late 1950s. A universal truism, if not a truth, is that virtually all humans have a built-in need for metaphorical narratives that give meaning to one's life and an explanation for why humans are born into this world.

> … the religious phenomenon is invariable at the base of social structure… throughout the immense variety of our myths, our religions, and philosophical ideologies, the same essential "form" always recurs… the explanations meant to give foundation to the law while assuaging man's anxiety are all narratives of past events, "stories" – or "histories"… that are "ontogenies" [These ontogenies represent an animist tradition that emerged with the first hominids, and are] the only form of explanation capable of putting the soul at ease … [by providing] meaning of man by assigning him a necessary place in nature's scheme .. [In order] to

195

appear genuine, meaningful, soothing, … [any] "explanation" must blend into the long animist tradition.[20]

Secondly, the "will-to-power," expressed as the will to objective, control of nature, also is innate in all humans, and produces in all civilized societies a patriarchal structure (not necessarily present in many primitive societies) in which the masculine persona representing objective control of nature dominates/represses the feminine persona representing subjective, participatory engagement with nature. The emergence of modern science in the third enlightenment brought this patriarchal structure to its peak of power by making the ethic of objective knowledge the *only* authentic source of truth. This ethic impels humans to thoroughly revise all traditional ethical premises and make a *total* break with the animist tradition that provides explanations for the meaning and purposes of the existence of humans.[21]

> Of all the great religions Judeo-Christianity is probably the most "primitive," [and the one most despised by modern science] since its strictly historicist structure is directly plotted upon the saga of a Bedouin tribe before enriched by a divine prophet.[22]

Thirdly, because of its "prodigious power of performance," that is, its usefulness, science, especially in the 20th century, has structured societies and given them "their wealth, their power, and the certitude that tomorrow far greater wealth and power will be [available to them if they so wish.][23] Fourthly, as a result of a fundamental opposition between traditional metaphorical narratives and the ethic of objective knowledge, all modern societies are fundamentally conflicted, and this internal conflict is becoming ever greater, in part because most people while addicted to science also fear and hate it. Fifthly, analogous to the classical Greek and medieval Thomistic enlightenments, the ethic of knowledge "freely chooses to make authentic scientific knowledge the supreme value, the measure and basis for all other values and choices.[24]

Sixthly, the ethic of knowledge distinguishes between authentic and inauthentic knowledge. *Authentic knowledge* is scientific constructivism that represents an empirical, metaphorical pattern by a logical, conceptual model that can be validated by empirical, objective criteria. This constructivism purifies the *inauthentic* metaphorical knowledge of its subjective aspects. It imposes possible validity onto inauthentic knowledge of nature rather than discovering truth in nature, and it converts ambiguous knowledge into non-ambiguous knowledge that may be useful.

---

[20] Jacques Monod, [trans. by Austryn Wainhuse], *Chance and Necessity*, (New York: Vantage Books, 1972), pp. 168-169.

[21] Jacques Monod, [trans. by Austryn Wainhuse], *Chance and Necessity*, (New York: Vantage Books, 1972), 168-171.

[22] Jacques Monod, [trans. by Austryn Wainhuse], *Chance and Necessity*, (New York: Vantage Books, 1972), p. 168.

[23] Jacques Monod, [trans. by Austryn Wainhuse], *Chance and Necessity*, (New York: Vantage Books, 1972), p. 170.

[24] Jacques Monod, [trans. by Austryn Wainhuse], *Chance and Necessity*, (New York: Vantage Books, 1972), p. 180.

Finally, the ethic of knowledge is *nihilistic idealism*. On the one hand, this ethic affirms the Manichaean vision. Logical, conceptual, scientific knowledge is mental and by definition of mental is spiritual in that it is totally independent of sensations and subjective, feeling perceptions or metaphorical representations of these perceptions. As a result, it is cold, austere, proposing no explanations of meaning or purpose, and it imposes an ascetic renunciation of all other spiritual visions. On the other hand, this ethic is rational and resolutely idealistic.

> It prescribes institutions dedicated to the defense, the extension, the enrichment of the transcendent kingdom of ideas, of knowledge, and of creation [human creation involving scientific constructivism] – a kingdom which is within man, where progressively freed from material constraints and from the deceitful servitudes of animism [representing traditional values of mythologies, philosophies, or religions], he could at last live authentically [as a result of authentic, scientific knowledge], protected by institutions [committed to scientific humanism] which, seeing in him the subject of the kingdom and at the same time its creator, could be designed to serve him in his unique and precious essence [which is to be the creator of authentic knowledge that structures societies to also be authentic].[25]

CONSEQUENCES OF REJECTING PARTICIPATORY SUBJECTIVITY

Rejection of Meaning

Rejection of traditional values. As described at the beginning of this chapter, in undermining subjective knowing science has generated justifiable fear that all traditional values will be swept aside.

The two-cultures double bind. The opposition between scientific culture and traditional humanistic culture has produced the following double bind. On the one hand, if we choose scientific culture we negate traditional humanistic culture, which always has been necessary for social cohesion; on the other hand, if we choose traditional humanistic culture, we negate scientific culture, which has become essential for human survival in our biosphere. Many thinkers, such as Willis Harman, acknowledge and see this opposition of subcultures as producing a crisis for all modern societies.

> Modern society has been attempting the impossible. We have been trying to manage our own lives, our societies, and our planet, on the basis of two mutually contradictory views of reality -- namely, the scientific-materialist and the spiritual. But both of these are limited. Within the spiritual view there is no evident consensus after one gets past the most general propositions [because of the

---

[25] Jacques Monod, trans. By Austryn Wainhuse, *Chance and Necessity*, (New York: Vantage Books, 1972), p.180.

diversity problem that led to the scientific revolution in the first place]. And the adequacy of the scientific view is basically questionable because of its systematic neglect of those deep inner experiences from which all societies have, throughout history, obtained their sense of ultimate meaning and guiding values.[26]

Summary of The Traditional Values Double Bind

1. With respect to science:
    a. Current postmodern societies absolutely depend on science and technology to give humans control over nature and social stability.
    b. Science and industrialism have guided an evolution of consciousness toward differentiation of radical individualism that: (1) is irreversibly replacing traditional individualism, and (2) is irreversibly destroying social cohesion.
2. With respect to social consciousness evolution:
    a. Evolution of social consciousness has led to postmodern societies characterized by radical individualism that is leading to a total breakdown.
    b. If we return to a society characterized by traditional individualism, we must recommit to patriarchal values that: (1) exploit women and various ethnic groups and (2) repress expressive individualism.
    c. If we don't return to a society characterized by traditional individualism, current postmodern societies will self-destruct.

Knowledge Fragmentation

Dialogue between two scientists about knowledge fragmentation. In this dialogue David Bohm points out that scientific analytical thinking splits off problems into specialized disciplines, which ignore the wider context of these problems that would show their interconnections. This type of thinking has been very successful in predicting, controlling, and manipulating things so as to produce short-term solutions to problems. This divide and conquer type of thinking now [1987] is becoming our culture's general approach to life as a whole. David Peat agrees and comments that the success of science solving more and more problems has a cost, which is more and more specialization and fragmentation to the point where rational thinking in general is disconnected from any vision that would give it meaning. David Bohm responds by putting this trend toward fragmentation into a historical context. A century ago the benefits of science generally outweighed any negative effects, but now in the late 20th century, we have passed a point of no return. We have progressed to where we have unlimited powers of destruction. David Peat responds by claiming that we need wisdom in order to transcend this science-generated fragmentary attitude to life. "It is lack of wisdom that is causing most of our serious problems rather than lack of knowledge."[27] By wisdom I think Peat means a spiritual perspective that relates knowing to the meaning of life for humans.

---

[26] Willis Harman, "Redefining the Possible: the Need for a Restructuring of Science" *The Quest*, 7

[27] David Bohm and F.D. Peat. *Science, Order, and Creativity* (New York: Bantam Books, 1987), pp.11-14.

Interdisciplinary double binds. The fragmentation in science has produced many diverse sub-disciplines. Each sub-discipline, guided by the scientific method, develops its own language. That is, each sub-discipline creates its foundational, operational terms that incorporate the mathematical and instrumental tools germane to its approach to posing and solving problems in its area. Esoteric theoretical terms are created and defined in relation to these equally esoteric foundational terms. The net result, of course, is a language that only people trained in the specialty, or in a closely related specialty, can understand. Note, this language problem is not just many terms that are esoteric to people outside the specialty, but the language embodies a *way of thinking* that is geared to a particular specialty and for the most part is not shared by other specialties. The scientific method has infected the social sciences and many aspects of the humanities, with the result that there is an analogous fragmentation of these areas in which each knowledge fragment has its own language. Clifford Geertz (1983) nicely summarizes the problem in this way:

> In particular, the hard dying hope that there can again be ... an
> integrated high culture, anchored in the educated classes and
> setting a general intellectual norm for the society as a whole, has to
> be abandoned in favor of the much more modest sort of ambition
> that scholars, artists, scientists, professionals, and ... administrators
> who are radically different, not just in their opinions, or even in
> their passions, but in the very foundations of their experience, can
> begin to find something circumstantial to say to one another
> again.... All we can hope for, which if it were to happen would be
> that rarest of phenomena, a useful miracle, is that we can devise
> ways to gain access to one another's vocational lives.[28]

The double bind, then, is: on the one hand, scientists (and scholars) continue to choose specialization and pay the price of cultural fragmentation which leads to the public's neither understanding what any specialists is doing nor appreciating what manner he/she is contributing to the welfare of society; on the other hand, if the scientists (and scholars) pull back from specialization, they will totally undermine the scientific approach to solving problems in their specialty.

Science education double binds. Specialized education today, for example, an education leading to a professional scientist or science related profession, such as a physician or an engineer, is partially analogous to a fertilized egg (a zygote) dividing billions of times and differentiating to a complex, multi-cellular baby. In an early stage of development (blastula) all the 100 or more cells are similar, but at the next stage, (the gastrula), three different cell lines (germ layers) are established, and each differentiates into many different types of cells. However, as the thousand or more different type cells are formed, they all are coordinated to form a unified whole body. Moreover, we now appreciate that these diverse cell types participate in elaborate communication networks that keep the cell functions coordinated with one another. However, here is where the analogy breaks down, because science-technology education

---

[28] Clifford Geertz, *Local Knowledge: Further Essays in Interpretive Anthropology* (New York: Basic Books, Inc., 1983), p. 160

produces thousands of different *autonomous non-coordinated specialties*. The intellectual capacity of most people is such that they can be fluent in only one or a few specialties and familiar with several others. In the last half of the 20th century many theoretical and practical problems are interdisciplinary, requiring that one see how many different disciplines interrelate to one another. However, it takes many years of intense study to become proficient in a particular specialty. This produces a double bind. On the one hand, in pursuing specialized education, one has no time for an overview, and thus cannot participate in solving interdisciplinary problems. On the other hand, in pursuing an overview of many diverse disciplines, one does not have sufficient time to master some specialty and thus cannot contribute to solving any technical problem.

College education double binds. Double binds associated with college education stem from science dominated post-secondary education and from postmodernism that emerged from the hippie revolution of the 1960s:

> Piagetian theories have had a profound influence on American education, primarily through Jerome Bruner's introduction of Piaget's work at the prestigious Woods Hole conference on education in 1958. This conference, which framed our nation's educational response to the Sputnik challenge, found in Piaget's developmental theory the ideal framework for emphasizing and rationalizing the techno-scientific dominance of education. In the race to regain the lead over the Soviets in technology supremacy, the educational system embraced *scientific thinking* [italics mine] as the way of knowing. Nurtured by federal largesse, educational institutions were transformed into technological institutions where scientific inquiry prospered and other approaches to knowing -- the affective expression of the arts, the metaphysical reflection of philosophy and religion, and the integrative ideals of liberal education -- all atrophied.[29]

The double bind of educators is similar to that of scientists discussed earlier. Educators cannot transcend the opposition between objectivity and a subjectivity that could bring insights that would integrate diverse disciplines. This is so because they do not know how to do this but more especially because if they were to accomplish this vision, they would lose money, status, and power.

Call for a Transformation of Science.

The previous sections pointed out that scientists rejecting participatory subjectivity has led to rejection of meaning, knowledge fragmentation, and various cultural double binds. In response to these negative consequences, David Bohm and David Peat call for a "revisioning" of

---

[29] D. A. Kolb, *Experiential Learning: Experience as the Source of Learning and Development* (Englewood Cliffs, New Jersey: Prentice-Hall, Inc., 1984), pp. 139-140

science.[30] First, they acknowledge that modern science that emerged between 1500-1700 had to break from the dogmatism of the Catholic Church, see chapter 16. But then, after science became institutionalized and embedded in Western culture, it developed its own kind of dogmatism called positivism. This dogmatic perspective claims that even though no one can attain to absolutely true knowledge, any knowledge gotten by means other than the scientific method is invalid, that is, not legitimate. But, according to Bohm and Peat, the scientific dogmatism does not stop there. Most scientists, at a level just below ordinary consciousness, cling to the hope that scientific research is bringing humans toward absolute truth. As a result of this subconscious attitude, they strongly defend against any attempt to point out these limitations of science so that this way of knowing can be incorporated into a holistic spiritual vision.

Bohm and Peat lament that any attempt to defeat scientific dogmatism in the present climate of science dominating American culture will fail. Then they proclaim:

> What is needed is some new overall approach, a creative surge ...
> that goes beyond the tacit and unconscious ideas that have come to
> dominate science. Such a novel approach would, however, involve
> questions about the nature of creativity and what, if anything, will
> help to foster it.[31]

Systems Science Is Not a Solution

Ervin Laszlo (1987) proposed that the emerging evolutionary paradigm overcomes the inadequacies of gross mechanistic science, which undermines human meaning and promotes fragmentation of knowledge. Laszlo proclaimed that in the 1980s there has emerged a new evolutionary perspective representing patterns of change studied by all the natural sciences. Its core concepts include systems persistence, chaos, and transformations that apply across time to the Big Bang energy point origin of the universe that underwent sequences of transformations to the hierarchal and complex universe of today. Now humans are applying this perspective to the social sciences, psychology, and the humanities so that it is the most interdisciplinary theory ever created by humans. It marks a new era in scientific thinking, which is equivalent to the call by Bohm and Peat for a revisioning of science. In this new way of thinking the creative process of universal evolution is becoming conscious of itself as expressed in individual humans and in human societies.[32]

Thus, Laszlo proposes that this systems conception of evolution as a universal creative process also provides a new approach to science. However, while this new approach rejects gross mechanistic reductionism, it retains scientific positivism in the form of subtle reductionism. This, of course, means that the end results of systems speculation are totally objective theories. Laszlo went on to proclaim that in the 1980s the evolutionary perspective (paradigm) is a positivistic, scientific theory in that it is totally based on empirical observations

---

[30] David Bohm and F.D. Peat. *Science, Order, and Creativity* (New York: Bantam Books, 1987), pp.24-25.

[31] D. Bohm and F. D. Peat, *Science, Order, and Creativity* (New York: Bantam Books, 1987), pp. 24-25

[32] E. Laszlo, *Evolution, the Grand Synthesis* (Boston: Shambhala, 1987), pp. 9-10

rather than introspection and the subjectivity of philosophy, and it proposes a logical, conceptual model that is testable and that describes the same pattern expressed in all conscious individuation, described in chapter 12 and in chemical individuation, described in chapter 2. Laszlo's view contrasts with the perspective of this text, which proposes that the pattern of transformation is similar but not exactly the same in different realms of this hierarchal universe. This is why one must describe patterns of transformation by metaphorical, conceptual statements rather than by logical, conceptual statements. In opposition to the metaphorical conceptual understanding of scientific models (narrative scientific constructivism), Laszlo went on to insist: "It is now possible to advance a grand evolutionary synthesis (GES) based on unitary and mutually consistent concepts derived from the empirical sciences."[33]

Chapter 19
AMERICAN SCIENTIFIC INDUSTRIAL CONSCIOUSNESS

The postmodern era, which began with the 1960s counter cultural movement, emerged from a destabilized social consciousness of the Industrial Age. There are four major aspects that characterize this industrial consciousness: traditional values ideology, traditional values-utilitarianism alliance, utilitarianism, and pluralization of social life worlds.

TRADITIONAL VALUES IDEOLOGY

The traditional values ideology was manifested as biblical individualism, for example, today's Christian coalition, republican individualism such as today's far right conservatives, or some mix of these two types of individualism. One of the major social expressions of this ideology is what Yankelovich calls the *old giving/getting compact*. This compact presupposes the patriarchal ideology, see chapter 12, that prescribes stereotype gender identities and rigidly defined social roles. More specifically, the compact is based on the three-fold familial success: marriage-family, material well-being, and respectability. The compact may be paraphrased this way. I can plan on having a nice house, a good job, and an ever-growing standard of living. My values are that I will also have a satisfying, loving family with a devoted spouse and decent kids who will take care of us in our old age. All this leads to receiving respect from my friends and neighbors and having a sense of accomplishment for making something of my life. The catch is that in order to fulfill this dream I must give work, loyalty and steadfastness. This giving will require that I swallow any frustrations and suppress my desires to do those activities that bring me enjoyment. All and all I must do what is expected of me, put the needs of others ahead of my own, which amounts to never putting myself first. But, this giving/getting compact leads to the realization that: "as an American I am proud to be a citizen of the finest country in the world.[34]

Moreover, according to Yankelovich the work ethic involving self denial is the core value for the old giving/getting compact. This giving, which involved the self-denial, was necessary for the country's economic success. At the same time, the rewards from this giving were necessary to reinforce the discipline imposed by self-denial. Thus, Max Weber's insight applies to America. The Protestant work ethic based on self-denial led to the great capitalistic expansion

---

[33] E. Laszlo, *Evolution, the Grand Synthesis* (Boston: Shambhala, 1987), p. 18.
[34] Daniel Yankelovich, *New Rules, Searching for Self-Fulfillment in a World Turned Upside Down* (New York: Random House, 1981), p. 9.

of the American economy following World War II.[35] The 1960s counter-cultural revolution attacked the legitimacy of this ethic, and by 1973, according to Yankelovich, the ethic lost its dominance in American culture.

TRADITIONAL VALUES-UTILITARIANISM ALLIANCE

Bureaucracy and Consciousness

Bureaucracy in the 20th century tends to depersonalize control consciousness to the point where it almost totally represses emotional intelligence. It encourages the mentality of problem-solving that "fixes" a current situation rather than create a new set of rules or an exception to the rules for dealing with a problem. Its repression of social, emotional competencies produces *anonymity* in which bureaucratic procedures applied to individual humans treat them as abstract categories that interact in the bureaucratic process. The individual characteristics, peculiarities, or eccentricities of both the bureaucrat and his/her client are irrelevant to the bureaucratic process.[36] As a result of this rationalized interaction, bureaucrats and clients, such as teachers and students in today's bureaucratic educational institutions, are not engaged in common tasks. Therefore, they cannot identify with, that is be empathetic to, each other's roles. This leads to the client's sense of impotence. Bureaucracy has a high degree of arbitrariness, that is, it sticks to its definitions and procedures even when these are shown to be irrelevant, non-achievable, or even counter-productive to the goals of the institution. For example, college institutions continue "business as usual" in spite of evidence of over two decades [as of 2005] that they are becoming progressively dysfunctional. Bureaucracy's systematic repression of emotional intelligence has produced *moralized anonymity*. Especially in the 1950s and early 1960s anonymity was morally legitimated as a principle of social relations.

Modern Rational Humanism

This Western version of humanism is an expansion of the Greek co-emergence of individualism and competitiveness, see chapter 10, to the modern scientific and free market mentality. This mentality claims that human existence is like an overriding game made up of component games. The believer can choose what games he will play and choose not to be discouraged if he loses; there always will be other games he can enter that he may win. One of the modern dogmas of this mentality is that if a person is patient and tries hard enough, he eventually will win. Of course a game is only possible in a civilized society that establishes and maintains the rules of the contest. In like manner, belief in human existence being like a game only is possible in a just society. Thus, spectators at the OLYMPIAN games were participating in a religious celebration of Greek humanism. In like manner but to an even greater extent, spectator sports are religious celebrations of a shared belief in a just society that sets up conditions for its citizens to be able to compete in the games of life, just as the players compete in sport contests. Participating in a competition either as players or as spectators is a ceremonial reaffirmation of our collective

---

[35] Daniel Yankelovich, *New Rules, Searching for Self-Fulfillment in a World Turned Upside Down* (New York: Random House, 1981), p.172.

[36] Peter Berger, *Homeless Mind: Modernization and* Consciousness (New York: Random House, Inc., 1974), p. 47.

commitment to rationality and its associated values such as individualism, justice, knowability (intelligibility) of the world, adventure, and rational hope for a good outcome.

If I ask another person to play a game with me, I don't first inquire or even care what political affiliation or philosophical-religious commitments he/she has. All I need care about is that the potential player has sufficient mastery of rationality to understand any game -- which is a metaphor for analytical-synthetic rational thinking. He also must have a moral commitment to obeying the rules of the game, and a willingness to enter into a mutual relationship with me wherein the relationship is defined by the game. Any game transcends all the human categories that tend to separate people from one another, such as sex, race, creed, philosophy, nationality, longstanding animosity, even centuries-old mutual hatred. Thus, the world OLYMPIC games is a religious celebration by world citizens of a transnational, world centric humanism. Even at the height of the cold war coinciding with great tensions between Russia and China, and the U.S, and European nations, these nations along with many others participated in the OLYMPICS.

The OLYMPICS is an embodiment of a world culture's commitment to the idea that human existence in the world is like a game, and the name of this game is *global economics*. Nuclear world war is unthinkable unless world consciousness becomes obsessed with global self-destruction. In order for the human species to survive, the contests among nations must shift -- and I believe the shift already is in place -- from military conquest for greater autonomous power to economic wars for greater material quality of life. Economics, like any game, is not discovered; it is created by humans. Again, this is one of the great realizations of the scientific Enlightenment. I wish to extend this realization to the idea that, in creating sport contests and economics, humans simultaneously create a type of rational mutuality. Thus, every time any of us participate in a game either as a player or a spectator, we participate in the mutuality defined by the game. If by reflection we become aware of this mutuality, then this personal knowledge can be a metaphor for the mutuality in any economic exchange that in today's trans-national marketplace seems beyond our experience. Modernism, which includes the scientific Enlightenment and *avant-garde* artists/philosophers' critique of it, has undermined social cohesion in any society influenced by it and has alienated individuals from one another and from nature. However, economic games have evolved from local, through national, to transnational marketplaces. This implies that an ethic of mutuality also has evolved to an emerging world culture. There is a basis for rational hope, which we can experience by reflecting on our participation in competitive sports (or any other type of games).

In competitive sports there is one winner and one or several losers. When competition in sports is used as a metaphor for competition in economic markets, there are several winners but a very large number of losers. Robert Frank and Philip Cook call these "those-near-the-top-get-a-disproportionate-share markets," *winner-take-all markets*. According to these authors, more and more segments of the American economy are taking on the characteristics of the winner-take-all markets. The forces that created these markets in the 1970s have intensified in the 1990s so as to bring about dramatic changes. Some of these changes benefit consumers. For example, modern technology allows the most talented people to serve wider audiences. A renowned author's manuscript may be made available at a relatively low price to a great number of people. However, there are many negative consequences of winner-take-all markets. (1) They have increased the gap between rich and poor causing a disruption of community. (2) They have lured some very talented citizens into careers that make a lot of money but are not self-full-filling and ultimately are destructive to themselves and others. (3) They have fostered wasteful patterns of investment and consumption because of a short term, bottom line perspective. [These last two points are particularly relevant to the September, 2008 economic collapse.] (4) Indirectly they have made it more difficult for "late bloomers" to have enough material success to develop their talents and contribute to society. "The runaway salaries of top performers ... has stemmed largely from the growing prevalence of winner-take-all markets, which ... is tied closely to the growth of competitive forces.[37] Overall these markets have caused American culture to degenerate because the focus is totally on winning, as defined by wealth, rather than allowing some time to develop value commitments.

The Typographic Mind

This phrase was coined by Neil Postman. According to Postman, evidence mostly from signatures imply that between 1640 and 1700 the literacy for men in Massachusetts and Connecticut was between 89 and 95 percent. From 1681 to 1697 the literacy for women in these colonies was up to 62 percent.[38] Beginning in the sixteenth century there was a shift to logical, conceptual knowing, especially because scientific humanism coevolved with the invention of the printing press that made books and articles about it available to the public.[39] America in the 19th century became a print-based culture in all regions. Also, scores of Englishmen who came to America in the 19th century were greatly impressed with the high level of literacy that was extended to all classes.[40]

As a result of this extensive literacy, the American social consciousness differentiated a scientific, utilitarian mentality. In the 18th and 19th centuries logical, conceptual thinking became the basic organizing principle of society as expressed in public discourse and in printed communication. The Age of Reason involving the scientific and industrial revolutions co-evolved with the growth of a print culture, first in Europe and then in America. The spread of this print culture, which Postman calls typography, kindled an arrogant projection that the mysteries of the universe could at last be comprehended, predicted, and controlled,[41] as, for example, described by Laszlo, see chapter 18.

Postman amplifies this analysis by saying that the print culture became the *Age of Exposition*, which is a mode of thought, method of learning and a means of expression involving logical, conceptual knowing. In the Age of Exposition this type of thinking characterizes the cultural definition of maturity that is expressed as control individualism, which opposes or dominates participatory individualism as described in chapter 12. Typography, which has a strong bias toward exposition, amplifies this definition of maturity. Exposition involves: (1) a sophisticated ability to think conceptually, deductively, and sequentially; (2) valuing reason and order over feeling and chaos; (3) an abhorrence of contradiction and correspondingly, an

---

[37] R. Frank and P. Cook, *The Winner-Take-All Society* (New York: Martin Kessler Books, The Free Press, 1995), pp. 4-6.

[38] N. Postman, *Amusing Ourselves to Death (Public Discourse in the Age of Show Business)* (New York: [Viking Penguin, Inc.(1985)] Penguin Books, 1986), pp. 31-32.

[39] N. Postman, *Amusing Ourselves to Death (Public Discourse in the Age of Show Business)* (New York: [Viking Penguin, Inc.(1985)] Penguin Books, 1986), p. 33.

[40] N. Postman, *Amusing Ourselves to Death (Public Discourse in the Age of Show Business)* (New York: [Viking Penguin, Inc.(1985)] Penguin Books, 1986), pp. 38-39.

[41] N. Postman, *Amusing Ourselves to Death (Public Discourse in the Age of Show Business)* (New York: [Viking Penguin, Inc.(1985)] Penguin Books, 1986), pp. 51-52.

abhorrence of paradox and double binds, which produce chaos that can initiate the non-rational process of creativity; (4) a great capacity for detachment and objectivity resulting from repression of the empathy competencies of participatory individualism, see Tables in chapters 12 & 14, and (5) a tolerance for delayed responses.[42]

UTILITARIANISM

Scientific Positivism

Chapters 16, 17, and 18 described how modern science from its very beginning but especially in the late 19th century and in the 20th century, reaffirmed an ongoing consensual rejection of subjective knowing. This consensus was and is expressed in many forms, such as., gross reductionism and logical positivism. The most fundamental and most common form is that of *scientific positivism*, which was summarized in chapter17. Scientific positivism (scientific humanism) not only is embedded in the practice of science by scientists, but in the practice of all technologies based on science. Especially since World War II, science-based technology has permeated virtually all aspects of American culture. As a result, the everyday consciousness of ordinary people who use these technologies have a tacit commitment to scientific positivism in the same way they have a tacit understanding of how to drive a car. This commitment is constantly reinforced by the media, that is, advertisements that claim that some statement is true because it is "scientifically tested." Moreover, especially since 1958 (in response to the Sputnik challenge), educational institutions embraced the scientific mentality with its embedded scientific positivism. Thus, Americans are oriented toward a positivistic attitude, even though they may have very little technical understanding of science-technology and no awareness of the idea of scientific positivism.

According to the positivistic attitude, subjective insights are legitimate only when converted into scientifically valid objective knowledge. This means that one's personal feelings and attendant evaluations must be kept separate and distinct from factual analysis. Daniel Yankelovich refers to this outcome of scientific positivism as the *fact/value split*. My interpretation of Yankelovich's analysis is to represent the fact/value split as the contrast between control individualism emphasizing objectivity and participatory individualism emphasizing subjectivity, see chapter 14 and table 14.1, which list 28 contrasts. Yankelovich points out that facts, corresponding to objectivity of control individualism expresses objective knowing associated with rationalized feelings. He contrasts this with values expressed as feelings and preferences, which I call subjective knowing associated with experiential feelings. Thus, the fact/value split may be reformulated as control objectivity versus participatory subjectivity. Up to the mid to late 1960s American culture was dominated by the patriarchal ideology described in chapter 12. In this ideology two generalizations apply to the present discussion. One, control objectivity defines the masculine persona and participatory subjectivity defines the feminine persona. Table 12.1, which lists nine contrasts between these two types of persona, indicates that participatory subjectivity produces a legitimate type of knowing that was subordinated to the objective knowing that defines the masculine persona. After the Civil War, scientific positivism began to have ever-greater influence on American culture evolving to

---

[42] N. Postman, *Amusing Ourselves to Death (Public Discourse in the Age of Show Business)* (New York: [Viking Penguin, Inc.(1985)] Penguin Books, 1986), p. 63.

dominance by the 1950s. The positivistic view maintains that only scientific knowledge is legitimate; other types of objective knowing, such as philosophy and the non-scientific aspects of the social sciences, are less legitimate. The subjectivity associated with value does not produce knowledge. Rather, as Yankelovich points out, in American culture;

> value-laden perspectives aren't considered to be knowledge and so
> are not taken seriously when policy is being shaped...
> values belong in the world of feelings and preferences that have
> nothing to do with knowledge.[43]

By the 20th century scientific positivism began to dominate the view of education held by colleges and universities. The result of this is that the patriarchal, scientific, positivistic version of the fact/value split

> is an important part of our elites' intellectual heritage and
> professional competence. Facts are automatically categorized as
> knowledge, and values are categorized as feelings, beliefs, and
> convictions that get in the way of knowledge.... The positivistic
> fact/value split remains strong in the general culture. Indeed, it is
> difficult to exaggerate how pervasive it still is, how much
> influence it has on our culture even today [1999].[44]

Yankelovich goes on to point out the positive and negative aspects of this fact/value split. This split, as modified by patriarchal, scientific positivism, has the good consequences of making it easy to find technical fixes for problems. This cultural perspective has led to a great increase of material well being of society. The bad consequence is that it represses or even undermines subjective knowing, which provides the grounds for commitment to communal and ultimate values.

> As a result, we are becoming technical giants, and sociological
> midgets [and]... our civic virtues of mutual respect, trust, concern,
> neighborliness, community, love, and caring are slowly eroding.[45]

The solution involves what I call the ethic of mutuality of knowledge described in chapter 27.

Another significant aspect of scientific positivism is that it is the highest level of differentiation of the patriarchal ideology that first emerged after the classical period of Greek culture, see chapter 12. According to this ideology, humans can obtain (create) objective knowledge that gives them some control over their lives. The mythic-rational patriarchal view reduced diverse forms of knowing to a hierarchy of knowledge in which some type of rational

---

[43] D. Yankelovich, *The Magic of Dialogue. Transforming Conflict into Cooperation* (New York: Simon & Schuster, 1999), pp. 188-190.

[44] D. Yankelovich, *The Magic of Dialogue. Transforming Conflict into Cooperation* (New York: Simon & Schuster, 1999), pp. 188-190.

[45] D. Yankelovich, *The Magic of Dialogue. Transforming Conflict into Cooperation* (New York: Simon & Schuster, 1999), p. 191.

knowledge, such as, metaphysics, theology, mathematics, is the highest form. Scientific positivism elevated utilitarian knowing based on science to this highest position. Though scientific positivism became a pervasive cultural perspective in the late 19th century, the idea of a hierarchy of knowledge emerged and became embedded in Western culture 2,500 years ago. Thus, in the West (and in any culture structured by the patriarchal perspective) there is a longstanding tradition of dividing society into elites versus the general public. The elites are those persons with specialized knowledge directly or indirectly validated via the criteria of scientific positivism. Indirect validation primarily consists of educational credentials, that is, if one received an M.D. degree he/she has valid specialized knowledge in some area of medicine. A majority of people always will be numbered among the non-elites because they do not have specialized training. However, all members of the elites will be considered non-elites in many (most) circumstances where they are dependent on specialized knowledge very different from their own field of expertise.

Under all circumstances the communication between elites and non-elites will be governed by this scientific positivistic version of the hierarchy of knowledge. The elites assume that they have important information to transmit to the non-elites – which are all of us at one time or another – who must passively receive the knowledge. The non-elites cannot make any significant contribution to decisions based on this knowledge because both elites and non-elites assume that only the elites have knowledge appropriate to these discussions. This assumption is everybody's blind spot because the 2,500-year-old hierarchy of knowledge and the 300+ year old scientific positivism are deeply embedded in the American industrial consciousness. This blind spot produces a breakdown of mutuality among all of us in the work world. This breakdown of social cohesion represents the fragmentation associated with radical, utilitarian individualism as described in chapter 17.

Yankelovich described his understanding and experience of this blind spot – what he calls the *elites' blind spot*. The blind spot is associated with a hierarchy of knowledge that came into prominence in classical Greece. Patriarchy was well established wherein not only did male citizens of the city-state dominate women, they considered women less than human. At this time, though masculine men dominated feminine women, there was an acknowledgement of subjective insights that both men and women can have as a legitimate valuable type of knowing. The metaphorical, conceptual, "poetic" knowing of the pre-Socratic philosophers implied the mutuality of subjective, experiential, metaphorical knowing and objective knowing. Beginning with Socrates, Plato, and Aristotle, human maturity was defined in terms of masculine persona enhanced by logical, conceptual thinking of philosophy and the Greek idea of diverse sciences. Subjective, experiential insights were considered a type of knowing associated with the feminine persona. As women were subordinate to men, subjective knowing was subordinate to objective knowing. When the religion of the Jesus cult became institutionalized by the Catholic Church, subjective knowing became divided into spiritual insights and ordinary feeling insights. Spiritual insights were the means by which Saints and mystics ascended to union with God whereas feeling insights were the ordinary types of knowing possessed by the non-educated, non-elite common people. The logical, conceptual interpretations of the literally true theological dogmas became the highest form of knowing. In the modern version of the fact/value split, scientific objective knowing is at the top, professional and other types of objective knowing are of secondary importance, and subjective insights are demoted to mere feeling expressions, which become a type of knowing only when they are converted into scientific theories.

Yankelovich describes his education at an Eastern elite university in the 1950s, which indoctrinated him to believe the modern version of the fact/value split. After several years of studying the opinions of the public, he realized with a shock that feeling insights are different from but not inferior to the scientific, objective knowledge of the elites. He lamented that he could never convince elites to remove their blind spot to this mutuality of subjective and objective knowing. The elites' blind spot was protected by their minimal contact with the public.[46]

Work/Private Life Split (Private/Public Split).

In modern technological societies where most of the population is located in cities, each individual's private sphere of consciousness is surrounded by job-related institutions. For example, when America consisted of a high percentage of farmers, such individuals had time for their own thoughts and to reflect on their values – and collectively celebrate them in religious services. In modern America even housewives who don't go to work nevertheless are constantly interacting with a host of mechanistic, bureaucratic institutions. One such institution, our educational system, not only structures the consciousness of our youth, it, via lifelong learning programs, continually renews and reinforces the industrial consciousness of adults. Thus, the work life is of paramount importance to most people in the society of the 1970s. As described earlier, the work life world is structured by the scientific positivistic version of the hierarchy of knowledge. This usually unacknowledged perspective segregates knowledge and cognitive style of work from other ways of knowing, especially those involving subjective insights and value commitments. This blind spot that we all have leads to the segregation of work from private life producing the p*rivate/public split.*

Cognition/Feelings Split.

The same forces that produced the private/public split, that is, scientific positivism, also produced a cognition/feelings split in the work world. We may summarize the effects of this split in terms of three types of dichotomies involving respectively, creativity, social relations, and self-identity. The *creativity dichotomy* segregates *adaptation creativity* from other types of creativity. Adaptation creativity involves problem solving, inventiveness, and a tinkering attitude to fix things or machines or humans to "work better." This approach often worked well in the work world and at least by the 1950s permeated the education system. It was a major force for undermining liberal education, which required a different kind of creativity. Such liberalizing creativity embraces abstract knowing as intrinsically valuable. It also involves understanding and appreciating values expressed in the classics and seeing how diverse forms of knowledge are interrelated. I experienced this opposition between science and liberal education in 1951 when I decided to major in chemistry at the University of St. Louis. This university was a leading center of Thomistic philosophy (which provided core principles for a Catholic version of liberal education) and, of course, required all their college graduates to obtain a liberal education with some area of specialization. When I told my chemistry professor I wanted to major in chemistry because I wanted to integrate this with other kinds of knowledge in order to

---

[46] D. Yankelovich, *The Magic of Dialogue. Transforming Conflict into Cooperation* (New York: Simon & Schuster, 1999), pp. 194-196.

better understand the universe, he was shocked at my naiveté. "Our chemistry graduates [rather than pre-medical school chemistry majors]," he said, "can get well-paying jobs in industry. That's why you should major in chemistry."

The other types of creativity also include personal transformations such as described in chapter 12, that is, transformation from childhood to adolescence and from adolescence to adulthood. Then there is intimate conversation and dialogue, which will be described in chapter 28 & 29, and contemplation and appreciation of nature and forms of art. These other kinds of creativity involve the simple enjoyment of people, nature, and art.

The *social relations dichotomy* results from scientific cognition that represses social emotional intelligence. In the work world one applies the problem-solving and tinkering attitude to personal relationships. Yankelovich describes this malady as "a mind set … that treats people as objects to be manipulated."[47] Yankelovich sees this emphasis on control of people, circumstances, and life itself as a deep-rooted distortion of American culture. Such a social distortion produces what Berger calls anonymous social relations. This produces the related dichotomy: people in our private life, such as friends and family, are concrete, individual persons with whom we have an *I-Thou* relationship versus people in the work world are anonymous functionaries with whom we have an *I-It* relationship. The I-It relationships bring people to work together in spite of personal likes and dislikes and in spite of various prejudices. This attitude may be carried over into private life relationships as well. When this happens, the person must confront the self-identity.

The *self-identity dichotomy* results from scientific cognition that represses personal emotional intelligence. In the work world people experience themselves as:

> …components [that] are continuously interdependent in a rational,
> controllable, and predictable way.[48]

The worker may not even be aware of the overall goal of the project he/she is working on, but sees himself/herself only as a part of the means for accomplishing some goal that only the elite decision-makers know about. Thus, in the public world, the worker sees "self" as an it, a mere component of some process. This produces the *self-identity dichotomy* of: the self in the private life sphere is a concrete person, versus the self in the public life sphere is an *it*, an anonymous entity. This mechanized social consciousness influences people in a capitalistic democracy that celebrates individualism to respond to their mechanized social consciousness by becoming disconnected from, that is, not as committed to, the legitimacy of public life. The public person, for example, the worker or people influenced by the work world, defines the individual self as more real than the "work-anonymous self." People perform their work roles as a necessary evil to satisfy their individual needs and desires.

The self-identity dichotomy in 20th century America manifests a fundamental paradox of life, a theme described by Hegel and other "religious philosophers," such as, Buddhists. On the one hand, 17th century mechanistic science and 19th century industrialization led to the emergence of traditional democratic individualism that further differentiated into radical,

---

[47] D. Yankelovich, *The Magic of Dialogue. Transforming Conflict into Cooperation* (New York: Simon & Schuster, 1999),p. 157.

[48] Peter Berger, *Homeless Mind: Modernization and Consciousness* (New York: Random House, Inc., 1974), p. 27

utilitarian individualism of the 20th century. On the other hand, this evolution to greater individualism led to the opposite of individualism in some individuals who showed self-alienation, apathy, and anomie. In still others it led to the opposite of utilitarianism. These *avant-garde* radical expressive individualists, that is, the 1950s beatniks and the 1960 hippies revolted against the scientific repression of subjective individuality. The revolt was at the fringe of culture that mainstream society could partially quell and ignore. But then, two major institutions of industrialization, mass education and mass communication, became carriers of the self-identity dichotomy and brought the revolt against this dichotomy to the mainstream society. Berger notes that these themes became independent of the primary carriers, which are those processes and institutions that are directly concerned with technological production. Then, the secondary carriers incorporated both themes into the late 20th century, that is from 1970 to 2000.

The first theme, alienation, apathy, anomie, played itself out in the following way. Some individuals experienced partial or total alienation. The partial alienation occurred when a person could not find any self-identity in work roles or in any of his/her anonymized roles. This led individuals to affirm self-subjectivity, cut off from the public world where subjectivity is denied. Total alienation occurred when the individual sought refuge in the very anonymity of his/her work situation because he/she found the non-anonymous relations of private life intolerable. Such individuals became like mechanical robots and correspondingly experienced apathy.

Another aspect of alienation is psychological engineering that produces, according to Berger, a specific mode of emotionality. In the workplace, individuals are expected to be low keyed, cool, and controlled, that is, repress all passionate engagement. For example, in 1965, the research lab I worked at from 1962 to 1964 (The American Foundation for Biological Research) was going to be shut down. I wrote a well-reasoned but passionate plea that this decision be reversed. The two people I wrote to – a Jesuit scientist who also was the provincial of his order in North America and a highly placed scientist in the Red Cross – were shocked by the passion of my plea. In their view, reasoned argumentation should be kept separate from one's feeling values. Thus, even if one's primary nature governed by emotional intelligence is integrated with rationality, it must be repressed and replaced via "emotional management" with a secondary, objectified, it-nature suitable to the mechanized work place. Of course, with the repression of one's feelings associated with subjective insights and value commitments, there is the threat of meaninglessness, disidentification and the experience of anomie.

PLURALIZATION OF SOCIAL LIFE WORLDS

According to Berger, the social life world

> …is social both in its origins and in its ongoing maintenance: the
> meaningful order it provides for human lives has been established
> collectively and is kept going by collective consent.[49]

Throughout most of human history the social life world in which an individual lived was more or less unified as a result of a single, dominant philosophical-religious perspective, for example., medieval Christianity and the 17th to mid-19th century Christian perspective in America.

---

[49] Peter Berger, *Homeless Mind: Modernization and Consciousness* (New York: Random House, Inc., 1974), p. 63

However, as the result of the emergence and differentiation of utilitarianism from the late 19th century to the present, the modern social life world became very diverse. In this pluralized social life world, the typical situation of the modern person was to experience various sectors of everyday life that were vastly different. The differences usually involved contradictory perspectives on the meaning of life and the nature of the common good of society. In the 1940s and 1950s the Catholic Church attempted to minimize these discrepant experiences by admonishing young people to only attend Catholic schools and, of course, socialize only with fellow Catholics as did their parents who lived in Catholic neighborhoods. Other religions attempted the same strategy, and the same sort of thing happened since the turn of the century with the partition of large cities into ethnic ghettoes. However, especially since the end of World War II, various factors broke down these social partitions. Some of these factors were the economic boom centered on the typical American family becoming a two car household located in the suburbs, the GI Bill enabling many more people to attend college, which in turn, enhanced social mobility, that is, moving from a lower economically defined class to a higher one, and the civil rights movement of the late 1950s.

Diversity and fragmentation went hand-in-hand with the differentiation of radical, utilitarian individualism. Yankelovich describes three major, interconnected fault lines of which the first and third were described earlier in this chapter. The first one is the I-It mentality and the third one is the hierarchy of knowledge with utilitarian, scientific knowledge at the top and the subjective insights and values at the bottom – the so-called fact/value split. Yankelovich calls the second fault line "the silo effect" in which culture fragments into subcultures that because of specialized language are unable to communicate with one another. The inability of people to accept subjective feeling insights prevents them from participating in any dialogue across disciplinary lines, that is, non dialogue among the silos. This debilitating mind set is prevalent in American colleges and universities in which academic specialties are so isolated from one another that these schools become an aggregate of silos. Yankelovich goes so far as to say that to make connections between subcultures, one "must resort to what anthropologists call 'cross-cultural communication,' [which is] the kind of semantic bridge building needed when people literally come from different cultures."[50]

Pluralization of social life worlds, that is, fragmentation, has occurred on many levels. As described earlier, the public and private spheres are segregated from one another, and within each of these spheres there is further pluralization. In the public sphere bureaucracies, which are rational, static institutions, are segregated from technological production, which involves rational, dynamic institutions that must continually, creatively adapt to a changing marketplace. The private life sphere also is pluralized due to expanding suburbia, Americans relocating in different places around the country, mixed marriages, and the intermingling of diverse people in the public educational system.

The private/public split also leads to what Berger calls the modern identity crisis. The post-World War II economic expansion leading to greater social mobility coupled with new opportunities for personal growth led to greater vulnerability and uncertainty. A modern human is "peculiarly unfinished" because he/she now can objectively express the innate capacity for transformation of identity in later life. As a result, more and more people began to make life plans for career development and for moving up into higher brackets of status and money.

---

[50] D. Yankelovich, *The Magic of Dialogue. Transforming Conflict into Cooperation* (New York: Simon & Schuster, 1999), pp. 152 – 153.

Increasingly intense competition made the outcome of life plans uncertain and correspondingly, each part of one's life world was experienced as relatively unstable and unreliable. Of course, no one individual life world is the "right" one or the "best" one; they are all relative. Amidst this uncertainty and relativity many individuals began to seek experience of an Inner Self that is more real and certain than one's experience of success or failure in the objective, social life world.

Thus, we have the modern human's permanent identity crisis. On the one hand, modern identity is defined by success in one's chosen career, but one's achievement is transitory and liable to random changes in the environment, such as the current economic crisis that became explicit in September, 2008. One's success is uncertain, which also makes one's identity uncertain. On the other hand, one may define his/her identity in terms of looking to the Inner Self that generates individual autonomy and individual rights of which one is to freely choose a life plan. The individuality based on the Inner Self gives modern identity a certainty even if one does not achieve success as judged by society.[51]

Pluralization of religion led to the privatization of religion, which is the religious aspect of the private/public split. Berger argues that one can make the subjective commitment to a religion that defines the meaning of life, its cognitive structure, and prescribes right and wrong behavior, its normative structure. Thus, even though one feels alienated and uncertain in the public world, the choice of committing to a religion makes one feel at home in the universe.[52] Scientific positivism undermines the cognitive validity and normative legitimacy of any religion. However, humans have the ability to live with contradiction by compartmentalizing opposing perspectives and then never allowing themselves to be aware of both points of view at the same time. This is what most white Americans did with regard to the core values of democracy and the supposed legitimacy of slavery of African-Americans up until the Civil War. Thus, I believe, the opposition of scientific positivism to any religion was not the major cause of its decline in the latter half of the 20th century.

However, the scientific mentality did contribute to the declining influence of religious thinking on 20th century Americans. In this regard Berger notes that it is questionable whether committing to the assumptions of modern science and technology is opposed to all religions. Berger's point is that a large number of people, who are not aware of the philosophical assumptions of scientific positivism, do not perceive modern science and technology as: "intrinsically and inevitably inimical to religion. [One can practice science as I did for 10 years without committing oneself to scientific positivism] ... [Nevertheless] The modern rationalization of consciousness has undermined the plausibility of religious definitions of reality."[53]

The three major interrelated factors for the decline of religion are: the private/public split, life planning, and diversity. Religious institutions always have been important determinants of the public mind-set. However, by the 1970s, most Americans had adjusted to the alienation produced by the industrial mind-set by creating the private/public split. As a result, the various religions that use to inform, via its teachings and its ceremonies, a person's subjective value

---

[51] Peter Berger, *Homeless Mind: Modernization and Consciousness* (New York: Random House, Inc., 1974), pp. 78-79.

[52] Peter Berger, *Homeless Mind: Modernization and Consciousness* (New York: Random House, Inc., 1974), p. 79.

[53] Peter Berger, *Homeless Mind: Modernization and Consciousness* (New York: Random House, Inc., 1974), p. 82.

commitments now became less able to do this. As Berger puts this, the religious definition of reality became less certain and correspondingly, less convincing. For many, religious faith was no longer socially given; rather, faith must be individually achieved.

Furthermore, by the 1970s most Americans experienced one another as having different philosophical and religious perspectives. At least some of the time, each person experienced some other person as being likable, sincere, morally admirable and as having other virtues along with that person having a radically different philosophic-religious commitment. Experiences like these led to greater tolerance and sometimes to a more relativistic understanding, that is, less dogmatic commitment to one's own religion. Correspondingly, religious commitment became a more individuated formulation of ideas and values, rather than a blind acceptance of the teachings of the religious community of which one happened to be a member.

Finally, life planning associated with the 1970s industrial consciousness disposed a person to be open to transformation of perspectives and value commitments later in life. This led to young people's choosing a religion for themselves or choosing not to be a member of any religion, rather than allowing themselves to be indoctrinated to the religion of their parents. Also, of course, with the rise of mixed marriages and the rise of divorces, young people were thrown on their own to make religious commitments. Likewise, older adults became more disposed to allow transformations of themselves either away from one religion to another or away from any religion, especially the one of their childhood with which they associated many bad experiences.

As with other aspects of the differentiation of industrial consciousness of the 1970s, mass communication and public education facilitated this development. The Church became less of a focus of community life.

> The public sphere, by contrast, [had] … come more and more to be
> dominated by civic creeds and ideologies with only vague religious
> content or sometimes no such content at all.[54]

Thus, while persons were seeking their own individuated religious commitments or rejecting any religious commitments, we all in the 1970s suffered and now suffer from a deepening condition of homelessness that religion used to help create.

---

[54] Peter Berger, *Homeless Mind: Modernization and Consciousness* (New York: Random House, Inc., 1974), p. 81

UNIT VII EMERGENCE OF THE FOURTH ENLIGHTENMENT
Chapter 20
OVERVIEW OF THE EMERGENCE OF THE FOURTH ENLIGHTENMENT

TRANSCENDENTAL PERSPECTIVE

If one is embedded in a particular culture, then the only way one can imagine a culturally contradictory perspective as possibly valid is to adopt a *transcendental perspective.* Such a view allows one in some manner to be able to remain committed to one's culturally defined identity and at the same time consider the plausibility of a contradictory or contrary identity. If one is absolutely committed to a Manichaean view of the world as True/False, Good/Evil, Right/Wrong, then, in absence of any personal transformation, a transcendental perspective is never possible. However, a personal transformation superimposed on a maintained Manichaean view can produce a limited transcendental perspective. For example, a male with a Manichaean view who adopts over several years a masculine persona, can never accept feminine traits as defining his conscious self-identity. However, as described in chapter 14, pp. 159-160, if he falls romantically in love with a woman who represents to him his unconscious feminine aspect, he may absorb this femininity into his conscious self-identity. Then, while remaining committed to his masculine persona, he can accept and even further develop the feminine aspects of his personality. That is, he has adopted a limited transcendental perspective. Many other personal transformations can lead to an ever-expanding transcendental perspective to include diverse ethnic groups, religions, sexual orientations, philosophical perspectives, and political affiliations.

The third enlightenment involving the emergence of scientific constructivism initially produced a conflict between a transcendental perspective and scientific humanism. The third enlightenment's attack on religious/philosophical dogma led in the American colonies to a limited, transcendental perspective of religious and philosophical diversity. Even though most people did not consciously adopt this attitude toward diversity, citizens directly or indirectly voted for a Constitution that enforced behavior that accommodated this diversity. Scientific humanism rejects participatory subjectivity that enables one to commit to values, see chapter 18. Fundamentalist religions, especially diverse forms of Christianity, modulated the influence of scientific humanism so that the US remained primarily a Christian nation committed to traditional Christian values though interpreted in diverse ways, see Bellah's analysis later in this chapter. This remained the case until the 20th century when science collaborated with industrialization to enhance the influence of scientific humanism. Especially in the late 1940s, 1950s, and early 1960s, as documented in chapter 19, this influence combined with the patriarchal ideology and the free market ideology to produce paradoxical results. It greatly increased America's standard of living – the middle class greatly increased in size during this time – but it totally undermined the core values of democracy as manifested by the US investing 1000s of lives and huge sums of money in waging the Viet Nam War.

During the post-World War II "cold war" with Russia, the American high school and college institutions taught two traditions that totally oppose one anther. The conflict between them was and still is manifested by contradictory attitudes toward abortion. Though there are diverse (and sometimes contradictory) interpretations of *liberal education*, all of them have roots in classical Greek rationalism that claims that one can have absolutely true knowledge of a static universe and that knowledge is good in itself. All these interpretations also commit to the medieval Thomistic idea of *natural law*. This philosophical ideology, which often is associated with dogmatic religions, is associated with the mantra, *right to life* that insists that abortion is murder. A diverse group of perspectives that includes scientific humanism is associated with the mantra, *right to choice*, which claims that abortion is not murder. Therefore, a pregnant woman has the right to choose abortion or having her baby. Scientific humanism is what totally opposes liberal education. Moreover, the liberal education taught at the elite colleges and at Catholic colleges was a patriarchal, dogmatic, and very demanding curriculum that diverted time and effort away from specialization.

Perhaps analogous to the abrupt emergence of "an interior life" in monks in the 11th century of the Middle Ages, see chapter 14, a *countercultural revolution* erupted at the elite colleges in the US. This revolution spread to or independently emerged in Europe and evolved in the US into a "hippie movement" among mostly young people, ages 13 to 25. Initially it was a rebellion against the patriarchal dogmatism of both liberal education and scientific humanism. Based on the analysis of the sequence of three enlightenments described previously, chapters 9 – 19, I propose that the countercultural revolution was the nihilistic phase of a transformation to a new enlightenment involving a postmodern, transcendental perspective that chapter 24 describes as narrative constructivism.

NIHILISM OF TRANSFORMATIONS

Physical and Self-Conscious Nihilism

Physical nihilism may be defined in terms of Eros-chaos as described in chapters 1 and 2. That is, *physical nihilism* results when Eros-chaos disturbs some aspect of a system so that it becomes destabilized. A *homeostatic system* is one that has mechanisms that bring the system back to or near its original stabilized condition. A system that cannot return to its stabilized condition either becomes extinct or transforms to a new stabilized condition, see end of chapter 2. A mature adult human is a self-conscious, homeostatic system that is adapted to its environment. It is able to return to a stabilized condition when disturbed by physical, individual psychological, or social-cultural factors. A factor becomes self-consciously nihilistic when it disturbs a human in such a way that he/she cannot return to a stabilized condition without undergoing a personal transformation, for example, puberty, see chapter 12.

Social-Cultural Nihilism and Enlightenments

Individualism is a core ideal of Western culture and may be defined as one's identity that is not determined by personas. Self-identity includes personal desires that may conflict with one or more other humans' self-identity. A collection of humans each expressing a self-identity is potentially very unstable because of the conflict of desires. The way ancient civilizations maintained stability is by mythologies believed to be literally true that defined personas and prevented most people in society developing a significant level of individualism. As a result, the mythical world view produced a tension between individual's non-persona self-identity and social stability. The more dogmatic and all-encompassing the mythical world view was, the more it opposed individualism. Correspondingly, as the occurrence and degree of individualism in a society increased, it produced greater loss of cohesion leading to social instability.

The evolutionary history of Western culture indicates a progressive increase in occurrence and degree of individualism that sometimes eventually produced a new vision that increased social cohesion, for example, Athens adopting in 508 BC direct democracy. This increase in individualism always led to an increase in diversity. Society tolerated this diversity as a result of a social transcendental perspective analogous to the individual transcendental perspective described at the beginning of this chapter. Correspondingly, the social transcendental perspective always co-emerged with a social conscious transformation. Therefore, the progressive increase in individualism in a human community that remains stable is an order, chaos, new order transformation process. Order is a particular level of individualism expressed in a stable society. When a new type or degree of individualism emerges, it destabilizes the society. The resulting chaos may lead to the extermination of the source of the new individualism; for example, the Persians crushed the new Athenian model of democracy adopted by city-states in Ionia. Alternatively, it may lead to transcending the chaos, as the Athenians did in adopting direct democracy in order to transcend the chaos of the conflict between the aristocracy and ordinary, poor, citizen farmers . Transcending chaos produces an institutionalized new world view that guides creating self-identities that produce social cohesion. The emergence of such a new world view is an *enlightenment*. The history of Western culture may be described, somewhat arbitrarily, as a sequence of the three main enlightenments described so far in this book. Each emerged as a result of nihilism that negated some fundamental aspects of the established world view thus opening up "cultural space" for greater individualism. Sometimes the resulting chaos spearheaded institutional innovations that produced a new kind of society.

Greek Enlightenment

The initial "Greek nihilism" was the disintegration of the Mycenaean, monarchial societies by 1,100 BC that led to the dark ages in which people lost the ability to read and write. The Ionians, one of the four Greek tribes, settled in Athens as well as founding several communities in Ionia in Asia Minor and in the island of Samos. These scattered communities on the mainland and in Asia Minor became Greek city-states. This social-cultural-political innovation emerged from the evolution of epic poetry that celebrated the humanistic ideal of heroic individualism and generated the Olympian religion and the Olympic Games. These innovations evolved to pre-Socratic arete expressed as metaphorical, conceptual, "narrative" philosophy such as first created by Thales (624-546 BC), and Parmenides (501-450 BC) and Heraclitus (535-475 BC). During the reign of Pisistratus, which was from 560 to his death in 527 BC, the center of Greek culture moved from Ionia to Athens. In 547 BC Cyrus' general Harpagus, brutally subdued all the city-states in Ionia. During this time, Greek humanism led to individualism that destabilized societies in all city-states as a result of the opposition between the aristocracy and ordinary citizens. The first stage of the Greek enlightenment ended on a short term positive note of direct democracy that extended voting equality to all Greek citizens. This institutionalized individualism spread like a virus to all city-states including those in Ionia, but the Persians that controlled that part of the world totally crushed it. The democratic individualism that took root on the mainland provided new social cohesion and inspirational energy that enabled the Greeks led by the Athenians to repel the Persian invasions even though they always were out numbered sometimes by as much as ten to one. The long term results of the nihilism of the first stage produced skepticism and relativity of all knowing, analogous to present day postmodernism, by the sophists. It also produced the tyranny of the masses that caused both Plato and Aristotle to hate democracy, and it contributed to fomenting wars between city-states, especially between Athens and Sparta.

The nihilism of the first stage also led to the emergence of logical, conceptual thinking of Socrates, Plato, and Aristotle representing the second stage of the Greek enlightenment. This classical Greek rationalism generated the nihilism of total rejection of metaphorical, conceptual thinking of pre-Socratic arete and rejection of the Olympian religion that provided cohesion to classical Greek culture. The dominance of this culture ended when Philip II established Macedonian power in mainland Greece and in 338 BC defeated the Athenians. In the mean time, the Hebrews over several centuries developed a metaphorical, conceptual narrative of a monotheistic spiritual vision integrated with a detailed moral and cultural code of behavior. Because the Mosaic law epitomizes logical, conceptual thinking, the Jewish, Old Testament tradition represents the possible collaboration between metaphorical and rational thinking though there always was a tension between the two modes of knowing. Often enough, Biblical narratives expressing mystical inspirations were reduced to literal mythologies dictating behavior. During the lifetime of Jesus, this facilitated the Pharisees, such as Paul before his transformation experience, to stifle religious, participatory individualism of many Jews. This patriarchal legalism as well as the possible collaboration between participatory subjectivity and control objectivity had a profound influence on the third stage of the Greek enlightenment.

The first phase of the third stage of the Greek enlightenment, called Hellenism, began in 323 BC with the death of Alexander the Great and ended in 31 BC when Rome defeated Ptolemaic Egypt. Classical Greek rationalism generated and spread Neo-Platonism and other philosophies (Stoicism and Epicureanism) to the over 70 city-states founded by Alexander on or near the Mediterranean. These philosophies, analogous to Jewish rationalism, created a new partnership between metaphorical and rational thinking. However, the partnership from the very beginning became one-sided because of the emergence of the Greek-Jewish patriarchal ideology. In this ideology control objectivity dominates participatory subjectivity. The educated elite in the Hellenic world developed rational maturity that presents one with choosing either control objectivity individualism or participatory subjectivity individualism, see chapter 12, table 12.1. Because of the patriarchal ideology, most adults, who almost always were males, choose control objectivity. Thus, the enlightened emergence of rational individualism had the nihilistic consequence of further repressing participatory subjectivity.

The second phase of the third stage of the Greek enlightenment began with the founding of the Roman Empire in 27 BC and lasted until the fall of Rome in 476 AD. The enlightenment aspects of this phase involved the emergence of the Jesus Christ mystery cult that interacted with the Jewish, Greek, and Roman traditions as it became institutionalized as a universal, Christian religion. Christianity provided the spiritual vision of personal salvation that masses of people hungered for but felt they could not receive from pagan mythologies, intellectual religions, such as Manichaeanism, Greek intellectualism, Judaism, and mystery cults other than the Jesus cult. St. Paul converted the provincial Jesus cult – a Jewish heresy – into a universal religion that with his aid spread all over the Roman Empire. While this was happening, many educated elites transformed to full, adult control individualism. Philo, a Jew (20 BC-50 AD) developed a Neo-Platonic justification for Judaism that (probably) had a profound influence on the Church's formulation of the doctrine of the Trinity, especially the idea of Christ being the "Divine Son" of

God, see chapter 13. Many Christian thinkers, in particular Augustine (354-430 AD) used Neo-Platonic philosophy, especially that of Plotinus, to guide one to develop a Christian, rational individualism. The nihilistic aspects of this phase was the Catholic Church expanding the patriarchal ideology that combined with Roman practicality to further repress participatory subjectivity.

Medieval Enlightenment

Augustine was the last Hellenistic man and the first medieval man. He promoted the dogmatic and patriarchal ideological aspects of the Church and toward the end of his life totally rejected the Plotinus version of Christian rational individualism. This rejection along with other factors brought about the medieval dark ages. This lasted until about 1050 when some monks began to develop an interior life. This occurred as a result of the Sufi, the avant garde of the medieval enlightenment. Just their presence and writings infected the culture of the Middle Ages with the ideas of romantic love, erotic mysticism, Spanish illuminism, Spanish theosophy, and the joyous, participatory engagement with nature all of which stimulated personal introspection. Though this could not overcome the dogmatic, patriarchal ideology of the Church, it did inspire Thomas Aquinas, through Albert Magnus, to create an intellectual, participatory engagement with reality by means of his idea of the analogy of being. The Thomistic philosophy-theology integrated into Church dogma imposed an extroverted, Aristotelian scholasticism onto the medieval understanding of nature. This was partially balanced by Church rituals and sacraments that facilitated a religious, participatory engagement with everyday life and Faith/Hope for salvation.

Scientific Enlightenment

The 15th and 16th century, Renaissance humanists "inherited" from a degenerate Thomistic scholasticism the intellectual diversity problem and an exclusively extroverted attitude toward reality rather than the introspective, Neo-Platonic intellectualism, which Augustine rejected. Augustine's apparent nihilistic views of Greek rationalism unintentionally prepared the way for the late medieval cultural break from the Catholic Church, see chapter 14 for the reasons for making this statement. These humanists also took in the magical and alchemist aspects of the Hermetic tradition unintentionally maintained by the rituals of the Catholic Church. These nihilistic trends combined with the constructivism of Ptolemy, Plotinus, and Copernicus to produce scientific constructivism integrated with systematic experimentation. This scientific enlightenment combined with laissez-faire capitalism in the context of England dominating the American colonies to generate the Declaration of Independence and Constitutional Democracy. On the one hand, this extension of the third enlightenment to democracy was the greatest and most extensive increase in human individualism in human history. For over two centuries it has increased societal cohesion not only in the US but in hundreds of societies throughout the world. It also increased the life expectancy, health, and standard of living for millions of people. On the other hand, by the 20th century it greatly decreased participatory subjectivity for millions of people and undermined all religions and mystical traditions including the spiritual vision of liberal, constitutional democracies.

RESURGENCE OF SUBJECTIVITY

Nihilistic Trends

Subjective constructivism of avant garde modernism. Avant garde modernism embraced the scientific view that even though reality is unknowable, one can construct objective, valid knowledge of it. However, this new version of modernity that emerged in the late 1800s turned scientific individualism upside down. On the one hand, the new modernity rejected the positivistic dogma that objective scientific knowing should discard all subjective insights that cannot be converted into valid scientific theories. On the other hand, it celebrated introspection that produced subjective insights that discarded or subordinated all types of valid objective knowing to subjective constructions, such as avant garde modernist art. This subjective constructivism became the new modernist dogma. Only subjectivity enabled one to be authentic, especially if it flied in the face of consensual objectivity, which made one inauthentic.

Postmodern absolutized diversity/relativity. Postmodernism affirmed the systems perspective, which included ecology and evolution implying one another, that no thing can be understood independent of its interactions with other things; that is, each thing has a process aspect. Conversely, no process can be understood independent of the things that participate in interactions that help define the process. For example, contact force cannot be imagined independent of the two things that interact to produce an *action* at a point in space. Likewise, a "field of electrical forces" cannot be imagined independent of each "force point" of the field potentially interacting with a test charge. There are no autonomous things or autonomous processes; there only are "things-processes." This is a philosophical generalization of the idea of energy, as described in chapter 1, representing both Order, analogous to a thing, and Chaos resulting from energy flux. (This postmodern insight also is analogous to the mutuality of the object that is known and the process of knowing an object; that is, Being representing metaphysics implies epistemology and epistemology implies metaphysics.)

Postmodern thinkers expanded this profound insight to *absolutized diversity* and correspondingly *absolutized relativity*. This absolutized diversity and relativity is but the further evolution of scientific positivism. For the positivistic scientist one's subjective insights are not connected to a deeper reality, and, in particular, the insights are not the products of an Inner Self, which is manifested by an ultimate SOURCE. The subjective insights only present possibilities for creating scientifically validated theories. For the avant-garde thinkers, subjective insights come from participatory subjectivity interacting with archetypes of the unconscious, and they provide the basis for autonomous, that is, authentic, personal self-expression. Such authentic self-expression is achieved by a *Will-to-meaning* that has no connection to the inner witness. According to this view, God is dead, the universe is absurd and has meaning only because I as a "truly authentic person" *will* the universe to have meaning in accordance to my personal experiences. For the postmodernist, participatory subjectivity not only interacts with archetypes of one's unconscious, but is determined – embedded in – a network of meanings.

This social aspect of postmodernism, according to Wilber, is that each individual born into a society is socialized to adopt the meaning embedded in the collective social consciousness. That is, just as a biological organism expresses traits determined by genetic programs interacting with a particular environment, a human expresses meaning determined by social perspectives modified by vast networks of background contexts. Thus, one does not initially fashion meaning by oneself. Rather, social meanings fashion the individual. Society partially determines the meanings an individual expresses, and he/she is not at all or is only vaguely conscious of the contexts of these social meanings. This is why postmodernists are correct in affirming that *any meaning is context-dependent* in contrast to the gross mechanistic, scientific theories that are claimed to be valid independent of social context.

> This is also why individual states of consciousness must to some degree be interpreted within a cultural context, and why any truly postmodern view should attempt to move toward an *all-context sensitivity* [in contrast to avant garde autonomous personal interpretations]…Not only is meaning in many important ways dependent upon … context … these contexts are in principle *endless or boundless*. Thus, there is no way finally to master and control meaning once and for all … [though an autonomous modern science would attempt to do so].[1]

Thus, in the postmodern perspective there are an indefinite number of meanings, and an indefinite number of interconnections among these meanings. When any particular meaning is specified to be true (or false) or valid (or invalid) by means of truth criteria or validity criteria, that specified meaning is merely another meaning in the indefinite collection of meanings. In other words, any set of truth criteria or validity criteria is *absolutely relative,* which means that no set of criteria can be said to be more true or more valid than any other set. Finally, all these relative meanings only are *surface meanings* because none of them is connected to ultimate Reality, which generates ultimate meaning. When this *nihilistic insight* is applied to humans, it prescribes that any human developmental-evolution, that is, any human individuation is an illusion. One may think that he/she is moving toward or has moved to a higher level of consciousness or a higher level of maturity, but this is an illusion. This specified higher level is just another meaning among an indefinite number of meanings.

How could anyone authentically claim to believe such a statement about the relativity of individuation? I answer: in exactly the same way that Einstein talked about time. According to Einstein, one may experience his/her life and events in general as irreversible. But one must reject this subjective intuition. Subjective experiences in themselves cannot even be acknowledged to contain validity or truth because they cannot be judged by the scientific method. An autonomous *Will-to-control* causes some scientists to reject the subjective intuition that time is irreversible. Likewise, an autonomous Will-to-control causes many (most?) postmodern thinkers to reject the subjective experiences of maturing to higher levels of consciousness in going from a childhood mentality to an adulthood mentality. Thus, postmodernism absolutizes diversity by claiming that there is no unity and there cannot be a unity among infinite diversity.

Wilber's description of this "bad news" of postmodernism emphasizes the superficiality of this perspective. Scientific modernity privileges the perspective of positivistic science and thus discards or reduces all subjective insights to scientific knowing. Postmodernism created a higher form of reason represented by aperspectivism and contextualism. *Aperspectivism* is the postmodern idea that there is no perspective that is the one true or most valid perspective. But postmodernists absolutized diversity by rejecting any experience of ultimate Reality, such as a

---

1. Ken Wilber, *Integral Psychology* (Boston: Shambhala, 2000), p. 166

Creator God, that generates all aspects of the universe including human knowing,. The rejection of any experience of ultimate Reality implies that all knowledge statements are relative; there are no higher types of knowing resulting from higher types of subjective insights. This postmodern perspective reduces all knowing representing nature to a two-dimensional plane, which Wilber calls flatland holism.

> In fact, most postmodernism would eventually go to extraordinary lengths to *deny depth* in general. Postmodernism came to embrace surfaces, champion surfaces, glorify surfaces, and surfaces alone....
> Extreme postmodernism thus went from the noble insight that all perspectives need to be given a fair hearing, to the self-contradictory belief that no perspective is better than any other (self-contradictory because their own belief is held to be much better than the alternatives). Thus, under the intense gravity of flatland, integral-aperspectival awareness became simply *aperspectival madness*, a total paralysis of thought, will, and action in the face of a million perspectives all given exactly the same depth, namely, zero.[2]

Bellah's analysis of US marketplace individualism. In *Habits of the Heart* Bellah et al describe how "marketplace individualism" developed and now is expressed in American culture. The foundation and dominant theme of American culture up to the Civil War (1865) was *traditional individualism* conditioned by three characteristics. (1) Most people lived in face-to-face communities of towns and small cities. (2) There was a patriarchal, dominant social structure governed by an interlocking network of stereotype social roles, rational morality, and shared values about the nature of the good person, the good life, and the good society, and (3) Society was sustained by what Bellah called "a moral ecology shaped by women," which is akin to what Carol Gilligan describes as "an ethic of care" found primarily in women.

This early American tradition was expressed in two distinct but complementary themes which Bellah et al designate as biblical and republican individualism. John Winthrop (1588 - 1649) and later Abraham Lincoln were exemplars of *biblical individualism*, which espoused hope for a just and compassionate society in which a genuinely ethical and spiritual life could be lived. The biblical tradition encourages moderation and moral freedom defined as liberty to do "the good." George Washington, John Adams, and especially Thomas Jefferson were exemplars of *republican individualism*, which elevated to a universal principle the dignity and equality of all humans. Correspondingly, each human has the God-given right of freedom to participate in contributing to the common good of the republic.

Between the Civil War and World War I, as the United States became industrialized, a *radical individualism* emerged. Its social basis was bureaucratic consumer capitalism and its spiritual center was the autonomous individual who was hostile to any traditional moral order that would limit a person's autonomy. People began to choose work commitments and modify social roles based on the criterion of life-effectiveness as the individual judges it. A calculating managerial style replaced the ethic of care. Morality and truth were considered relative so that

---

2. Ken Wilber, *Integral Psychology* (Boston: Shambhala, 2000), pp. 169-170

normative commitments were viewed as alternative strategies of self-fulfillment. "In the absence of any objectifiable criteria of right and wrong, good and evil, the self and its feelings became our only moral guide."[3] Utility replaced duty; self-expression replaced authority, and feeling good replaced being good.

This modern trend to radical individualism differentiated into two complementary, though at times contrary themes: utilitarian and expressive individualism. In the 20th century the exemplars of *utilitarian individualism* are persons totally focused on careerism, entrepreneurs, and the professional manager. A middle-class mentality emphasizing rationality and technical rules demanded technical, specialized education, bureaucratic occupational hierarchies, and a market economy. To succeed, a person must develop the virtues of discipline, hard work, careful calculation, self-reliance, autonomy, and rational self-improvement. The good life defined in this way allows little room for love, feelings, and appreciation of aesthetics and transcendental meanings.

Bellah lists Walt Whitman (1819 - 1892) and the therapist as exemplars of *expressive individualism*. I would add to this list avant garde modernist and post-modernist artists, existentialists, and new-wave thinkers. There is an openness to diversity of ideas, personal appearance, and behaviors coupled with a rejection of stereotype roles and expectations. Experience and feeling awareness rather than rational understanding predominate, thus leading to the "source of life," which is the expansive and deeply expressive self. Consistent with an earlier tradition of romantic individualism, one seeks to express the self rather than submit to external authority, cultural tradition, and social institutions.

Bellah summarizes the crisis of American democracy in the last decades of the 20th century as follows: (1) Exclusive radical individualism would lead to destruction of a free democratic society, (2) Our society still is intact because there is a remnant of biblical and republican individualism that sustains the common good; (3) traditional individualism continues to decline. 4) If we are unable to revitalize biblical and republican individualism, society will collapse and some form of totalitarianism will fill the void. Thus, Bellah proposes "...that our most important task today is the recovery of the insights of the older biblical and republican traditions."[4]

Competing visions: free market versus civil society. Yankelovich reformulates the crisis of American, radical individualism in terms more appropriate to the late 1990s. In 1999 Yankelovich proposed that the US society is in a struggle for the soul of America. The struggle is between the perspective of *laissez-faire capitalism*, which Yankelovich called the *Vision of the Free Market* and the *Vision of the Civil Society* defined in terms of what Bellah called the biblical and traditional republican individualism.[5] The Vision of the Free Market is another name for radical, utilitarian individualism and the Vision of Civil Society is another name for the integration of traditional biblical and republican individualism. But in 2000 there was no way Americans could "return to the noble and profoundly traditional dream of America as a City on a Hill." For one thing, a modified version of the self-fulfillment ethic was still alive and growing in intensity as we moved into the new millennium. This ethic plus all the "deconstructions" of

---

3. Robert Bellah, et al, *Habits of the Heart* (New York: Perennial Library, 1985), p. 76

4. Robert Bellah, et al, *Habits of the Heart* (New York: Perennial Library, 1985), p. 303.

5 Daniel Yankelovich. *The Magic of Dialogue* (New York: Simon & Schuster, 1999), pp.202-203.

the 1980s, which produced the civil rights movement and the feminist movement, totally oppose the return to the American dream, which not only was racist and sexist but represented a lower-level stage of individuation, and most people recognized it as such.

Yankelovich recognized that the struggle between the two visions is lopsided; the leaning toward one pole is greater than the other:

> The Vision of Civil Society is a major source of the moral values that must contain and support the thrust of our market economy. *But it limps along far behind the Vision of the Free Market in vitality, enthusiasm, and power* [italics mine].[6]

The emergence of American industrial consciousness occurred in association with the business world - especially big business – in many ways in opposition to traditional American individualism. Expressive individualism, which emerged from the countercultural movement of the 1960s, focused its attack on "the military-industrial complex," that is, government contracts with big business, as well as an attack on traditional values. Economic realities in the last two decades of the 20th century, that is, the recession of the late 1970s and the one from 1989 to 1992, the move to a global economy, and the increase in the winner-take-all markets, virtually eliminated the emphasis on *radical, subjective, control individualism*, see chapter 22 dealing with the countercultural revolution. All this helped to produce a modified, more socially oriented ethic of self-fulfillment.

The middle class social mentality from the middle to late 1990s combined the ethic of self-fulfillment, see chapter 22, with laissez-faire capitalism to produce a utopian understanding of the Free Market Vision. This new idealized version of capitalism seemingly overpowered both the traditional and radical expressive types of individualism. Writing in 1998-1999, Yankelovich proclaimed that:

> Many of America's leading business executives and political leaders believe that the free market has moral virtues over and above its pragmatic advantages in allocating resources efficiently. It is these moral virtues that give the Vision of the Free Market its ideological and political power.
>
> The most probable scenario for the near future is that some form of this free-market vision will prevail. At the moment, it dominates the American climate of opinion....
>
> [T]here is little doubt that our American technology-driven high-entrepreneurship market economy is the model for the rest of the world and is destined to shape our lives for decades, if not centuries, to come....
>
> The prospect of greater material well-being serves a purpose that goes far beyond materialism: it helps to validate the American Dream.... Material well-being is indispensable to our system of upward mobility. The American Dream depends on an

---

6. Daniel Yankelovich, *The Magic of Dialogue* (New York: Simon & Schuster, 1999), pp. 211-212.

economy that can deliver rewards for hard work and self-improvement.

But the Vision of the Free Market has its dark side.[7]

The dark side of the Vision of the Free Market consists of all the negative aspects of the industrial consciousness as described in chapter 19. The 1990s produced some "new wrinkles" of these negative aspects such as radical disruption of jobs, skills, and older enterprises, the income gap between winners and many more losers, downsizing, and reengineering. Moreover,

> [The business world depends on] Economic ideas [that] always come embedded in a matrix of social values. The long-term success of the economy as well as the well-being of the larger society depends utterly on these values. But the values themselves do not come from the market economy, and they are not self-sustaining. They need constant reinforcement. If they don't get it form some source other than the market, they wither and die.[8]

Just a few years later this bubble of hope burst. Not only did a recession officially occur March, 2001 (exactly tens years after the longest post-World War II business expansion which began in March, 1991), but also at about the same time Americans and the rest of the world became aware of the widespread, rampant greed in the business community that produced professional misconduct and gross and unlawful immorality. The recession officially ended in November, 2001, but as of December, 2003, there were reports of widespread, unethical business transactions, such as, the mutual funds scandal. In the late 1990s Yankelovich noted that the only way the conflict between the Vision of Civil Society and the Vision of the Free Market can be reconciled pragmatically and realistically is through participatory dialogue, as will be described in chapter 28.

> My studies of the public reveal an immense pool of good will and good faith all over the country. Americans are hungry for enhanced quality of life, for deeper community, [and] for endowing our communal life with spiritual significance.... They are ready to accept truths over and above those of science and technical expertise without discarding their immense contributions. They are ready... [to engage in participatory dialogue] in order to endow their own lives and those of others with a larger meaning.... What we [Americans] don't know very well, and where we are surprisingly awkward and not at all adept, is in the arts of listening with empathy, setting aside status differences, and examining with open minds the assumptions that underlie all the old scripts we all live by – in a word, dialogue.... I believe that the greater mastery

---

7. Daniel Yankelovich, *The Magic of Dialogue* (New York: Simon & Schuster, 1999), pp. 202-203.

8. Daniel Yankelovich, *The Magic of Dialogue* (New York: Simon & Schuster, 1999), p. 205.

of dialogue will advance our civility – and our civilization – a giant step forward.[9]

Compassionate conservative free market vision. The Bush-Cheney presidency represents itself as having a *compassionate conservative free market vision.* This vision is of the Baby Boomer idealist generation (born 1943-1960) came to political dominance over the previous civic generation. Each of these archetypal generations attempts to use the political process to achieve or defend deeply held personal values and social commitments. The civic generation consisting of the GI generation (born 1901-1924), which elected Franklin Roosevelt in 1932, supported big government enacting social programs, such as social security, and regulating big business. These progressive ideas were very different from the laissez-faire economic policies that dominated the idealist generation that brought about a political realignment with the election of William McKinley in 1896. The Roosevelt civic agenda also differed from the Baby Boomer idealist realignment since 1968.[10] Under the leadership of President Reagan the politics of the Baby Boomer generation deregulated big business over two decades that, in turn, led to shrinking the middle class, reducing the spending power of most Americans, and greatly increasing the gap between the wealthy and most Americans. The 62% disillusionment of all Americans with the bush-Cheney presidency in 2007 and 82% in May of 2008 indicated a disillusionment with scientific capitalism that became total after the economic collapse in September, 2008. The current recession eventually will end, but the old style free market vision is unsustainable, and a new, watered down version of it will not by itself be satisfying.

Radical ego constructivism. The combination of science modernity and the new postmodern modernity affirms that reality is not intelligible and therefore unknowable, but one can construct valid interpretations of it. On the one hand, this *radical constructivism* is a rejection of the ego illusion of having power based on absolutely true knowledge of reality. Correspondingly, it rejects dogmatism that gives a few people, who supposedly have true knowledge, power over other people who need the guidance that the true knowledge can give. Thus, one rejects ego power based on the illusion of true knowledge and on dogmatism. On the other hand, a radical constructivist may choose his/her self-validating ego to replace any higher power, such as God or some socially validated hierarchy of power such as that defined by the U.S. Constitution. When radical constructivism leads one to reject any limitation to one's self-validating ego, then it becomes what I call radical, ego constructivism. This produces even greater mind self and ego inflation than power hierarchies based on dogmatism associated with absolutely true knowledge.

Guiding Principle: Collaboration between Eros-Chaos and Eros-Order

The fourth enlightenment involves the resurgence of participatory subjectivity but in a balanced way. This can be achieved by the collaboration between Eros-chaos associated with participatory subjectivity and Eros-order associated with control objectivity. All the themes of

---

9. Daniel Yankelovich, *The Magic of Dialogue* (New York: Simon & Schuster, 1999), pp. 217-218.
10. Morley Winograd and Michael D. Hais. *Millennial Makeover* (New Brunswick, New Jersey: Yale Uni. Press, 2008).

the fourth enlightenment described in Unit VIII involve this collaboration along with the commitment to an ultimate SOURCE.

Chapter 21
AVANT GARDE MODERNISM

ROMANTICISM BECOMES NIHILISTIC SELF-EXPRESSION

In the early to mid 19th century, science-technology and romanticism were integrated in the ideology of scientific humanism, but as science became more positivistic, romanticism became isolated from any objective, rational foundations. The romantic poet was confronted with his own personal response to a universe devoid of meaning and to a "life-world" given meaning by an emerging bourgeois society that came under a radical critique by Right and Left Hegelians. Both these camps agreed that traditional philosophies and the religions wedded to them were dead. These young Hegelians created a modern philosophical discourse (following the lead of Hegel) to overcome the positivism of science. The Left Hegelians, especially Marx, unmasked bourgeois society as driven by utilitarian values that destroyed communal relations and contradicted traditional philosophical-religious beliefs. Marxism became one example of the *avant-garde* that defined itself by its opposition to all traditional value systems and institutions. However, Marxism clung to the belief that a revolutionary form of scientific humanism could transform bourgeois society into a non-alienated, ideal community. The Right Hegelians chose a strong state government to save bourgeois society by superimposing on it traditional moral values even though the evolution of science wrenched these values from their religious and metaphysical foundations.

Thus, the passion of romantic poets originally drawn out to the creative energy in nature and to a rationally enlightened humanity now became dissipated in the Void of a meaningless universe and in a superficial, fragmented, and alienated bourgeois society. The romantic passion brought the mind self to confront the Void and to replace the earlier belief in progress with the will to create authentic experiences purified of all past or present cultural ties:

> [F]rom about 1885, when Modernist thought was shaping itself in sociopolitical and psychological areas as well as in the arts, the avant-gardeist was creating unique instants; revelations and privileged moments. The idea of progress is, of course, connected to the future becoming "now";... we are the future. The avant-garde artist partakes of that meaning, but he embodies in himself a significant extension of the idea, the sense that utopia has arrived purified of pastness and historical arguments. His work, whatever shape it takes as poetry, and fiction, or as musical composition, painting, and dance, is the utopian moment. It is far more than the "future is now," and it has leaped beyond progress....
> The avant-gardeist, in his desire to purify, is always seeking a still moment or a spiritualized moment in which time and space are disconnected from past, present, future.[11]

11. Frederick R. Karl, *Modern and Modernism* (New York: Athenaeum, 1985), p. 164.

To confront the Void is to experience a kind of death, which tempts one to suicide, and to negate all social connections. This is to court madness. Indeed, many avant garde artists and thinkers did attempt and sometimes succeeded in committing suicide, or they became insane. However, this romantic *leap beyond progress* frees up energies from the soul to create new patterns and establish alternate modes of perception. To do this the avant gardeist had to create new languages that were different from the representational languages of science and modern philosophy or the symbolism of traditional art forms, including romanticism. Subjective experiences became occasions for the artist to express his inner self by means of his personal, non-rational, non-symbolic language, that is, an expression of soul. In opposition to all traditional values the work of art was proclaimed to have value in itself, beyond all objective criteria. While it may have shocked all those who perceived it, the work invited them to enter into it and experience its wholeness, integrity, and authenticity. Thus, through his personal language manifested in his work, the artist created an authentic experience not only for himself but also for all those who chose to enter into it.

The leap beyond progress is a metaphor not only for modernism as a cultural movement but also for the experience of scattered avant garde poets such as Charles Baudelaire (1821-1867), Mallarme (1842-1898), Lautreamont (1845-1870), Rimbaud (1854-1891), and Paul Verlaine (1844-1896) who bridged the gap between romanticism and avant garde modernism. This metaphor embodies rushing toward transcending death and madness with

> ...strategies [that] involve exalting the world as "nihil" or "nil,"
> "nada," nothing and yet proceeding as if that "nil" can be overcome
> through self, senses, sensations, yearning for absolutes within.[12]

Avant garde artists evolving from romanticism took this leap in contradistinction to autonomous reason absolutized to the pure instrumentality of logical positivism. Consequently, the ineffable insights resulting from this kind of leap are rightly designated by various thinkers as aesthetic experiences of the "autonomous self" (autonomous feeling body ego) *absolutized to the other of reason.*

The enlightenment equated with modern culture is fundamentally paradoxical: It produced two kinds of individualism, traditional and radical, which are diametrically opposed and bent on destroying one another. For example, thinkers committed to the ethic of objective knowledge define individualism associated with this ethic as authentic in contrast to individualism associated with traditional values seen as authentic. These two "individualisms" present in 20th century Western cultures also are present in many persons. The resulting internal war is producing societal-cultural and individual-psychic stress. Speaking from the perspective of the emerging avant garde artists, Frederick Karl notes:

> This movement toward... autonomy of self was in reality a move
> toward freedom, functionlessness, anarchy. That is , the individual
> for the first time might cohere as a whole, but he cohered at the
> expense of community and society. Here we have one of the many
> paradoxes of the Modern movement: that striving toward

---

12. Frederick R. Karl, *Modern and Modernism* (New York: Athenaeum, 1985), pp. 42-43.

individual wholeness based on freedom, choice, liberation and expression of self [radical subjective individualism]; and its reciprocal, an ever growing anarchy in social/political terms, which can be contained only by regimentation and prohibition of Modern Art.... At one end was total freedom (unachievable) [a kind of mind self generated participatory consciousness] and at the other end total control (also unachievable) [mind self generated control consciousness][13]

AESTHETIC EXPERIENCE ABSOLUTIZED TO THE *OTHER OF REASON*

Aesthetic experience absolutized to the other of reason is self-expression that is nihilistic on two counts: (1) It embodies a kind of death and madness, and (2) it disrupts all social interactions. However, just as positivism representing a core theme of the scientific Enlightenment is a higher level of consciousness with nihilistic side effects so also with "the other of reason," in which its nihilistic side effects are counter balanced by three positive outcomes. First of all, the leap of an avant garde artist and the work of art that embodies his insights resulting from that leap are analogous to embracing death and the consequent rebirth (chaos and new order) of any human creative process. The nihilistic side effect is a necessary aspect of the creative process leading one to a higher level of consciousness. Secondly, as described above, the leap is toward a utopian moment, a spiritualized moment analogous to the ecstasy of a mystic and toward personal wholeness. Thirdly, a particular aesthetic experience may be a *shattering, willed discontinuity*.

This third positive aspect of "aesthetic experience absolutized to the other of reason" may be pictured by the metaphor of Bohr's quantum mechanic understanding of the hydrogen atom. The electron is moving in a continuous elliptical path around the hydrogen nucleus when in a "non-instant" it is moving in a new elliptical path farther from the nucleus. The old pattern (path) is shattered when the electron receives light energy from the sun and uses it "to will itself" to move in the new pattern. The "willed non-instant" is an "existential point" of discontinuity: It is beyond time and is absolutely indeterminate. The event is not a "present" but rather is a *gap* between the past and the future; nothing in the pattern of the past leads up to this event, and nothing in the current pattern points to a past pattern from which it emerged. Henceforth, I will refer to the aesthetic experience understood to mean a "*shattering, willed discontinuity*" as the *gap experience*. As will be explained below, the gap experience is beyond truth-falsehood or good-evil, but it is positive in that it simultaneously marks the end of an old pattern and the beginning of a radically new pattern.

Though the gap experience became manifest in avant garde artists, it is by no means only encompassed by a work of art. St. Paul's conversion on his way to Damascus was a gap experience; so also was the Buddha's enlightenment, and so also was Job's "enlightenment" when he was "touched by God," that is, moved by Grace. It also is *samanya* of Hindu Buddhism or satori of Zen Buddhism. The avant garde artists of the scientific enlightenment discovered a "secular gap experience"; before that time it always had been associated with a mystical tradition of some religion. A truism of this tradition always has been that the gap experience is a kind of death that may be a leap into madness; hence, mysticism was only for the few who could be

---

13. Frederick R. Karl, *Modern and Modernism* (New York: Athenaeum, 1985), pp. 41-42.

protected from madness by sufficient mental and physical discipline and by the guidance of a spiritual director or a guru. Avant garde artists eschewed discipline and, cut off from any mystical tradition, they eschewed spiritual guidance as well. It is not surprising, then, that many of them pursued self-destructive activities that often led to death or insanity.

The gap experience does not necessarily produce a "morally better" person, but it does give one more power. For example, the mostly male avant garde moderns were as sexists as the Nazis were racist.

> We note a profound paradox: that while Modernism stressed the individual achievement and the unique artist, its chief male figures and many of its theorists denied individuality to women....
>
> But coequal with Nietzsche and Freud was the large number of poets, novelists, painters, playwrights, and composers who perceived woman as Earth Mother, on one hand, as destructive Eve, on the other.[14]

Following the arguments of James Rhodes and Fritz Stern, Morris Berman notes that Fascism was a gnostic phenomenon (a gap experience). The Nazi experiment was an attempt to bring back together the secular and sacred worlds split by the scientific enlightenment.

> It was a demonic attempt to reenchant the world. The language of National Socialism was that of transcendence. Hitler recognized, instinctively, a religious need on the part of the masses and he responded with a gnostic political program, though the gnosis was well disguised. The real story here is one of ecstasy and ascent, salvation and redemption.[15]

Berman goes on to describe the gap experience of Hitler:

> The "character of Hitler's compulsive power over men's minds can only be understood in religious terms," writes Joachim Fest; and we have already noted how the Abel documents reveal an odyssey of redemption, an *imitatio Christi*, with Hitler taking the place of Jesus and *Mein Kampf* replacing the Gospels....
>
> ... Hitler lost his sight as a result of the gas attack, which was only slowly restored at Pasewalk, and this was accompanied by depression and mental instability... As Hitler lay in despair on his cot, recovering from Germany's defeat and surrender, he had a "supernatural vision" (this quotation is from the *OSS Hitler Source Book*, cited in note 67). The blindness lifted; in an ecstatic trance, or "inner rapture," Hitler heard voices summoning him to save Germany, to deliver her from defeat....

---

14. Frederick R. Karl, *Modern and Modernism* (New York: Athenaeum, 1985), p. 146.

15. Morris Berman, *Coming to Our Senses* (New York: Bantam Books, 1989), p. 269.

Hitler emerged from the event [the *gap experience*] as a charismatic with a politico-religious agenda. He changed markedly after this, which is, of course, a common feature of ascent experience [*gap experience*]. Hitler had not been a talented speaker before the war's end, yet by the spring of 1919 a member of the DAP (German Workers' Party, which Hitler would later join) heard him address a soldiers' council in Munich and noted "his almost occult power of suggestion over the assemblage." [note 73] Hitler went from someone who was inward and bookish to being a powerful orator.... Hitler's effect was repeatedly described as hypnotic; leading Nazis described their moment of allegiance to Hitler as a kind of religious conversion, and this included Speer, Heydrich, Hess, Hans Frank, Julius Streicher, Goring, Goebbels, Hanfstaengl, Ribbentrop, and many others. In his public addresses in particular, the magic was the message; Speer said that the effect of a Hitler speech went far deeper than the content. The ascent experience got translated into a national, political context, with awesome results.[16]

FRAGMENTATION OF SOCIAL COMMUNICATION LEADING TO DIVERSITY

The integration in the early to mid-19th century of science-technology, capitalistic institutions, and romanticism began to fall apart as Western Europe moved to cultural modernism, which began around 1885. The three "kinds of passion" intersected to produce the ideology of scientific humanism. The trajectories of these three passions continued on from this point of intersection to produce what Max Weber called three irreconcilable cultural value spheres.[17] Jurgen Habermas modified Weber's theory of societal rationalization in the following way. In modern societies people, "actors," are better able to learn about and evaluate their activities when these are put into three categories called *life worlds*, which provide the "multidimensional learning potential of modernity." This learning potential has progressively become enhanced by these life worlds becoming crystallized into separate, specialized forms of argumentation, which simultaneously become institutionalized into corresponding cultural spheres of action. Each cultural sphere has its own inner logic and corresponding validity claim and leads to a general structure of social consciousness, which produces a particular type of knowledge as shown in table 21.1.

---

16. Morris Berman, *Coming to Our Senses* (New York: Bantam Books, 1989), pp.281-282.

17. These are: 1) cognitive involving science/knowledge and economy/wealth; 2) normative involving religion/morality and politics/power; 3) aesthetic involving art/taste and counterculture/love. see: David Ingram, *Habermas and the Dialectic of Reason* (New Haven, CN: Yale University Press, 1987) p. 52.

TABLE 21.1
Three Spheres of Knowledge Corresponding to Cultural Spheres of Action and Life Worlds

LIFE WORLD	CULTURAL SPHERE OF ACTION	TYPE OF KNOWLEDGE
Natural	Science-technology	Cognitive
Social	Morality-law	Normative
Subjective	Art & Literature	Aesthetic

Though on many counts I disagree with Habermas's theory of societal rationalization, it does provide a convenient terminology for describing the fragmentation of social communication.

This fragmentation is nihilistic because many people locate themselves primarily in the natural or subjective life world, and each of these has its own brand of deep alienation. Moreover, each has had a profoundly destructive effect on the social life world. The quotes in chapter 18 of Jacques Monod give eloquent testimony to the alienation of scientists. Though I disagree with Allan Bloom's analysis of and solution to the problem of higher education in America today, his assessment of the destructive effect of the natural sciences fits my experience as a graduate student for five years and as a science professor for 45 years (June, 2009). In the 1980s and still today, the positivistic sciences (what Bloom calls the natural sciences) dominates American culture by dominating academia, which, among other tasks, trains teachers for grades 1 through 12. The natural sciences studies humans only in so far as they are like other living organisms or, for the strict mechanistic scientists, as like physical systems. The social sciences try to study humans, individually and as social units, as if they are objects that are predictable and controllable just as physical objects are. Some in the humanities influenced by idealistic philosophy beginning with Kant, who sought to overcome the dehumanizing effect of science, treat humans as if they primarily are spiritual minds residing in a body. Postmodern thought in the humanities but in a different way de-emphasize the objective, material aspects of humans. Thus, the positivistic sciences, and the further differentiation of the social sciences and the humanities have produced a culture maintained by an education system that is unable to understand humans as the paradoxical integration of body and soul.[18]

Let us now look at the reflections about science of Charles Darwin, who lived at the time when the three life worlds were splitting from each other:

> Up to the age of thirty, or beyond it, poetry of many kinds, such as
> the work of Milton, Gray, Byron, Wordsworth, Coleridge, and
> Shelley, gave me great pleasure, and even as a schoolboy I took
> intense delight in Shakespeare, especially in the historical plays. I
> have also said that formerly pictures gave me considerable

---

18. Allan Bloom, *The Closing of the American Mind* (New York: Simon and Schuster, 1987), pp. 356-359.

[pleasure], and music very great delight. But now for many years I cannot endure to read a line of poetry; I have tried lately to read Shakespeare, and have found it so intolerably dull that it nauseated me. I have also lost my taste for pictures and music. I retain some taste for fine scenery, but it does not cause me the exquisite delight which it formerly did....

My mind seems to have become a kind of machine, for grinding general laws out of large collections of facts, but why this should have caused the atrophy of that part of the brain alone, on which the higher tastes depend, I cannot conceive. A man with a mind more highly organized or better constituted than mine, would not, I suppose, have thus suffered; and if I had to live my life again, I would have made a rule to read some poetry and listen to some music at least once every week; for perhaps parts of my brain now atrophied would thus have been kept active through use. The loss of these tastes is a loss of happiness, and may be injurious to the intellect, and more probably to the moral character, by enfeebling the emotional part of our nature.[19]

I have known many scientists with "Darwin's malady," with the only difference being they do not even lament the lack or loss of taste. As Eugene Hargrove notes:

Though this loss of taste is a mystery to Darwin, it need be no mystery to us. It is a natural consequence of his attempt to be scientific, to deal with the facts alone. [*exclusively stay with the inner logic and form of argumentation in the life world of nature*] This experience is uncommon today only because scientists are now usually so little exposed to the humanities in their insensitivity and even aversion to literature, poetry, art, philosophy, music, religion, and ethics.[20]

The nihilism of the natural life world comes from control objectivity totally suppressing participatory subjectivity. In the subjective life world, participatory subjectivity totally displaces control objectivity. The previous section indicated the rise to dominance of participatory subjectivity and how this, for example, the gap experience of avant garde artists, often led to madness or suicide or short of these, to extremely self-destructive life patterns. The stability and even the future survival of the social life world have been severely challenged by the natural life world which has "invaded" and undermined traditional values in the social life world. So also has this stability been challenged by the subjective life worlds. Each according to its "inner

19. Francis Darwin, ed., *The Autobiography of Charles Darwin and Selected Letters* (New York: Dover Publications, 1958) pp.53-54.

20. Eugene C. Hargrove, *Foundations of Environmental Ethics* (Englewood Cliffs, New Jersey: Prentice Hall, 1989), p. 42.

logic" and "specialized form of argumentation" has undermined religion and metaphysics that formerly brought cohesion to society. Social cohesion is further undermined by the ongoing war between cognitive and aesthetic "types of knowledge," that is, the war between control objectivity and participatory subjectivity.

HEIDEGGER'S MODIFICATION OF NIETZSCHE'S NIHILISM

Heidegger claimed that the nihilism of Nietzsche carried traditional philosophy to a new, higher level of understanding. Nietzsche extended the nihilism of the ethic of objective knowledge to a nihilism of all conceptual knowing both traditional explanations and scientific descriptions. Nietzsche also extended the validity of the subjective affirmation of the ethic of objective knowledge to the subjective knowing involved in metaphorical, conceptual explanations that science totally rejects except as descriptions of empirical patterns that can be represented by logical, conceptual models that can be empirically validated. He was inspired by the dynamism of classical Greek art, particularly the plays written before and during the pre-Socratic philosophy period. I surmise that he interpreted these plays as giving meaning then and now to all those people who experience them. But, according to Nietzsche, the interpretations of meanings elicited by these Greek plays are illusions because, in agreement with the ethic of science and with respect to human knowing, reality has no meaning. How can anyone know this? Nietzsche may have known this to be subjectively true as a result of one or many gap experiences. He philosophically justified such a proclamation by claiming that he, the "last disciple" of Dionysus, the god who does philosophy, understands Heraclitus to say that all is change and therefore there are no eternal truths that arise from a static structure of reality. Thus, by a subjective will to power, analogous to the will to power that justifies the ethic of objective knowledge, Nietzsche imposes meaning onto any situation. However, unlike the will to power of science that has objective, utilitarian value, Nietzsche's aesthetics absolutized to the "other of reason" by definition of aesthetics has no utilitarian value; it supposedly only has value in itself.

While Heidegger agreed that Nietzsche shocked Western philosophy into a radically new direction, Heidegger chose to "correct Nietzsche" by incorporating modified aspects of traditional philosophy and modern science into this new vision. First of all, Heidegger converted the most fundamental postulate of both the ethic of objective knowledge and Nietzsche's nihilistic philosophy into an *ethic of collaboration*. The postulate of objective knowledge is the will to objective control of nature, represented by the phrase, "will to power," and Nietzsche's will to power dominated the will to subjective, participatory engagement with nature. Heidegger's transformed postulate is that the will to participatory engagement with nature is primary but should collaborate with the will to power. The will to power of both modern science and Nietzsche's philosophy affirms a fundamental nihilism which is: participatory engagement with nature will never produce subjective, true knowing that can be converted into objective, true knowledge. Heidegger agrees with this nihilism but then purports to transcend it. Reality does not have a static structure that humans can know. Being manifests the reality that humans experience, but Being can never be even approximately comprehended by the human mind (in agreement with all the great world mystics). Nevertheless, just as Being manifests reality, it also manifests knowledge in the human mind. It does this by "talking" to humans through their sensations leading to perceptions of reality. These perceptions become the starting point for humans interpreting their experiences of reality. The collective interpretations generated by various earlier civilizations became metaphorical, conceptual narratives represented by myths, religions, and in classical Greek society, the poetic philosophy of the pre-Socratics. Starting with Socrates, Plato, and Aristotle, metaphorical narrative explanations were rejected and then replaced by metaphysical, logical, conceptual models thought to be absolutely true. The rejection of the possibility of absolutely true, objective knowledge is the authentic nihilism of the ethic of knowledge and of Nietzsche's philosophy.

However, Heidegger's interpretation of the pre-Socratic philosophers, Heraclitus and Parmenides, see chapter 10, convinced him that valid but never absolutely true theories are legitimate ways of interpreting one's experience of reality. These provisional interpretations of experience replace the real nihilism of absolute metaphysics that claims to provide an absolutely true representation of reality or of science that claims to provide a provisional but progressively more approximately valid representation of the world that excludes all other interpretations.

> By contrast, to press inquiry into being explicitly to the limits of nothingness, to draw nothingness into the question of being [which means that Being never can be totally captured by any particular representation of it] this is the first and only fruitful step toward a true transcending of nihilism.[21]

Thus, we must acknowledge that we cannot know being in itself, but we can distinguish being from its *essent*, and instead of claiming to understand the essent we should recognize:

> In accordance with the hidden message of the beginning, man should be understood, within the question of being, as the site which being requires in order to disclose itself. Man is the site of openness, the there. The essent juts into this there and is fulfilled. Hence we say that man's being is in the strict sense of the word "being there." The perspective for the opening of being must be grounded originally in the essence of being-there as such a site for the disclosure of being.[22]

But in order for a human to be "the site of openness," and therefore "being-there as such a site for the disclosure of being," one first must have participatory engagement with nature. Collectively, such engagements provide empirical, metaphorical, conceptual patterns as the basis for constructing logical, conceptual representations of these patterns. Constructing objective knowledge in this way is humans' will to power. Thus, objective knowledge comes from the collaboration between the will to subjective, participatory engagement with nature and the will to objective control of nature (the will to power). In his later writings Heidegger further developed this insight by distinguishing between:

> ...two kinds of thinking, each justified and needed in its own way: calculative thinking and meditative thinking.... I would call this

---

21. M. Heidegger, *An Introduction to Metaphysics* [R. Manheim, trans.] (New Haven: Yale Uni. Press, 1959), p. 203.
22. M. Heidegger, *An Introduction to Metaphysics* [R. Manheim, trans.] (New Haven: Yale Uni. Press, 1959), 205.

comportment toward technology which expresses "yes" and at the same time "no," by an old word, *releasement toward things.*

Having this comportment we no longer view things only in a technical way....

That which shows itself and at the same time withdraws is the essential trait of what we call the mystery. I call the comportment that enables us to keep open to the meaning hidden in technology, *openness to the mystery.*

Releasement toward things and openness to the mystery belong together. They grant us the possibility of dwelling in the world in a totally different way [than merely controlling nature]. They promise us a new ground and foundation upon which we can stand and endure in the world of technology without being imperiled by it. [23]

Calculative thinking comes from social and individual, control objectivity and meditative thinking comes from individual, participatory subjectivity. Thus, Heidegger is expanding Nietzsche's strictly patriarchal point of view to a non-patriarchal perspective that includes the patriarchal one as a limited special case. Nietzsche proposes that each of us can obtain foundational understanding by a patriarchal "will-to-power" that de-individualizes, that is, causes one to regress to the polar self with mythic consciousness which in turn regresses to the action self with magic consciousness that brings one into an archaic union with nature. The farther away one is from this archaic union as a result of individuation, the farther away he/she is from foundational knowledge. According to Nietzsche, this foundational knowledge provides the basis for a critique of rational knowing. Heidegger corrects Nietzsche by insisting that we may retain the fruits of the scientific enlightenment and celebrate our radical individualism, but we must balance this with a type of knowing that comes from participatory subjectivity that he calls meditative thinking.

Secondly, Heidegger redefined the meaning of authentic knowledge. According to him, inauthentic knowledge comes in many forms including (1) the claim of absolute metaphysics, such as Thomistic philosophy, (2) the claim of modern science representing the ethic of objective knowledge, and (3) the claim of Nietzsche's will to power that imposes onto one's experience of nature subjective, mystical insights coming from the inner self. Authentic, objective knowledge comes from the collaboration of the will to participatory engagement with nature and the will to power. The idea of inauthentic versus authentic knowledge is linked to nihilism and the postmodern idea of deconstruction. I view Heidegger's ideas of authentic nihilism as also implying an analogous distinction between inauthentic deconstruction and authentic deconstruction. *Inauthentic deconstruction* is associated with inauthentic nihilism, whereas *authentic deconstruction* is associated with authentic nihilism. What makes it authentic is that the deconstruction does not totally destroy or otherwise totally negate some human creation. Rather it holds up an idea as being neither good nor bad, true nor false, and thereby allows the idea to reveal new possible mutualities with other ideas even if they are contradictory or contrary to it. Alternatively, authentic deconstruction destabilizes a system so that it exhibits new

---

23. M. Heidegger, *Discourse on Thinking.* [J. M. Anderson and E. H. Freund, trans.] (New York: Harper Torch Books, 1966), p.46

possibilities for restructuring itself in conjunction with one or more other systems, even if these other systems oppose it or seek its demise. Thus, authentic deconstruction is the chaos phase – the death phase – of the creative process represented as Order, Chaos – deconstruction – New Order including modified aspects of the old order. That is, it is the process of Life, Death, Rebirth, which describes the developmental-evolutionary transformations of humans.

Thirdly, Heidegger transformed the meaning of the doctrine of "the eternal return of the same situation." Heidegger's version is: A*ny created meaning represents an aspect of Being; it* may replace an old meaning or it may complement that meaning. In any case, *meanings are only partially created by a knowing subject; they also are manifestations of Being to which the person has given himself over in order for Being to produce meaning in him.* Being, itself, though experienced in this way, can never be known. Heidegger "deconstructs" all traditional metaphysics stemming from Plato and Aristotle. At the same time, he affirms that humans "create meaning" only as a result of Being manifesting itself in the human knower. The human must give himself over to Being in order to participate in the creation of meaning. Heidegger's "creative person," analogous to Nietzsche's "superman," is the human who chooses "being there," that is, to engage in meditative thinking which is "releasement toward things." This implies that the rational models driving calculative thinking have "evolved from" meditative thinking. Traditional philosophers and modern scientists are destructively nihilistic precisely because they have "forgotten Being." They have denied the existential and metaphorical-narrative knowing from which their rational models emerged.

Nietzsche's version is associated with his non-evolutionary expression of mystical, heroic creativity. Heidegger interpreted Nietzsche's version as the *evolutionary* process of creating hierarchal, new patterns. As a result, *if we separate Heidegger's philosophy from his political-social commitment to Nazism,*[24] the mystical, heroic creativity that produces these hierarchies of patterns is the evolutionary process of "no-knowledge that leads to no-self," see chapter 13. My further interpretation of this is that all patriarchal philosophies and dogmatic religions including the Judeo-Christian religions, which many scientists and Nietzsche hated, should not be despised but rather should be brought into the dialogue of all points of view in the context of evolutionary, mystical, heroic creativity.

Fourthly, if, again, we separate Heidegger's philosophy from his Nazism, his philosophy may be interpreted to imply the *individualism of creative dialogue* in which there is the collaboration of participatory, subjective individualism with control, objective individualism that includes scientists' radical, objective control individualism and Nietzsche's radical, subjective, control individualism. This new kind of individualism involving creative dialogue is disposed to transcend conflicts of different points of view rather than, by a will to power, impose one point of view as a replacement for all other points of view, see chapter 28.

Fifthly, Heidegger acknowledged in agreement with the ethic of knowledge, see chapter 18, and with Nietzsche's philosophy that modern societies have become addicted to utilitarian science. These societies also have become infected with the will to power of avant garde modernism. As result, there is an intense internal conflict between values associated with these two types of will to power and traditional values. Heidegger sought to transcend this conflict by applying his philosophy to facilitate his commitment to Hitler and the greatness of National

---

24. Victor Farias, *Heidegger and Nazism* (Philadelphia: Temple University Press, 1989)

Socialism.[25] But again if one separates Heidegger's philosophy from his commitment to Hitler and National Socialism, his ideas of a new definition of authentic knowing and what may be interpreted to be the individualism of creative dialogue implied by his philosophy form the basis of an emerging *new enlightenment*. Creative dialogue then becomes one of its major theses that has supreme value and a basis for committing to other values and choices. It represents an idealism to transcend the conflicts in modern societies and produce better societies, and, indeed, a better world society.

Chapter 22
THE COUNTERCULTURAL REVOLUTION OF THE 1960s

AVANT GARDE MODERNIST'S FAILED ATTEMPT AT INTEGRATION

The evolution to nihilism initiated by the scientific enlightenment prepared the ground for a differentiation of a new kind of modernist culture beginning around 1885. Thus, aspects of the enlightenment-generated nihilism which include innovations of the avant garde artists described in chapter 21 make up the cultural-social innovations of modernism. These are: (1) the gap experience, (2) diversity of personal and social *evolutionary* differentiation involving unpredictable transformations as opposed to *epigenetic* differentiation involving predictable transformations implied by the patriarchal perspective, and (3) fragmentation into life spheres. Modernism failed to integrate these innovations into society because of what Habermas calls "the paradigm of [self] consciousness" and its associated "philosophy of the subject." That is, the avant garde artists and thinkers such as Nietzsche separated themselves from society but then in patriarchal fashion attempted to impose their insights onto its social institutions. The avant-gardeists, like priests or prophets speaking from on high, believed they could instruct or better yet shock people into changing their way of thinking, thus changing the way they live, which would transform society.

> [T]he idea of a fusion between radical art and radical politics, of art as a direct means of social subversion and reconstruction, has haunted the *avant-garde* since Courbet's time [1819-1877]. On the face of it, it has a kind of logic. By changing the language of art, you affect the modes of thought; and by changing thought, you change life. The history of the *avant-garde* up to 1930 was suffused with various, ultimately futile calls to revolutionary action and moral renewal, all formed by the belief that painting and sculpture [and poetry-literature-philosophy and music] were still the primary, dominant forms of social speech they had been eighty years before. In uttering them, some of the most brilliant talents of the *avant-garde* condemned themselves to a permanent self-deception about the limits of their own art [and philosophy].

---

25. Victor Farias, *Heidegger and Nazism* (Philadelphia: Temple University Press, 1989), Foreword by Tom Rockmore and Joseph Margolis, pp. x – xi.

Though it hardly alters their aesthetic achievement, it makes the legend of their deeds seem inflated.[26]

Ironically, the bourgeoisie, whose values the avant garde artist opposed, promulgated this transforming view of art and simultaneously brought the period of avant garde to an end:

Over the last half-century, the Museum has supplanted the Church as the main focus of civic pride in American cities. (At the same time, European churches were busy converting themselves, for survival, into museums.)... But the later nineteenth century, in America as elsewhere, saw the rise of that intense belief in the reformative and refining powers of art which was, in itself, one of the taproots of the *avant-garde*. Paintings ... were conceived as vehicles of moral instruction, and the museum, in assembling them, performed some of the functions associated with a religious gathering place.... The idea of social improvement through art struck a responsive chord in the American rich, who proceeded to pour millions upon millions of dollars into the construction and endowment of museums and the getting of collections that would eventually fill them....[For example, the Toledo Museum of Art in Toledo, Ohio.]

But the great change came in 1929, when the Museum of Modern Art (MOMA) was founded in New York.

Until then, the words "museum" and "modern art" had seemed, to most people, incompatible....

'The values propagated by MOMA were blown through the American educational system, from high school level upwards. (They also filtered downwards to the kindergartens, considerably raising the status of "creativity" and "self-expression" in primary education.) ... Moreover, with the colossal enlargement of the museum audience... the old distance between the coterie and the mass audience was swiftly being abolished. The work of art no longer had a silence in which its resources could develop. It had to bear the stress of immediate consumption....

...the good will of museums, universities, and other institutions of liberal thought had deprived modern art of its function as an *avant-garde*.[27]

By the mid-1960s the period of avant garde art was over.

The years from 1965 on saw the eclipse of art movements. The very word "movement" had become a curator's device for

---

26. Robert Hughes, *The Shock of the New* (New York: Alfred A. Knopf, 1982), p. 371.

27 Robert Hughes, *The Shock of the New* (New York: Alfred A. Knopf, 1982), pp. 391-394.

generalization... yet it remained useful as a dealer's gimmick for prodding the market along by lending pseudo-historical weight to new art.[28]

...by 1976 "*avant-garde*" was a useless concept: social reality and actual behavior had rendered it obsolete. The ideal -- social renewal by cultural challenge -- had lasted a hundred years, and its vanishing marked the end of an entire relationship -- eagerly sought but not attained -- of art to life.[29]

ETHIC OF SELF-FULFILLMENT

Three Stages of Its Development

The self-fulfillment movement was a response to the negative aspects of industrial consciousness. The movement emerged and differentiated in three stages. The <u>first stage</u> is the separation of the private and public spheres, the private/public split. Separating the private sphere was a balancing mechanism to provide meanings to compensate for the discontents brought about by structures of the industrial consciousness. In the private world one is allowed to express feelings repressed in the work world. A person develops a specific, private identity as a shelter from threats of anonymity experienced in the work world, see chapter 19.

The <u>second stage </u>is the incorporation of three major themes of the 1960s countercultural revolution into postmodern industrial consciousness. One theme is the intense opposition to the private/public split. This dichotomy is perceived and condemned as hypocrisy and as pathology. The individual should be at home in all sectors of his/her social experience. This means that he/she should be able to develop and express his/her individuated self not only at home or with intimate friends but also while participating in bureaucracies and in the workplace. The second theme is nihilism, expressed as opposition to many aspects of the industrial consciousness. A list of the things that are opposed includes: functional rationality, the rational dominance of one's personal feelings and subjective insights; traditional repression of sexuality, objectification of women, that is, women seen merely as sex objects; and mindless exploitation of nature rather than reverence for her. Furthermore, it opposes calculated life planning and delayed gratifications. The countercultural people, especially the youth who can neither plan nor wait, rejected: achievement, which is an aspect of life planning, the "Protestant ethic" of hard work, sobriety, saving, and ambition to achieve status, wealth and power. Instead, one should "hang loose, turn on to drugs, and "work the system." It also opposes organizations and bureaucracies that are inimical to community life and correspondingly all stereotype roles and all aspects of order that are resistant to modification or transformation. The third theme is the reaffirmation of participatory consciousness that spilled over into the feminist movement, the ecology movement, and into the resurgence of occultism, magic and mystical religions. It also is a disposition to greater receptivity-openness, that is, surrender, letting go, an essentially passive stance toward the world that some may describe as feminine.

---

28. Robert Hughes, *The Shock of the New* (New York: Alfred A. Knopf, 1982), p. 385.

29. Robert Hughes, *The Shock of the New* (New York: Alfred A. Knopf, 1982), p. 366.

The <u>third stage</u> is what Yankelovich designates as the *self-fulfillment ethic* that entails the search for the full, rich life that is ripe with leisure, new experiences, and enjoyment as a substitute for the orderly, work-centered ways of earlier decades. This cultural shift – the hippie revolution – in the 1960s did not touch the lives of the majority of Americans. But by the 1970s Yankelovich's surveys showed 72% of Americans were beginning to be more preoccupied with satisfying the inner needs of the self.[30] The two major themes of the self-fulfillment ethic are the sacred/expressive aspects of life and radical subjective, control individualism.

The sacred/expressive aspects of life. "Sacred" here is used as a sociological concept not opposite to secular or profane but rather opposite to functional rationality, that is, instrumentalism. "Expressive" also is opposed to functional rationality. People and activities have value in their own right, things such as myths, art, poetry, monuments, story telling, song, dance, customs, architecture, ritual, harmonics of nature. These activities stem from the self-fulfillment search for personal meaning that resides in the sacred/expressive aspects.[31]

In the 1970s the American society began to have an ambivalent attitude toward competition. The self-fulfillment ethic sought to compensate for the negative aspects of the competition in the work world. In this world success is defined in terms of a competitive zero-sum game. Whenever a few people win, other people, usually a large number, lose even though they are intelligent, work hard, and are passionately intent on winning. As a result, these losers may feel personally responsible for their failure and therefore feel bad about themselves. The ethic of self-fulfillment enables them to shift from an exclusive commitment to the laissez-faire capitalism game and to judge personal success according to one's internalized standards.[32]

Yankelovich quotes the results of a study he did with Skelly and White. In the 1970s, 17 percent of the workers they studied valued personal self-fulfillment high above all other concerns, such as money, security, performing well or working at a satisfying job. However, these workers, which Yankelovich designates as "strong self-fulfillment seekers," the "strong formers," feel that the new rules of the 1970s give people the go ahead to break loose from old life patterns. So long as people are not antisocial rebels, they now can dismantle and remake their lives, such as getting a divorce, and still be accepted by traditional society.[33]

These "strong self-fulfillment seekers" exhibit several characteristics. They are ill at ease with their heritage of moral values and in particular, they believe that the self-denial virtue once thought to be good in itself now makes no sense. They are convinced that they are free to choose life goals and the means to achieve them without the guidance of tradition, custom, and conventional rules.[34] These "strong formers" are in the mainstream of the American tradition of self-improvement, but this now has a meaning very different from its traditional definition. Along with that they have a streak of earnestness and great faith in education. Finally, they

---

30. Daniel Yankelovich, *New Rules ,Searching for Self-fulfillment in a World Turned Upside Down* (New York: Random House, 1981), p. 5.

31. Daniel Yankelovich, *New Rules ,Searching for Self-fulfillment in a World Turned Upside Down* (New York: Random House, 1981), p.5.

32. Daniel Yankelovich, *New Rules ,Searching for Self-fulfillment in a World Turned Upside Down* (New York: Random House, 1981), p.7.

33. Daniel Yankelovich, *New Rules ,Searching for Self-fulfillment in a World Turned Upside Down* (New York: Random House, 1981), p.7

34. Daniel Yankelovich, *New Rules ,Searching for Self-fulfillment in a World Turned Upside Down* (New York: Random House, 1981), p.7.

cherish creativity, which underlies all the other aspects of the strong search for self-fulfillment. At the same time, most of them are not prepared to deal with the conflict between creativity and achieving material affluence.[35]

This somewhat unrealistic attitude toward creativity blends in with a shift in priorities of a much larger portion of the population in relation to the traditional "giving/get-contract." The risks of failure are greater and the rewards of success are somewhat less appealing because of the sacrifice of leisure time and time with one's family one has to make. The majority of Americans in the 1970s tried to integrate achieving the goals of economic success with the satisfaction of marriage and family life. Americans wanted to reduce the duties of self-denial by increasing individual freedoms.[36]

In the 1970s the average American still closely identified "getting ahead" with the goals, values, and meanings of familial success. The strong seekers of self-fulfillment replaced the ethic of self-denial with the duty to self, and this destabilized the familial triangle. As the economy began to worsen in the late 1970s, some "strong formers" as well as many other Americans began to seek alternatives to the "rat race." They sought ways of being less competitive so that it required less effort. They sought types of work or life situations where they would not be "… manipulated like a laboratory animal regardless of one's own needs and well-being."[37] Nevertheless, according to Yankelovich, competition still enjoyed the place of honor in the rules of American life.

Radical subjective, control individualism. Yankelovich describes this second major theme of the self-fulfillment ethic as the expectation of acquiring more of everything as a matter of personal entitlement rather than from desire and hard work. This attitude turns the self-denial ethic upside down. Instead of moral concern for others, one has a "duty to self" in which all personal desires are satisfied. Yankelovich summarizes this world-view in its most extreme and naïve form as:

> I am entitled to more; I owe it to myself to get more. To do so, I
> need to learn who I really am [in terms of what I want] and then
> assert myself.[38]

POSTMODERNIST TRANSFORMATION OF SOCIETY

In the 1960s, as the avant garde was fading, a *counterculture* was emerging. Many factors such as the museums promulgating "creativity" and "self-expression" and indulgent parents in the upper middle-class affluent society pampering their children, led to a small group absorbing without much reflection the innovations of avant garde modernism. This small minority swelled to include students and mentors from colleges and universities throughout the country rather than

---

35. Daniel Yankelovich, *New Rules ,Searching for Self-fulfillment in a World Turned Upside Down* (New York: Random House, 1981), Ch.3.

36. Daniel Yankelovich, *New Rules ,Searching for Self-fulfillment in a World Turned Upside Down* (New York: Random House, 1981), p. 157

37. Daniel Yankelovich, *New Rules ,Searching for Self-fulfillment in a World Turned Upside Down* (New York: Random House, 1981), p. 152.

38. Daniel Yankelovich, *New Rules ,Searching for Self-fulfillment in a World Turned Upside Down* (New York: Random House, 1981), p. 157

from just the prestigious ones like U. of California at Berkeley, Harvard, Yale, and Columbia. Very quickly these groups of students captured the imagination of young people everywhere, leading to the full-blown "Hippie movement." After the conservative and liberal backlash producing events like the National Guard killing students that happened at Kent State, cooption by commercialism greatly aided by the media and economic hard times of the late 1970s, overtly, hippies and obvious signs to identify a growing counter culture had disappeared as quickly as they emerged. However, all classes of society including blue collar workers and the disenfranchised minority groups had adsorbed, again usually in an unreflective way, the values of the counter culture.

All institutions, but especially the educational system, which has replaced the function of organized religion in the United States, had become "infected" and some would say perversely transformed by counter cultural values. Allan Bloom and a host of conservative intellectuals and leaders who agree with him believe that our education system is irretrievably lost. Bloom states that "in 1955 no universities were better than the best American universities in the things that have to do with a liberal education and arousing in students the awareness of their intellectual needs.[39] But according to Yankelovich's idea of the fact/value split, under the influence of positivistic science, the non-Catholic American universities were degenerating to a perspective that undermined all religions and philosophical insights in general and liberal education in particular. In Jesuit universities, such as St. Louis University, which I attended from 1950 to 1954, students were required to minor in Thomistic philosophy, integrated with other required courses to generate a liberal education, but the diverse science departments, sometimes explicitly but more usually implicitly, taught ideas and values that undermined the ideal of a liberal education. For example, when I briefly majored in chemistry, I was encouraged to think of the utilitarian value of such a major rather than the joy of understanding the chemical aspect of nature. In the 1950s it was common knowledge in academia that, even at the Jesuit colleges and universities, positivistic science was to an ever greater extent making obtaining a liberal education less possible. This perception even was true at the University of Chicago where Bloom taught philosophy. It is not at all surprising to me that "in the mid-sixties the natives, in the guise of students [at the best American universities] attacked."[40] That is, these students rejected the old definition of authenticity associated with the ethic of objective knowledge and radical objective control individualism and embraced a new definition of authenticity associated with avant garde nihilism and radical, subjective, control individualism.

Bloom believed that the best non-Catholic universities were providing students with a secular scientific version of a liberal education, but these students also were being influenced by the "beatniks" in California and New York and by European philosophers, especially Sartre and Nietzsche, who provided an intellectual basis for existentialism and avant garde modernism respectively. Also, they directly attacked the positivistic science, fact/value split. The countercultural revolution did bring about "a collapse of the entire American educational structure.[41] The celebration of subjective insights and feeling, existential awareness did make

39. Allan Bloom, *The Closing of the American Mind* (New York: Simon and Schuster, 1987), pp. 323-324.

40. Allan Bloom, *The Closing of the American Mind* (New York: Simon and Schuster, 1987), pp. 323-324.
41. Allan Bloom, *The Closing of the American Mind* (New York: Simon and Schuster, 1987), pp. 320-321.

the reforms of university education to be without logical, conceptual content. These reforms were made for the inner directed person. Also, "the instinctive awareness of [logical, conceptual] meanings, as well as the stores of authentic learning that were in the heads of scholars,"[42] were lost. The traditional values underlying American democracy had their home in the American universities, "and the violation of that home was the crime of the sixties. [From Bloom's perspective of American education], so far as universities are concerned... [there was] nothing positive coming from that period [of the late 1960s]; it was an unmitigated disaster for them."[43]

As a result of the fragmentation of knowledge produced by scientific humanism, see table 21.1, combining with the countercultural revolution, , universities no longer had a unified vision of what it means to be educated. Nor were there a set of competing visions. According to Bloom, the question of the meaning of being educated "has disappeared, for to pose it would be a threat to the peace [of the current, scientific power structure of the universities.]"[44] The type of reason embodied in the scientific enlightenment in conjunction with the educational system that is by far the most important if not the exclusive social value-organ for the *reproduction* of the American *democratic* way of life provides the intellectual foundation for democracy. Once we fully acknowledge this then we must also accept that the current radical critique of reason, such as *deconstructionism* and the demise of the educational system, will inevitably lead to the extinction of American and indeed Western democracy as we now experience it.

In less than 30 years postmodernism has accomplished what modernism could not even begin to accomplish in 80 years. Of course modernism made the emergence of postmodernism possible, but how did this new arrival accomplish such a transformation? I think that this transformation is analogous to the emergence of world religions and to the establishment of American democracy. With the emergence of world religions such as Christianity, religious insights, moral responsibility, and salvation formally the prerogative of the spiritual and secular leaders were made available to all humans in a particular society. The mechanism for doing this, however, was to replace esoteric knowledge and rigorous discipline with easy to imagine, literal myths, rituals designed for the masses, and a general, straight forward moral code. Likewise, the founders of American democracy made the outrageous proposal that uneducated masses, whether intelligent or not, nevertheless should be allowed to make decisions via their votes on political and economic practices, as well as on who governs them. Values previously embodied in and guarded by a priestly class and/or an aristocracy, now were embodied in political principles and rational education of the populace. Of course making these values available to all citizens also made them vulnerable to being trivialized, and/or rejected because of self-interest, and/or ignored because of lack of diligence.

Postmodernism has undermined, that is, *deconstructed*, hierarchies and dominance relations such as leader-managers versus nonparticipating followers, males versus females, white Euro-Americans versus everyone else, parents versus children, teachers versus students, reason versus feeling, Judeo-Christian-Greek culture (liberal education) versus other cultures, high art

---

42. Allan Bloom, *The Closing of the American Mind* (New York: Simon and Schuster, 1987), pp. 320- 321.

43. Allan Bloom, *The Closing of the American Mind* (New York: Simon and Schuster, 1987), pp. 320- 321.

44. Allan Bloom, *The Closing of the American Mind* (New York: Simon and Schuster, 1987), p.337.

(for example, avant-garde art) versus low art, great books, such as the classics versus ordinary books, and so on. The result has been a rush toward anarchy, a breakdown of moral responsibility, a breakdown in rational argumentation, and an acceptance of the banal along with the sublime aesthetic experiences. However, modernist innovations purified of their elitism have been incorporated into society as a whole. The conservatives, like Bloom, are justified in fearing that this will lead to destruction of traditional cultural values and institutions. But, this destruction is the nihilistic chaos that precedes a new vision, which I propose is the fourth enlightenment.

1960s SEARCH FOR SELF-FULFILLMENT, A TRUE REVOLUTION

Yankelovich refers to Hannah Arendt's two criteria that define a true cultural revolution. First, it must start a new story in human affairs, that is, it must introduce a genuine novelty into the human adventure. Second, the revolution must advance the cause of human freedom, and this always must involve a larger community. The search for self-fulfillment satisfies these criteria. It was a new story in that it introduced new meaning into our culture. It lessened the influence of radical instrumentalism by revising the old giving/getting compact to accommodate the sacred/expressive yearnings. The old system involving the Protestant ethic was very successful but had many drawbacks, such as loss of community, alienation, and a rationalism that overshadowed our sense of mystery in nature and overshadowed spirituality in other humans and in ourselves – nature, other humans, our own selves became *Its* rather than *Thous*. Yankelovich observed that various critical thinking about modern industrial society sometimes conflicted with one another, but they all agree that the instrumental focus of this type of society, at least distracts and often prevents most people from appreciating and acknowledging non-instrumental aspects of living that leads one to see himself/herself as a Thou rather than an It. However, until the 1970s the majority of Americans rejected these criticisms explicitly or simply ignored them.[45] The benefits of political freedom and material progress outweighed the bad aspects of radical utilitarianism. Then in the 1970s Americans began to heed the many criticism of the scientific, industrial consciousness, see chapter 19.

Americans knew the flaws of this kind of consciousness, and therefore focused on the private/public split in order to make room for greater personal choice. The old giving/getting compact provided an acceptable trade off between sacrifices and benefits. But then in the 1970s, influenced by ideas coming out from the countercultural revolution in the 1960s, people began to question the trade off. People began to experiment with modifying the giving/getting compact in their private lives. But Yankelovich notes that millions of these individuals provided life experiments for transforming the public sphere. By the mid-seventies, a majority of Americans began to notice and then agree with the earlier critics of the industrial consciousness. Though people had little or no control over their public lives, directly or indirectly dominated by values of the work world, they could experiment with introducing into their own giving/getting compact community values, expressiveness, care for others, and renewed orientation toward people and

---

45. D.Yankelovich, *New Rules, Searching for Self-Fulfillment in a World Turned Upside Down* (New York: Random House, 1981), p. 228.

nature as sacred. That is, the compact was modified to include diverse versions of self-fulfillment.[46]

Yankelovich makes the point that the search for self-fulfillment is a cultural revolution that is similar to the one that started American democracy. The start of this original revolution proclaimed that a free people can, with sufficient will and energy, improve their material lot by the exercise of their political freedom. As of 1981, the date of the publication of Yankelovich's book, Americans still made this commitment, but as of the 1970s, most Americans launched a new cultural, revolutionary commitment similar to the old one. Then and now we believe that poverty is not destiny, but now (1981) we also believe that instrumentalism is not destiny. Then and now we believe that political freedom can coexist with material well-being and enhance it. The search for self-fulfillment, with all its fallacies, contradictions and ambiguities nevertheless directs one's personal, political freedom to shape his/her life while participating in the instrumentalism of modern technological society. The hope is that the individuated searches of millions of people will transform our fragmenting, traditional community into a new kind of community maintained from the bottom up rather than from the top down.

FLAWED BUT NECESSARY REBELLION

Writing about the 1970s both Berger and Yankelovich agree that the countercultural movement begun in the 1960s, though fundamentally flawed, was and continues to be a necessary fight against the dehumanizing forces of industrial consciousness. Berger commented that even though the countercultural revolution has its limits, a modern society totally dominated by a patriarchal, positivistic, scientific fact/value split would be a science-fiction nightmare.[47]

Writing in 1981 rather than 1974, Yankelovich comments that if ultimate religious values do not restrain the full expression of American industrial consciousness, then this mind set would produce the science-fiction nightmare predicted by Berger. Alternatively, the democratization rebellion against hierarchy would emphasize the pleasure principle, hedonism, and duty to self. This would undermine "the disciplined effort required to sustain a modern industrial society."[48]

POSTMODERNISM VERSUS HEIDEGGER

Heidegger virtually had little if any impact on the postmodern culture that emerged from the 1960s countercultural revolution. Postmodernism in conjunction with cybernetics and systems science that emerged in the 1940s and 1950s respectively, is a nihilistic ethic that is driving U.S. society and culture to the brink of collapse.[49][50] Beginning in the 1990s there are clear signs that

---

46. D.Yankelovich, *New Rules, Searching for Self-Fulfillment in a World Turned Upside Down* (New York: Random House, 1981), p. 232.

47. Peter Berger. *Homeless Mind: Modernization and Consciousness* (New York: Random House, 1974), p.229.

48. D. Yankelovich. *New Rules, Searching for Self-Fulfillment in a World Turned Upside Down* (New York: Random House, 1981), p.255.

49. Morris Berman, *The Twilight of American Culture* (New York: W.W. Norton & Co., 2000)

50. Robert Bly, *The Sibling Society* (New York: Vintage Books, 1977)

the collapse has begun.[51] U. S. society today is in a phase similar to the "hit-bottom event" of an alcoholic who either spirals down to self-destruction or finally looks for seeds of a new vision. The deconstruction aspect of postmodernism prepares society either to spiral down to total collapse or to take advantage of the weakening of conservative barriers to transformation so as to evolve to a new vision. In contradistinction to postmodernism absolutizing diversity, Heidegger's philosophy, *separated from his alliance with Nazism*, proposed that the constructivism approach to human knowing is a collaboration between Being and individual or collective human knowing. Being is the ultimate SOURCE manifesting reality that humans can experience. Heidegger agrees with all the nihilisms stemming from science to postmodern humanism that humans are unable to know reality, but he converted this skepticism into the transcendental value that reality is fundamentally mysterious. As a result of reality being a manifestation of ultimate SOURCE, it does express ultimate meanings. Therefore, from time to time some individuals in an existential moment may glimpse an aspect of this ultimate meaning and then construct a conceptual representation of this non-conceptual seeing. The conceptual representation is neither true nor false, but it may be valid according to a consensually agreed upon set of criteria. By means of an abstract set of value criteria, there may be a hierarchy of valid knowledge constructs. For example, Einstein's special theory of relativity may be reduced to Newton's theory of non-gravitational motion, but the converse is not true; so there is the hierarchy of the more general theory of relativity including the less general Newtonian theory. Heidegger's idea of Being collaborating with any human knower implies the necessary collaboration between subjective and objective knowing and the necessary mutuality between participatory subjectivity individualism and control objectivity individualism. In particular, this means that when an individual or a group of humans give themselves over to a subjective participatory engagement with nature, Being that manifests nature will generate in humans an interpretation of nature.

Heidegger's interpretation of Nietzsche's doctrine of the eternal return of the same situation affirms the mystery of reality manifested by unknowable Being and also affirms the qualitative, hierarchal creativity of human constructivism that mirrors the creativity of Being manifesting reality. Just as scientific constructivism, in response to repeated reinterpretations of nature, has generated a hierarchy of valid, progressively more abstract, more universal scientific theories, so also other types of constructivism, in response to repeated reinterpretations of nature, have generated a qualitative, hierarchy of interpretations. However, Heidegger's philosophy in itself did not and does not solve the fundamental problem of postmodern American democracy in the 21st century, which Bellah and his colleagues proposed in the 1980s and may be described as follows. *Radical individualism at the very beginning of American democracy has evolved to the self-destruction generated by diverse integral-aperspectival visions of postmodern democracy.* In order to act, an individual or a particular group of individuals must choose a particular integral-aperspectival vision. The U.S. Constitution provides guidelines for some of these decisions, but there are many situations, which the Constitution does not cover or there are diverse interpretations of the Constitution related to the situation in question. According to American postmodern individualism, when particular choices are not specified ahead of time, everyone gets to choose whatever he or she desires. In large, complex, postmodern societies, there are many situations for which no one knows what is the best way to contribute to the

---

51. Morris Berman, *Dark Ages America. The final Phase of Empire* (New York: W. W. Norton & Co., 2006)

common good. Alternatively, there are diverse interpretations of the best way to contribute to the common good. Bellah's formulation of this problem is: how do we maintain American postmodern individualism that continually and progressively disrupts the cohesion of American society? We have here a classic version of a double bind. If we choose individualism, the cohesiveness of society continues to decline toward self-destruction. If we choose cohesiveness of society, we must empower a central government to impose progressively greater limitations to the expression of individualism.[52] This double bind always has been with us but only now is the social paradox so intense that it is destroying the American empire.

Heidegger's philosophy did not confront and transcend this double bind because his commitment to Nazism eliminated the problem. According to Heidegger's Nazism, "the people of the land in Southern Germany" were a "master race" who by means of high-German language trace their roots directly to classical Greek language and to classical Greek culture. These people are ideally suited to creating a hierarchy of interpretations of reality. When a hero, like Adolph Hitler, comes along who is thought to represent this superior, master race, then all people of this "pure" race should give themselves over to the master Fhurer and superimpose this Nazi integral-aperspectival vision on all other peoples of the world. From the very beginning of his thinking – rooted in German Catholicism that included many Catholics who were very anti-Semitic and antidemocratic – Heidegger was totally against democracy that celebrated radical individualism. Heidegger's "philosophical Nazism" converted avant garde nihilism from subjective knowing separated from any ultimate SOURCE as being the source of authentic knowledge to the master race collaborating with Being as the ultimate source of authentic knowledge.

It is a great puzzle to many thinkers of how Heidegger's many mystical insights – analogous to those of the Buddha – could lead him to embrace "philosophical Nazism." Perhaps this degeneration to subjective absolutism and ego inflation is the danger of any mystical vision. However, mystical insights coupled with humility and Faith/Hope in an absolute SOURCE can point one toward a new vision. I believe that the nihilism of avant garde modernism and postmodernism have exposed new possibilities, such as the mystical insights of Heidegger. These and other insights over the last 125 years or so have generated the emergence of the fourth enlightenment.

52.- Robert Bellah, et al, *Habits of the Heart* (New York: Perennial Library, 1985)

UNIT VIII THEMES OF THE FOURTH ENLIGHTENMENT
Chapter 23
MODERN SCIENCE IS AN EVOLUTIONARY PROCESS

HISTORY OF MODERN SCIENCE IS ANALOGOUS TO BIOLOGICAL EVOLUTION

First Differentiation of Scientific Paradigms

The emergence of a human community expressing modern scientific thinking is analogous to the first emergence of a biological species. In the perspective of biological evolution a *species* is a population of organisms that share a set of traits that distinguish the population from any other species and from any collection of non-living entities. In like manner, a modern scientific community is a population of humans that share a set of epistemological (ways of knowing) traits and values that distinguish the community from other science communities and from non-science communities. The set of epistemological traits and values is what Kuhn called a *paradigm*, which later he also called a "disciplinary matrix."[1] The individuals in a particular scientific community have diverse subjective interpretations of the meaning of a particular paradigm, but they have a consensus on how the paradigm governs their scientific thinking. Using the ideas from Kuhn's model one may describe the evolution of the emergence of modern science in the following sequence of stages: 1. communities with non-scientific paradigms; 2. Crisis resulting from chaos in non-science communities; 3. emergence of *prescience communities* each with elements of scientific thinking but no consensus for one unifying paradigm; 4. *revolutionary stage* involving conflicts between non-scientific and scientific paradigms and conflicts among diverse scientific paradigms; 5. the emergence of a *normal science community* defined by a single paradigm. Chapter 16 gave an overview of the evolution from non-scientific knowing to scientific knowing governed by a *limited version of scientific constructivism* that may be defined as: the process of a community of scientists discovering empirical patterns, describing them by metaphorical, conceptual stories (myths), and then representing these stories by logical, conceptual models that can be empirically tested for validity. Chapter 16 also described systematic experimentation that, because of Newton, became integrated with scientific constructivism. Chapter 17 described Newton's thinking about gravity that illustrates: 1. the importance of metaphorical, conceptual thinking in constructing a scientific theory, pp. 189-190; 2. the emergence of scientific positivism, pp. 190-192; and 3. the emergence of Manichaean modernism, pp.193-194. Also, a paragraph on page 192 summarizes the evolution to modern science as an order, chaos, hierarchal, new order process.

This section of this chapter provides more details about the first prescience communities and the Newtonian, normal science that emerged from it. These new ideas will contribute to understanding narrative constructivism to be described in chapter 24. The limited version of scientific constructivism that emerged in the prescience stage confronted two opposing approaches to understanding nature: rationalism and empiricism. From a historical perspective rationalism is represented by Plato's philosophy, Neo-Platonism involving mathematical

---

[1] Thomas S. Kuhn. *The Structure of Scientific Revolutions* (Chicago: The Uni. of Chicago Press, 1st ed., 1962, 2nd ed. Enlarged, 1970).

constructivism (see chapter 16, Plotinus, Ptolemy, Copernicus) and the limited version of scientific constructivism. Empiricism is represented by the philosophy of nature as proposed by Aristotle and modified by Thomas Aquinas and by the experimentation of the early Renaissance magicians/scientists, see chapter 16.

More specifically, in the "age of Reason" (16th and 17th centuries) the rationalism of Rene Descartes (1596-1650) opposed the empiricism of Francis Bacon (1561-1626). Correspondingly, early 17th century rationalism became the *Cartesian world view* consisting of the following sequence of ideas: 1. God created the universe consisting of autonomous things (beings, entities) whose interactions only can be precisely described by mathematical models; 2. Scientists say to study nature is to intuit "self-evident," fundamental principles, for example, two parallel straight lines never meet when extended indefinitely, (the major metaphor for Descartes was geometry) and then deduce theorems that describe nature; 3. The resulting mathematical models lead to *certainty of scientific knowledge of nature*; 4. The first generalization stated above is expanded to a belief in scientific *reductionism* in which any complex phenomenon or problem must be "reduced" to its constituent, elementary parts or sub-problems; 5. The universe is a vast mechanical system[2] expressing deterministic laws that govern *how*, not *why*, material objects interact and move in space over time intervals; 6. The human mind that commits to reductionism and a mechanistic view of the universe is spiritual in contrast to all other aspects of the universe that are non-spiritual, that is material, and 7. Manichaeanism, see chapter 13, pp.131-132 and chapter 17, pp.193-194. Late 16th century empiricism became *Baconiaism* consisting of the following ideas: 1. Science must collect factual data about nature by a type of experimentation independent of the Hermetic tradition; 2. Scientists must develop scientific induction in which one performs experiments, draws general conclusions from them and then tests the conclusions by means of further experiments; 3. Experimentation should be associated with innovations in technology; 4. The goal of experimental science must be to dominate and control nature; 5. The exclusive value of experimental, scientific knowledge is that it is useful – the utilitarian value of scientific knowledge is more important than understanding the structure of the universe.[3]

First Emergence of Normal Science: Newton's Paradigm

Galileo Galilei (1564-1642), whose life span overlapped that of Francis Bacon, was the first to combine rationalism equal to the limited version of scientific constructivism with empiricism equal to systematic experimentation. He had the idea of graphical representation of a function such as v(t) where for every value of time starting with zero one could associate a particular value of velocity. Beginning with this and other mathematical assumptions he *speculated* that the distance any object that falls toward the earth is proportional to the square of the time that

---

[2] All the philosophers or historians of science that I know of claim that both Descartes and Newton considered the universe to be a machine that operates according to the mechanistic laws of nature, but as described in chapter 1, any machine, especially as governed by the second law of thermodynamics, contradicts the mechanistic perspective of both Descartes and Newton.

[3] In 1968 my NIH three year grant was temporarily suspended by order of President Lyndon Johnson who established the government guideline that all research supported by government money must be primarily if not exclusively utilitarian. From what my scientific colleagues tell me, this guideline still is in place.

elapsed during the fall; that is, let s = distance and t = time and g = a constant, then $s = gt^2$. Galileo's mathematical reasoning that led to this equation for free fall represented a special case of the fundamental theorem in calculus.[4] Isaac Barrow was the first to recognize this idea in Galileo's writings, which led Barrow to formulate and give a rigorous proof for the validity of this theorem. Based on these speculations Galileo designed ingenious experiments to test the validity of them. In agreement with Descartes' rationalism Galileo enjoined scientists to reduce one's perceptions of nature to material objects in motion that can be observed and measured. He believed his "empirical scientific constructivism" is the one path to an absolutely true understanding of nature in contrast to the Aristotelian-Thomistic scholasticism espoused by the Catholic Church. Also, in agreement with Baconiaism, Galileo maintained that scientific knowledge should be utilitarian and thus give humans control of nature and power in human societies. Galileo's paradigm "… represent[s] the final stage in [Western culture's] … development of nonparticipating consciousness."[5]

Newton greatly expanded as he incorporated Galileo's empirical scientific constructivism in his paradigm for mechanistic, modern science. Newton further differentiated the calculus as hinted at by Galileo and partially developed by Isaac Barrow. In contrast to Galilean physical laws that applied only to motion on the earth's surface, Newtonian physical laws applied to motion of objects anywhere in the universe. In the *Principia, the Mathematical Principles of Natural Philosophy*, published in 1686, Newton used his theories to reject all of Descartes' propositions about the natural world, but he validated the mechanistic and Manichaean core theme of the Cartesian world view,[6] which partially contained Galileo's physics. Cartesian thinkers argued that Newton's theory did not explain gravity but only stated its effects, and therefore, gravity was merely an occult property of the universe. As described in chapter 17, Newton's response was to introduce positivism as a dominant characteristic of his scientific paradigm. In his work, *Optics*, published in 1704, he described experiments that divided white light into separated color rays of light that then were recombined to form white light again. As a result of the acceptance of Newton's theory of light, his paradigm became the combination of *philosophical atomism*, positivism, systematic experimentation, and a Manichaean perspective that reinforced the implicit choice in Galileo's paradigm to totally reject subjective, participatory engagement with all aspects of nature.[7]

---

[4] Otto Toeplitz. *The Calculus: A Genetic Approach* (Chicago: The Uni. of Chicago Press, 1963), pp.128-129. Note: I received a preliminary copy of this book at an NSF summer institute for high school math teachers in the summer of 1959. With the aid of this book and vol. 1 of Courant's classic textbook on calculus that provided rigorous proofs of theorems and calculus problems, I taught myself calculus during my first year in graduate school. This gave me the background to take graduate courses in real number theory that relates to the theory of probability.

[5] Morris Berman. *The Reenchantment of the World* (Ithaca, New York: Cornell Uni. Press, 1981), p.39.

[6] Morris Berman. *The Reenchantment of the World* (Ithaca, New York: Cornell Uni. Press, 1981), p.42.

[7] Morris Berman. *The Reenchantment of the World* (Ithaca, New York: Cornell Uni. Press, 1981), pp.43-446.

BIOLOGICAL THEORY OF EVOLUTION

Free Market Capitalism as a Metaphor for Understanding Biological Evolution

Evolution is the process of transformation that also may be called individuation. Individuation begins when the Order of a system degenerates to Chaos that exposes new possibilities. Actuation of some of these possibilities leads to a New Order with emergent properties. An economic unit such as an individual human, a business, or a society, survives in a market, which is an environment for buying and selling. An economic unit exhibits Order when the market is *suitable* for the unit's survival and the unit is *adapted* to the market. For example, if a business is making a profit for an extended period of time, then two things are simultaneously true. First, the people running the business are doing what is necessary to be successful. In particular, they are making the necessary adjustments to changes in the market, such as changing their product or service to meet the needs of their customers (they are maintaining the homeostasis of the business). Second, the market is such as to enable a well-run business to survive. The most well-run, aggressive, automobile company, such as GM, cannot survive if it depended on selling expensive cars with poor fuel efficiency, for example, trucks and SUVs, to people during an economic crisis when it is difficult to barrow money, gas prices are or threaten to become very high (3 – 4 dollars a gallon), and there may be a shortage of gasoline. These two ideas represent what may be called the *mutuality of economic unit and the market*.

An analogous understanding of the same two ideas applied to biological evolution gives us the *mutuality of a species and its environment*. A species, which is a population of individuals with similar traits, reproduces from one generation to another a similar life pattern and participates in a network of interactions with other species and with its physical-chemical environment. This complex network of interactions is an ecosystem in which the species is analogous to a business that exhibits a two-fold mutuality with its environment; the environment is analogous to the market. In a stable ecosystem the environment is suitable for the survival of the species. This means that the species life pattern contains mechanisms by which it can survive on a moment-to-moment basis and reproduce its life pattern. Furthermore, the species life pattern exhibits an array of homeostatic mechanisms by which the species can adapt to changes in its environment. In summary, in a stable ecosystem, the environment is suitable to the continued survival of a species and the species is continually making adaptations to changes in the environment.

Chaos occurs when there is an irreversible disruption of the mutuality of the economic unit and the market. Continuing with the example of evolution of a business, we may conclude that the market has changed beyond the capacity of the business to adjust to this change; or for some reason – usually poor leadership – the business has decreased its effort to adapt to a changing market, for example in 2008-2009 both Chrysler and GM in contrast to Ford Motor Co. The Chaos of the disruption of the mutuality of a business and the market simultaneously leads to the loss of one niche and the emergence of new possible niches in the market. Some businesses, like the Ford Motor Company in the 1980s & in 2007 and IBM in 1991, will – if guided by CEOs with vision coupled to strategic plans that carryout the operationally defined goals of the vision – undergo extensive restructuring that enable them to occupy a new possible niche in the market. The restructuring involves trying strategies for occupying a particular new niche and selecting those continually modified strategies that lead to financial success in occupying that niche. Alternatively, the business will try strategies to fit into several different niches and select those that produce the greatest success. Other businesses will not make the necessary restructuring and therefore will not survive in the changed market. In effect, the *market selects* those businesses that will be successful and prosper over those that will go out of business. This idea of the market selecting those businesses that survive over the many other businesses that do not survive represents the second and third features of an evolutionary process, see later section in this chapter. Competition is a central theme in evolution. With respect to evolution of business, there will be a few winners and many losers.

Biological evolution begins with an irreversible disruption of the two-fold mutuality of a species and its environment. On the one hand, this disruption can lead to the species becoming extinct, but, on the other hand, a central idea of biological evolution is that chaos opens up new possible collaborations that can lead to the emergence of a new species life pattern. Reproduction of a species leads to mutations, that is, changes in genetic information, to some offspring in each generation. Most of these mutations have negative survival value but some have positive survival value that generates new species-environment mutualities in the changing environment. The occurrence of mutations in each generation in a sequence of several thousands of generations and the elimination of negative mutations represents the second feature of any evolutionary process. Some of these new mutations will confer onto individual organisms expressing them a greater chance to survive and reproduce the emerging new species life pattern containing them. In a long sequence of generations of a population of individuals representing a particular species, the number of individuals in each new generation having the superior species-environment mutuality will increase and the species will become more stable, and correspondingly, the number of inferior mutualities will progressively decrease. In this ongoing process of increasing superior mutualities (and decreasing inferior mutualities), which is called *natural selection*, the environment selects the new mutualities that progressively are incorporated into the species life pattern. This transformation to a new species life pattern more adapted to the environment is called *species adaptation*. The accumulation of many species adaptations in the evolving species eventually leads to a species life pattern that is radically different than the life pattern just before the species began to evolve. The difference is so great that the new pattern is classified as a new species, and the overall process of accumulating species adaptations is called *species transformation*. The species adaptations and species transformation represent the third feature of any evolutionary process.

Natural selection is the core idea of biological evolution. As many species become extinct, a few by natural selection over thousands of generations will accumulate traits that represent a "restructuring of the species life pattern." These restructured species survive in that they establish a New Order represented as a new, species-environment mutuality. With respect to biological evolution most species that have emerged in the history of the biosphere have become extinct. The current network of interacting species in the biosphere represents the winners.

Free Market Capitalism As Analogous to Biological Evolution Implies Negative Features

In my experience most biologists, like Darwin, are not interested in philosophy or theology. As a result, biologists who study evolution focus on a valid description of extensive empirical observations that clearly indicate: (1) many species change their life pattern over many generations and (2) many species become extinct while new species emerge. The modern theory of evolution provides the desired scientifically validated description of these empirical observations. The existence or non-existence of a creator God is irrelevant to such a description; there is no need for the "God hypothesis." In like manner, for those committed to social evolution *exclusively driven* by free market capitalism, there is no need for the "God hypothesis." The changes in the market are by and large unpredictable, just as is the case for each human life story. To pragmatic realists the changing market, not God, determines who will become the winners or losers. One tries to be a winner but accepts – perhaps begrudgingly – fate equal to randomness if he/she is a loser. However, fate equal to randomness also is coupled to the rule of law. In the biosphere when a species-environment mutuality is irreversibly destabilized, the laws of nature dictate what species most likely will become extinct. But this same rule of law, as described by the theory of evolution, dictates that natural selection of millions of mutations will lead to the emergence of one or more new species. In like manner, the evolution of free market capitalism depends upon the rule of law plus human free choice analogous to natural selection of random mutations producing new niches in the environment (new ecological niches). A democratic government is one that facilitates individuation producing individual free choices, but the rule of law plus a police force selects as acceptable those choices that either are neutral or do or can contribute to the survival of society. Thus, free market capitalistic evolution only occurs in the context of at least a minimal democratic society stabilized by the rule of law.

Creative evolution has profound negative aspects; it is profoundly nihilistic! According to the modern theory of evolution, any individual organism is subordinated to the species of which it is a member. The survival of any species is subordinated to the survival of the community of which it is a member. The survival of any community is subordinated to the survival of the ecosystem of which it is a member. The survival of any ecosystem, such as the biosphere, is subordinated to the process of evolution that has no purpose or intrinsic meaning or connection to any idea of an ultimate Reality such as God or Brahman. Thus, any unit of life, such as the human species, as well as the process of evolution has no purpose or meaning or connection to any idea of an ultimate Reality. Thus, *there is no need for commitment to purpose, meaning or connection to any idea of an ultimate SOURCE.*

In like manner, for those totally committed to free market capitalism independent of any other vision, such as an ethic of care for the poor, there is no need for commitment to purpose, meaning or connection to an ultimate SOURCE for any human individual or for any human society. Since democracy is reduced to free market capitalistic evolution, there is no need for any spiritual vision such as the spiritual vision of democracy of the fathers of the American Revolution. There is no need for giving priority to manifesting beauty, love, or compassion and care for the nation's citizens. There is no need to care for the many individuals and economic classes of individuals who work diligently but still fail in the competition of the market. This produces the *irony of American Democracy*. American democracy of the 1700s was based on a collective commitment to a belief in a creator God and to individualism associated with personal freedom that is expressed in one's choices of what to believe and of life style and of one's contribution to the common good of society. This spiritual vision of democracy led to the emergence and prosperity of postmodern America seemingly committed to unregulated free market capitalism. The irony is that this committing to laissez-faire capitalism, not modified by an ethic of care and a new, postmodern, spiritual vision of democracy, leads to rejection of any idea of an ultimate SOURCE, such as a creator God, rejection of all religions including Christian religions and all versions of the Bible, and to reject traditional rationality that evolved from Socrates, Plato, and Aristotle. Scientific rationality associated with an extreme form of radical skepticism embedded in a pragmatic understanding of nature replaced all traditional forms of rational individualism that evolved from the classical Greek enlightenment.

THREE DEFINING FEATURES OF ANY EVOLUTIONARY PROCESS

First Feature: Order, Chaos, Hierarchal New Order Process

The first feature is complex in that it requires a narrative, scientific, constructivism perspective on the one hand, and on the other hand, it applies to non-evolving systems such as any machine as well as "self-organizing," evolving systems, all of which will be described in chapter 24. The order of a potential in nature and the order of machine work may be represented by the same mathematical-physical symbols defined in terms of units of energy, but the orders are *qualitatively* different, see chapter 1. The autonomous potential in nature is not a potential to accomplish a particular task; it has no machine purpose. It only becomes a "task potential" when it collaborates with an appropriate energy coupler. This collaboration is neither determined by the physical laws that specify the potential in nature nor by the laws that specify the operation of an energy coupler. There are indefinite numbers of ways the collaboration can occur resulting in the efficiency of the machine varying from low to high. An autonomous potential in nature does not contain or in any way refer to this idea of collaboration. Therefore, order of a potential in nature and the order of machine work are fundamentally different. That is, a machine has an organization pattern that can convert flux into work done by a machine. Machine work, the new order created by a machine, implies a potential in nature, but, as stated earlier, the old order does not imply the new order. Thus, machine creativity is an order, chaos, hierarchal new order process. The word "hierarchal" in this statement means that the new order includes a modified old order. However, one cannot make this statement from a scientific constructivism perspective. This hierarchal creative process, which also involves the mutuality of Eros-chaos and Eros-order, only can be understood in this way from an objective, narrative, scientific constructive perspective, see chapter 1.

Second and Third Features

The second and third features are exhibited by evolutionary creativity but are not exhibited by machine creativity. Machines only produce information transformation *external* to the machine structure. Any mechanistic or machine understanding of aspects of nature, such as perceiving an ecosystem only as a homeostatic machine (one version of the Gaia hypothesis) or perceiving any organism only as a homeostatic machine, at best only gives an external, narrative, scientific constructivism understanding that implies that evolutionary creativity cannot occur. Thus, many practicing scientists or science teachers exclusively committed to mechanistic or machine theories of nature, consciously, for example, Albert Einstein, or unconsciously deny the idea of creative evolution. However, internal, narrative, scientific constructivism, see chapter 24, acknowledges the chaos phase of a system's narrative as breaking up some old collaborations thereby exposing new possible collaborations between parts of the system and/or new possible collaborations between the system and its environment. This leads to the *second feature* of any evolutionary process.

Second feature: trial-and-error coupled with a selection process. The selection process selects stable collaborations from the many possible collaborations that emerge during the chaos phase. The various possible collaborations are tried and criteria of stability intrinsic to nature select those collaborations that are sufficiently stable to survive and reject those collaborations that are not sufficiently stable.

Third feature: the context or environment in which the trial-and-error process is occurring further selects those stable collaborations that continue to survive in the environment of the evolutionary process. For example, as described at the end of chapter 2, a system consisting of positive hydrogen atoms and negative oxygen atoms near equilibrium will transform to hydrogen and oxygen molecules. The same system far from equilibrium will transform to water molecules.

Metaphorical Understanding of the Three Features of an Evolutionary Process

Metaphor for the first feature: DNA duplication. DNA, the molecule in cells that contains genetic information, may be thought of as consisting of two parallel linear molecules bonded to one another. Each linear molecule is like a string of beads where each bead (one of four types of nucleotides) contains a small complex molecule (purine or pyrimidine bases). The two parallel linear molecules are held together because of each bead in one linear molecule is bonded to the bead next to it in the other parallel linear molecule. The bond between each pair of beads is due to the small complex molecule in one bead bonding to the small complex molecule in the bead next to it in the other parallel straight molecule. The bonding between two small complex molecules is like two pieces of a puzzle fitting into one another. In this pairing, the one small complex molecule is said to be *complementary* to the other one where the relationship is analogous to a foot being complementary to a foot print. With these insights in mind one may imagine a DNA molecule as consisting of a *foot molecule* bonded to the complementary *foot-print molecule.* The current theory of DNA molecular duplication (called DNA *replication*) may be metaphorically visualized as an order, chaos, hierarchal new order process as follows. Order is the DNA molecule before duplication begins. This order goes to chaos when the foot molecule separates from the foot-print molecule in each DNA. During this chaos phase, the cell synthesizes "bean molecules" and each bean molecule making up a foot molecule bonds with a complementary, synthesized bean molecule. As a result, the foot molecule is a *template* for making its *replica*, which is a foot-print molecule. When this happens, the resulting foot – foot-print molecule is a new DNA molecule identical to the original DNA molecule. Likewise the foot-print molecule is a template for making a replica of itself, which is a foot molecule. When this happens, the resulting foot-print – foot molecule is a new DNA molecule identical to the original DNA molecule. Thus, the original DNA molecule is transformed into two identical DNA molecules. Each new DNA molecule contains a linear molecule from the original DNA molecule.

Metaphor for the second feature: remarriage. Order is two people bonded by marriage. Chaos is divorce where the old marriage bond is gone forever. Occasionally the divorced people remarry. Alternatively, one or both of the divorced people may choose to remain single. The more usual outcome is that at least one former partner engages in a *trial-and-error process* involving dating many different people and eventually choosing a new partner to form a new marriage.

Metaphor for the third feature: see previous section, "Free market capitalism as a metaphor for understanding biological evolution.

Three Features of the Evolution of Modern Science

First feature. Kuhn's model that gives central importance to competition resulting in a "scientific revolution" detracts from the fundamental idea recognized by Darwin that evolution and ecology imply one another. This mutuality is captured by acknowledging evolution as an order, chaos, hierarchal new order process as follows: With respect to science, any human community always is embedded in a network of communities that generates a particular human ecosystem. Crises in this ecosystem produces the chaos of a destabilized society, for example, June 20, 2009 rebellion in Iran that, in turn, produces the anguish that motivates humans to find ways of returning to a stabilized condition. The social chaos also opens up many possible ways of reestablishing order. Reestablishing order often is <u>not</u> the result of a revolution in which a new order replaces an old order. Rather, for example, with respect to science, normal science begins to have a unifying complex paradigm <u>but</u> this new paradigm includes modified aspects of the old paradigm. For example, the Newtonian paradigm not only included modified aspects of ideas of Bacon, Descartes, and Galileo, it also had roots in the whole of the Greek enlightenment (the pre-Socratics, especially Parmenides and Heraclitus, as well as Socrates, Plato and Aristotle) and in the medieval enlightenment (especially St. Augustine, neo-Platonism of Plotinus, the Manichaean heresy, and Aquinas). This is why it is a grave error to treat any major event, such as the "so-called scientific revolution," as an autonomous happening.

Moreover, the success of any newly emerged paradigm depends on its integration into a stabilized new ecology. For example, Newton's paradigm was experienced as complementary to the emerging new human individualism associated with the Protestant revolts, the differentiation of modern philosophy beginning with Descartes, and new political theories, especially that of John Locke. The new individualism became closely associated with industrialization that collaborated with the idea of laissez-faire capitalism – *The Wealth of Nations* by Adam Smith (1723-1790) was publish in 1776, the same year as the Declaration of Independence! All these interconnections became greatly entrenched in Western culture with the emergence of US Constitutional liberal democracy and its intense collaboration with the industrial revolution of the early to mid 1800s. This development of industrialization was guided by the newly emerged science of thermodynamics. The growing financial success of all these collaborations plus the spiritual vision underlying US democracy enabled diverse religious and philosophical communities not only to "live and let live" with one another but to tolerate the absolute opposition between science and values defined by the Judeo-Christian traditions, see chapter 18. Incidentally, scientists, for the most part, did not see the opposition between the Newtonian paradigm and thermodynamics wedded to the theory of evolution. This realization only occurred to <u>some</u> scientists in the mid to late 20th century. By the beginning of the 20th century the Newtonian paradigm not only was firmly entrenched in Western culture, it began to seriously undermine the West's spiritual, democratic vision, possibly in part resulting in World Wars I and II. Ironically, the American alliance's success of winning World War II reinforced the commitment to the Newtonian paradigm. But just beneath social consciousness other forces were mounting an attack that led to the countercultural revolution of the late 1960s, see chapter 17. However, even today [September, 2009] most people in power cannot see the inadequacy of the Newtonian paradigm.

Second feature. Paradoxically the core value of the unifying paradigm of normal science leads to it eventually being partially rejected. However, many of the practitioners of normal science are unable to see this paradox; or if they do see it, they cannot transcend it because they do not understand science as an evolutionary process. The radicalness and profound implications of Kuhn's model stems from his evolutionary perspective that transcends this paradox. *The core of the unifying paradigm of normal science is non-ideological pragmatism, which involves "trial-and-error" independent of ideology.* The emergence of the Newtonian paradigm from 1686 (publication of *Principia)* and its evolution to its unchallenged dominance in the late 1800s resulted in it becoming a scientific "value/fact." Many crises beginning in the late 1400s emerged in the European non-scientific, human ecosystem that could not be resolved or transcended by its dominant political-religious-philosophical paradigm. During the emergence of the Newtonian paradigm, diverse humans began to solve some of these problems by abandoning – at least in part – the old paradigm but without the guidance from a consensus about a new paradigm. Thus, they enacted a non-ideological pragmatism. Eventually with the insights of geniuses, such as Newton, aspects of these solutions coalesced into the Newtonian paradigm. As more and more thinkers became indoctrinated, because of personal success, into committing to this approach to understanding nature, it became the unifying paradigm of the Newtonian, mechanistic world view. However, the commitment of most practitioners of science to this "operationally defined view of nature" was not to any set of dogmatic statements representing it. Rather, scientists and their students digested and incorporated this scientific, non-ideological pragmatism by "learning" known puzzle-solutions to problems that exemplify this kind of pragmatism. Thus, any scientific paradigm governs not the subject matter of any study of nature but rather the way the study is carried out. This makes any scientific community radically different from diverse philosophical, religious or many spiritual/mystical communities. Any "so-called scientist" who has at one time or now practices science has *experiential faith expressed as belief* in scientific pragmatism of a particular paradigm. In radical contrast, members of non-scientific communities believe in specific dogmas about the universe that guides how one solves his/her life problems.

Because any scientific paradigm at its core is non-ideological pragmatism, it eventually leads to the chaos of revolutionary science. During this period, the tacit understanding (gut understanding) by many practicing scientists does not guide one to solve important problems that emerge. This leads to a particular paradigm being modified so that it guides solutions to these new problems. Just like in biological evolution to diverse species, normal science diverges into many variations of normal science by a process analogous to species adaptation (rather than species transformation). These "scientific adaptations" co-exist without producing anguish to the whole scientific community until they eventually coalesce into a world view that contradicts the world view of the dominant scientific paradigm. Then there is a shift to a radically new world view – analogous to the emergence of a new species. Because of the dramatic rate and extent of this shift, Kuhn referred to it as a scientific revolution. This is what happened in the shift from the Newtonian to the Einsteinian relativistic world view as well as the shift from the determinism of Newton or Einstein world views to the probabilistic, quantum mechanics world view of Bohr and others. The evolutionary history of modern science indicates many other shifts that are, perhaps, less dramatic. However, the postmodernism implied by Kuhn's model indicates that may diverse scientific paradigms can co-exist each guiding solutions to problems associated with diverse contexts or environments.

The diversity of these contextually validated scientific communities superficially is analogous to postmodern absolutized diversity that denies the possibility of discovering true or valid knowledge or finding meaning in one's life, see chapter 20, pp. 220-222 and chapter 22, pp.242-245. However, the non-ideological pragmatism of science transcends this postmodern absurdity. The practicing scientist confronted with diverse approaches to studying nature always embraces some integration of these diverse approaches. That is, at least some 21st century scientists are forced to choose an *integral-aperspectivism* rather than floundering in the absolutized diversity of mere aperspectivism. Nevertheless, just as biological evolution has no *intrinsic meaning* or *known goal*, modern science has no intrinsic meaning; nor does it evolve toward an ever greater approximately true knowledge of nature.

Third feature. The third feature implies understanding any unifying paradigm in normal science in diverse ways according to the society-culture in which one is embedded. Even though the core of any scientific paradigm is non-ideological pragmatism, science always is associated with ethical choices and ideologies as described in chapters 18 and 19. For example, when Newton confronted trying to explain the force of gravity as action at a distance, he influenced thousands of scientists who followed his lead to choose positivism. Alternatively, Newton could have chosen diverse ways of knowing nature. That is, in this postmodern point of view each scientific or non-scientific paradigm is valid with respect to a particular context. Such a postmodern perspective was partially operative from the first emergence of the Newtonian paradigm. Each practitioner of this or any other scientific paradigm can and did choose to integrate its pragmatism with any one of several ideologies. For example, my boss for 2½ years at a research institute where some of the research I was doing applied to my getting a PhD, epitomized a radical version of the empiricism of the Newtonian paradigm. Nevertheless, he was a practicing Catholic priest (he said Mass and herd Confession, though not mine in that I would have had to confess how much I disliked him) with a PhD in biology and a PhD in physics. Ironically, he also enjoyed arguing with me about different interpretations of Thomistic philosophy and Catholicism. More to the point, there are hundreds of religious universities, usually Catholic, that are approved to train and award PhDs in science. One major problem with the evolution of modern science is that many scientists adopted the positivism and/or utilitarian ideologies associated with a particular scientific paradigm as absolutely true or uniquely valid. As the Newtonian paradigm became more entrenched in Western culture, Western societies, as described by Monod, see chapter 18, became more and more internally divided by the opposition between scientific and non-scientific ideologies. This internal division is evolving to or already has become analogous to the "house divided against itself" that preceded the Civil War in the US.

At the same time, the non-ideological pragmatism of science progressively generated a subtle and more dangerous destabilizing force to any society. The collaboration between scientific and free market pragmatism has circular interactions (positive feedback) with materialism: the pragmatism increases material benefits to individuals and societies and the material benefits stimulates increased pragmatism to obtain greater material benefits, and so on. This circular dynamism began to enter the American psyche to an ever greater extent in the context of the free market ideology coupled with government deregulation that the Reagan presidency initiated in the 1980s. The *materialistic, circular, reinforced pragmatism* became an attack on all ideologies that progressively overtook American college education. As a teacher of science for non-majors, I noticed in the late 1980s that more and more students in my classes ranging from 100 to 300 members became totally closed to understanding any theory or ideology about the universe. By 2006 I estimate that 60 – 80% of my students totally resist understanding – in contrast to memorizing – any new ideas about nature. Several of my colleagues who teach introductory science to "science majors" report that 40 to 60% of students in their large classes reject learning any scientific ideas. Note, it is not that many students are incapable of understanding ideas; rather, they choose to reject being open to understanding anything! I believe this culturally destructive choice is due to the vicious cycle that produced materialistic pragmatism becoming dominant in American higher education.

Before writing this chapter I believed that a major crisis in American society is the conflict between positivistic science and some version of traditional values.[8] With my current understanding that non-ideological pragmatism is the core value of science in association with reflection on my teaching classroom experiences and participation in departmental and university faculty meetings, I now realize how totally wrong I was. The ideological conflict between positivistic science and traditional values is *irrelevant* to most practicing scientists and science teachers. It also is irrelevant to most academicians and to most students! Such ideological conflicts are relevant only to those who understand and are passionately committed to a particular world view. As the virus of materialistic pragmatism infected the American psyche, especially as the education system since the 1960s spread this virus, more and more people no longer understand or are passionately committed to any ideology. This generalization may seem too radical or not grounded on sufficient sociological data until one realizes that the capitalistic, science-generated, materialistic pragmatism is one version of postmodernism. This radical skepticism rooted in the 1960s countercultural revolution began to emerge in the 1970s and finally was acknowledged in the 1980s, see chapter 20, pp.220-222 and chapter 22, pp.242-245. Practicing scientists, of course, did transcend the absolute relativism of postmodernism in their pursuit of solving problems – they always created an integral-aperspectival solution. However, scientists and science-influenced thinkers and teachers did not transcend the absolute relativism of morality and other values. They, for the most part, did not explicitly denounce value-laden world views; they and their students simply ignored them.

Again, this connection with postmodernism may seem like a stretch beyond the data since most people including academicians and students do not know what postmodernism is; or if they do know about it, they have not thought about or care about its consequences. But, after (June, 2009) rereading Zakaria's book, *The Future of Freedom*,[9] especially chapters 5 and 6 and the Conclusion, I realize that materialistic pragmatism equal to postmodernism is *democratization*, which Zakaria often calls democracy in contradistinction to *constitutional liberalism*. With these connections in mind, the data supporting the generalization about the American psyche being infected with materialistic, non-ideological pragmatism is over whelming. The generalization also explains the coincidental similarity of Greek direct democracy and postmodern, late 20th century democracy. As described in chapter 10, in 508-507 BC Cleisthenes proposed direct democracy in Athens that was successfully adopted by many city-states on the Greek mainland. This inspired many Greeks, including the Spartans with their own militaristic version of democracy, to unite and defeat the Persian invasions thus saving Western culture from total annihilation. But, this democratization world view was embedded in pre-Socratic philosophy that had degenerated into the radical relativism and

---

[8] Donald Pribor. *Spiritual Constructivism: Basis of Postmodern Democracy* (Dubuque, Iowa: Kendall/Hunt Pub. Co., 2005), pp.21-24.
[9] Fareed Zakaria. *The Future of Freedom* (New York: W. W. Norton & Co., 2003).

skepticism of sophism that is similar to postmodernism. The Greek politicians trained by sophists converted rhetoric as arete to rhetoric as technique that led to amoral utilitarianism. Greek politicians gained power by persuading the masses to adopt their dictates. Thus, Greek politics had evolved from city-state tyrants to direct democracy and then to Pericles (495-429 BC), who basically was a "good" leader but nevertheless exemplified a new kind of tyranny where his power to win over public opinion enabled him to dictate policy decisions. The logical, conceptual thinking of Socrates, Plato, and Aristotle eliminated this sophist disaster but also eliminated the possibility of direct democracy in the subsequent stages of the Greek enlightenment and in the medieval enlightenment. That is, these enlightenments produced variations of rational individualism that undermined the possibility of democracy. The scientific enlightenment generated pragmatic individualism integrated with various versions of Judeo-Christian, rational individualism that led in the US to democracy combined with constitutional liberalism. However, beginning in the 1970s-1980s materialistic pragmatism had evolved to postmodernism that undermined all ideologies equivalent to the sophist disaster of Greek direct democracy.

In his book's conclusion Zakaria insists that in order for the US democracy to "work," we must reintegrate "… constitutional liberalism into the practice of democracy … [which] requires that those with immense power in our societies embrace their responsibilities, lead and set standards that are not only legal, but moral. Without this inner stuffing, democracy will become an empty shell, not simply inadequate but potentially dangerous, bringing with it the erosion of liberty, the manipulation of freedom, and the decay of a common life. This would be a tragedy because democracy, with all its flaws, represents the "last best hope" for people around the world."[10]

The democratization aspect of American democracy in the 1970s-1980s became associated with "radical, subjective control individualism," which was the Eros-chaos drive toward greater individual freedom expressed as the self-fulfillment ethic, see chapter 22, p.241. The constitutional liberalism aspect of American democracy is the Eros-order drive to limit the expression of individual freedom so that humans can not only co-exist but also collaborate to produce a stable society. Each person can choose to believe whatever he/she wants, but one's actions must be governed by laws generated by a Constitution and enforced by the police or army. America's current crisis, which generated the current economic downturn, is the imbalance of Eros-chaos overwhelming Eros-order. The democratization, countercultural revolution that was a search for self-fulfillment, was a true revolutionary change for the better, see chapter 22, pp.245-246. This flawed but necessary rebellion, see chapter 22, p.246, deconstructed many patriarchal hierarchal structures associated with traditional values, liberal education, and modern science. In doing so it opened up psychic space for enhancing the civil rights movement and for initiating the gay-rights movement in 1969, and the feminist movement in the 1970s as well as stimulating a flourish of creativity in all the arts. The fundamental flaw of this rebellion was that it also generated radical, ego constructivism, see chapter 22, p.226. This, in turn, produced the social and culturally unsustainable individualism in which the person chooses his/her self-validating ego to replace any higher power such as SOURCE (God) or some socially validated hierarchy of power, as, for example, defined by the US Constitution. This kind of radical individualism would emphasize the pleasure principle, hedonism, and duty to

---

[10] Fareed Zakaria. *The Future of Freedom* (New York: W. W. Norton & Co., 2003), p.256.

self that undermines "the disciplined effort required to sustain a modern industrial society."[11] Chapter 24 describes a perspective that incorporates a modified version of this radical individualism that transcends its socially destabilizing effects. *This transcendental vision emerges from the evolution of normal science first to narrative, scientific constructivism and then to the more universal narrative constructivism.* Narrative constructivism involves the collaboration between Eros-chaos and Eros-order and correspondingly, the collaboration between participatory subjectivity and control objectivity.

Chapter 24
EVOLUTION TO NARRATIVE CONSTRUCTIVISM

HIERARCHAL SCIENTIFIC CONSTRUCTIVISM

Anyone who has a pet knows that some animals consciously recognize patterns. The pet owner establishes certain routines, sometimes associated with words, and the dog or cat recognizes certain cues and then prepares to participate in the routine. For example, when my wife or I mention getting take out for dinner, our dogs, Goddy and Murray, start preparing to get into the back seat of my car. The difference between humans and other mammals is that humans are to some degree conscious of themselves being conscious of some pattern or an event. When a human becomes civilized, which in modern societies usually occurs in the seven to eight year old child, he/she can construct a conceptual, language representation of perceived patterns. A young person who has developed logical, conceptual thinking can understand or even construct logical, conceptual models of patterns that are thought to be objective, true representations of nature. At the still higher level of individuation, which is the scientific, autonomous self, one can understand or create a theory of some pattern in nature that is neither true nor false but can be valid in a particular context. This approach, which chapter 23 called a *limited, scientific constructivism*, may evolve to hierarchal levels of describing nature. For example, Newton's theory of motion is valid for objects moving far from the speed of light but must be replaced by Einstein's special theory of relativity for objects moving close to the speed of light. Newton's theory has aspects that not only contradict Einstein's theory but does not imply that theory; Newton's theory cannot be simplified to or reduced to Einstein's theory. Einstein's theory represents a hierarchal description of nature in that it is a complex theory that includes Newton's theory; that is, it can be reduced to Newton's much less complex description of motion. Thus, Einstein's theory can be reduced to Newton's theory, but Newton's theory cannot be reduced to Einstein's theory. This non-reciprocal relationship is why Einstein's theory, by definition, is a hierarchal description of nature. A two-level hierarchy, such as life and non-life, means that the higher includes the lower level, but the lower does not include the higher level.

SCIENTIFIC CONSTRUCTIVISM BREAK FROM TOTAL CONTROL OBJECTIVITY

A scientific constructivism description of the *Carnot ideal heat machine* provides the foundation for formulating thermodynamics that also exemplifies a hierarchal, scientific description of nature. The *first law of thermodynamics* expanded Newton's ideas to include the concept of

---

[11] D. Yankelovich. *New Rules Search for Self-Fulfillment in a World Turned Upside Down* (New York: Random House, 1981), p.255.

energy metaphorically understood as a substance that flows from a higher to a lower level into or out of a system. There are two kinds of energy; *mechanical energy* that describes motion and *thermal energy* that in flowing into or out of a system results in the system changing its temperature. A difference in level of energy was called potential energy that represents structure equal to order in nature. According to the mechanistic ideology underlying the first law, a loss of order in one place always shows up as an equal amount of order some place else. Also, in the perspective of the first law, time is analogous to a dimension in space and thus is reversible. In this extended mechanistic world every event is predictable, thus implying determinism, and there is no loss of order; that is, there is no chaos.

The formulation of the second law of thermodynamics radically broke from this mechanistic description of nature. In this new perspective associated with the second law, energy no longer can be thought of as a substance that flows. Rather, energy is paradoxical in that it has two aspects that seem to oppose one another. On the one hand, energy is a *potential* (not the same as potential energy because kinetic energy also is a potential to do work) for two types of events to occur: (1) heat event where there is a change in temperature and (2) a work event plus a heat event where there is a change in motion and a change in temperature. *There is no potential in nature for only a work event. Potential* represents the order aspect of nature. On the other hand, energy is *flux* in which a potential transforms into a heat event or a work event plus a heat event. As a result of flux, the original potential equal to order is gone forever, and so flux is associated with the loss of order, which is the same as chaos. Furthermore, according to the second law, this loss of order sometimes may lead to creating a new order elsewhere in nature, but the quantity of the new order, called *negative entropy*, is less than the quantity of old order; that is, the new order has greater entropy than the old order. The usual way of saying this is that *in any change from old order to new order, there always is a net increase in entropy that is to say, a net decrease in order equal to a net increase in chaos.*

The original definition of entropy is that it always is associated with flux and is a measure of the quantitative change in order of a system. That is to say, entropy of any system in flux is a measure of its chaos. As a result of any flux in nature, the net entropy always increases; that is, net chaos always increases meaning that the net order always decreases. This means that, since any event always occurs over time, any and every event is irreversible. There is no event in nature in which there is a net increase in order corresponding to a net decrease in chaos. This being the case humans subjectively experience time as irreversible contrary to what Einstein maintained as a result of his theories of relativity. *This entropy law implies that we acknowledge subjective perceptions as well as objective representations of these perceptions.* This law also implies indeterminacy and lack of knowing. During any flux, we can know the initial state before the flux and the final state of a system after the flux, but there are an indefinite number of ways for the transition to the final state to occur.

THEORY OF MACHINES GENERATES NARRATIVE, SCIENTIFIC CONSTRUCTIVISM

As described in chapter 1, the description of an ideal heat machine implies seven interrelated generalizations about the functioning of any machine. These generalizations lead to the understanding of the functioning of any machine as described below.

Machine Task Results from Machine Collaboration with a Potential in Nature

Every machine has one or several interconnected energy couplers; the set of interconnected energy couplers is equivalent to a single energy coupler. A potential in nature communicates with an energy coupler. This communication may be interpreted as an energy flux involving a transfer of information to the energy coupler resulting in it having a potential to accomplish a small machine task. The energy coupler then transfers information to the environment resulting in a small machine task. Repetition of the cycle of the energy coupler leads to the sum of many small tasks equal to the large task of the machine. Any machine as an autonomous system does not accomplish a machine task. Rather, the machine accomplishes a task only as a result of it collaborating with a potential in nature, see figure 2.1. The sequence of energy events of one cycle of an energy coupler in any machine is as follows: (1) As an order in nature goes to chaos, the resulting energy flux leads to a heat event plus the energy coupler acquiring the new order (order out of chaos) of a potential to do a small task; (2) as the task potential of the energy coupler goes to chaos, the resulting flux leads to a heat event plus a work task such as a small motion of a car; (3) some of the energy flux into the machine leads to the non-machine task of bringing the energy coupler back to its initial state so that it can start a new cycle; this non-machine task is why even the ideal heat machine is not 100% efficient.

Metaphorical, Conceptual Understanding of Order and Chaos

Each cycle of an energy coupler produces a hierarchal mutuality of chaos and *machine creativity*. The machine may be said to create a task order out of the chaos of the energy flux passing through it. The machine task work is a two-level hierarchy. The task work implies that there must have been a potential in nature to generate it (this is what the first law says), but this potential does not imply any particular task work. Thus, the higher level machine task order includes the lower level order that existed as a potential in nature. Scientific constructivism gives quantitative definitions of order, such as a quantitative definition of potential and chaos defined in terms of the mathematical definition of entropy. *When we realize and acknowledge that an energy coupler produces a two-level hierarchy, then the ideas of order and chaos must be understood metaphorically as qualities that have degrees of differences.* Machine task is qualitatively different than an order in nature represented by a potential; that is, machine task is a higher level order than is an energy potential. The quantitative definition of machine task equal to work is not different than the quantitative definition of energy potential defined as the ability to do work, but, as described above, there is a qualitative difference. Likewise there are degrees of chaos corresponding to degrees of order.

Collaboration between Eros-order and Eros-chaos

The energy flux into the energy coupler metaphorically may be understood as resulting from a kind of drive (representing one version of the second law of thermodynamics), which I call *Eros-chaos*. The creation of a new order of the energy coupler acquiring the potential to accomplish a small machine task metaphorically may be understood as resulting from a drive that converts energy flux into a new energy potential or a work event. I call this drive, *Eros-order*. The same ideas apply to energy flux out of the machine that leads to a small machine task. Machine creativity may be understood as the mutuality of Eros-chaos and Eros-order, see chapter 2, figure 2.1. This mutuality is the more complex version of the second law of thermodynamics proposed by systems scientists.

Machine Collaboration with Nature Implies Narrative, Scientific Constructivism

The description of machine creativity involves a narrative interpretation of an historical process. Each cycle of an energy coupler is a concrete event that occurs in a particular context in thermodynamic time equal to one's subjective experience of time. The main structure of the plot of this narrative is the collaboration between the machine and a potential in nature. The actors in this narrative are humans who collaborate with nature by means of an energy coupler to achieve some goal. This more complete description of a machine is *narrative, scientific constructivism* that points to a new, higher, more comprehensive way of understanding science. It is higher because narrative, scientific constructivism includes scientific constructivism, but for practical reasons, one can reduce the narrative involving subjective understanding of metaphorical concepts to the objective model for accomplishing a goal. It also is higher because it enables one to be open to see possible connections with other areas of science and with non-science areas of knowledge. This more comprehensive way of understanding nature overcomes fragmentation of knowledge and disposes one to participate in nature, rather than just control her; this can actuate a kind of love of nature, see chapter 1, pp.12-13.

SYSTEMS SCIENCE

Dynamic Systems Perspective of Creating Order

A machine is a dynamic, open system in the chaos of energy flow, that is, flux, through the system. Another way of saying this is that the system is open to and to some extent not in equilibrium with the environment. A *phase space* is a system confining an entity and its relationships. Attractors are the principles or controlling factors that put order into the functioning of a phase space. That is, attractors organize the processes that occur in the phase space into specific patterns. The machine is a phase space and its energy coupler is an attractor. Because the energy coupler goes through cycles, it is a *periodic attractor*. This periodic attractor organizes energy flux as it passes through the machine – the phase space – into cycles that accomplish small work tasks. As a phase space degenerates into chaos, its attractor becomes less able to organize the energy that is passing through it. Eventually the attractor becomes unable to organize energy flux into any semblance of the pattern that it once produced. In applying these ideas to a machine one can say that a degenerating machine will become less efficient and eventually will stop functioning – "machine death."

Modern theories of chaos have shown that some phase spaces *that are not machines* may degenerate to what is called a *bifurcation* rather than phase space death. A bifurcation is the process of a complex dynamic system degenerating into chaos, which begins to generate new possible patterns; that is, there is the possible emergence of new orders from the degenerative chaos. The continued energy flux that produced the degenerative chaos begins to become organized by what are called *strange attractors* into the emerging new patterns. As a result of strange attractors, the phase space undergoes a *transformation* to new patterns. The emergence of these new patterns are said to *transcend* the degenerating pattern. That is, instead of some internal or external control bringing the degenerating phase space back to the old pattern, strange attractors become released by the chaos to generate new patterns. A phase transition of a phase space is a degeneration to chaos that either ends in phase space dissolution or to phase space death that is followed by a transformation "rebirth" to new patterns. Thus, a dynamic systems perspective of a machine provides another defining characteristic of it. A machine eventually will degenerate to dissolution; it never undergoes phase space transformation.

Two Types of Narrative, Scientific Constructivism

External, narrative, scientific constructivism. The theory of machines leads to narrative, scientific constructivism that is a type of knowing that transcends the totally objective, mechanistic, scientific, constructivism knowing. The word "transcends" refers to the hierarchy in which narrative, scientific constructivism includes scientific constructivism but not the other way around. A systems perspective of a machine points to a fundamental inadequacy of narrative, machine, scientific constructivism. Machine constructivism leads to a hierarchy of environment information transformed into machine work organization, but this perspective does not allow for the idea of any systems transformation. Machines have an intrinsic unity that may be called a *machine self.* A machine self can transform incoming information but it cannot transform itself. Rather the machine functions over time, but eventually it "self-destructs." In radical contrast some systems in the universe become configured into a unitary pattern that designates each of them as a system's self. Under some conditions a system's self can transform to one or more transcendental system's selfs. The word "transcendental" again refers to a hierarchy in which a newly emerged system's self includes aspects of an old system's self. Everywhere one looks at the biosphere there are instances of these system's self transformations. For example, subatomic particles become atoms, simple atoms become complex atoms, atoms become molecules, and so on. This points to a fundamental distinction between machine transformation leading to machine work hierarchies and systems transformations leading to system's self hierarchies. With this distinction in mind one may specify that machine transformations only produce information transformations in reference to an unchanging machine self. As the machine self degenerates, so also the quality of the information that it produces. The machine information transformation always is external to the machine self; it never can flow inward to become incorporated into the machine self thereby transforming it into a new machine self. Thus, any mechanistic or machine understanding of Nature is at best an *external, narrative, scientific constructivism* that never can describe a systems transformation.

Internal, narrative, scientific constructivism. A complex system's self remains stable as a result of one or a set of attractors that maintain it. When one attractor or the collaboration of attractors begins to breakdown, the system's self begins to degenerate. The resulting chaos opens up hidden potentials within the system that enables intra-communications (within system communications) or communications with potentials in its environment. Collaboration among some of these communications produces information transformations that are incorporated into a degenerating old self. This internalization of information leads to the transformation of the old self into a new self. For example, degeneration of a system of hydrogen and oxygen gases leads to the chaotic systems state of some hydrogen positive atoms and some negative oxygen atoms. If this system is near the attractor called equilibrium, then intra-system communications will regenerate the equilibrium state of hydrogen and oxygen molecules. However, if the gas system is far from equilibrium, then regeneration is not possible. This facilitates intra-molecular communications that collaborate to internalize information in water molecules making up the new system's self. That is, the positive hydrogen atoms and negative oxygen atoms do not recombine to form hydrogen and oxygen gas molecules. Rather, they collaborate to form water molecules. This understanding of systems transformation results from *internal, narrative, scientific constructivism*.

Origin of Life: A Uroboric Puzzle that Represents System Self-Organization

Max Delbruck described the problem of the origin of life as a Uroboros Puzzle.

> **The Uroboros Puzzle** What is the fundamental problem in biology? Max Delbruck answers as follows: "Thus there is a clear case for the transition on Earth from no-life to life. How this happened is a fundamental, perhaps *the* fundamental question of biology." [in Xlll the Nobel Conference, Gustavus Adolphus College, Oct (1977). <u>Nature of Life</u>, edited by William H. Heidcamp, Uni. Park Press, Baltimore, 1978] The problem of the initiation of life is beautifully embodied in the ancient idea of the uroboros. The uroboros, symbolized by a serpent with its tail in its mouth ..., represents an entity that is self-generating and self-sustaining.
>
> Life itself is a self-generating and self-sustaining system that has evolved into a state of being in which its origins are no longer discernable. Organisms use proteins in energy-transducing structures for obtaining the energy to make the proteins needed for energy transduction. The uroboros puzzle of the transition of no-life to life is: how could this have begun?[12]

Major biological organic molecules of life are large polymers composed of monomers: (1) Proteins consist of many thousands of amino acids; (2) complex sugars consist of many simple sugars joined together, that is, glycogen in animals and starch in plants are made up of glucose subunits; (3) nucleic acids consist of nucleotides joined together.

Uroboros puzzle restated. Organisms have complex structures stemming from bonding among polymers.

> [W]hat organisms must do to achieve these structures involves energy. The energy requirement exists because biological polymers are surrounded by water, which tends to degrade them by hydrolysis. Organisms acquire this energy through the catalytic activity of enzymes, special proteins that the organisms manufacture from instructions coded in genes (DNA). This is the essence of the uroboros problem: to make polymers, polymers are needed. Alternatively, this may be expressed: to make the energy needed for synthesis of polymers, energy for synthesis is needed.

---

[12] Ronald F. Fox. *Energy and the Evolution of Life*. (New York: W.H. Freeman and Co.1988), pp 3 & 5.

The transition from monomers to polymers poses the first real difficulty.[13]

The environment of the earth before the emergence of life thermodynamically favored the formation of water as well as siliceous crystal rock (energy from the sun plus a high chemical potential [large negative free energy of formation] drives these reactions). In this environment, polymers that form by chance tend to immediately degrade by *hydrolysis* into their constituent monomers. We know that living cells are able to take in and use chemical or light energy to activate monomers to form polymers more rapidly than hydrolysis degrades them; so polymers accumulate over time. Likewise, any *primordial form of life* must be able to do the same thing, but the problem is that already existing polymers are needed to activate monomers to form more polymers. Somehow what must emerge is a primitive Uroboric polymer system, that is, a "self-begetting" polymer system.

Summary of Uroboric puzzles.

1. A biochemical version:
    a. In the non-living world, polymers that spontaneously form also break down more rapidly than they are formed.
    b. A living system consists of polymers that are "self-sustaining," that is, the polymers capture energy from the environment and use this energy to make new polymers more rapidly than the spontaneous breakdown of newly formed polymers.
    c. How can "non-self-sustaining" polymers become "self-sustaining" polymers; that is, how can a living system evolve from the non-living world?
2. Two general ways of stating the Uroboric puzzle with respect to the emergence of life:
    a. If any living organism has properties that a non-living system does not have, then how can a primitive living system evolve from non-living matter?
    b. Life is a self-generating and self-sustaining system that evolved from non-self-generating and non-self-sustaining systems. How can this occur?
3. Two human analogies for a Uroboric Puzzle:
    a. *Learning to ride a bike.*
        1) In order to ride a bike, one must have a "feel" for riding a bike.
        2) In order to have a "feel" for riding a bike, one must get it by riding a bike.
    b. *Learning to solve a type of mathematical problem*
        1) In order to solve a type of mathematical problem, one must have a "feel" for solving these kinds of problems.
        2) In order to get a "feel" for solving these kinds of problems, one must experience solving these kinds of problems.

Systems Science Solution to the Paradox of Self-Organization

The systems solution to paradoxes of self-organization is rejecting gross reductionism[14] and proposing a theory of creativity involving the interconnected ideas of random events and emergent properties.

---

[13] Ronald F. Fox. *Energy and the Evolution of Life.* (New York: W.H. Freeman and Co.,1988), pp 34-35.

General description. From time to time the Order in the universe is disturbed by random events that produce local Chaos. As a result of one Order being destroyed, new structures become possible; that is, many new potential structures arise and conflict with one another. Sometimes the context of these conflicts is appropriate for some of these potentials to "self-organize" into a semi-stable new system. If this new system is "protected by circumstances" long enough to stabilize, it becomes the "rebirth from the death of the old order" and exhibits emergent properties. Besides not being present in the old order, these new properties cannot be predicted to occur from only knowing all about the old system. For example, as described at the end of chapter 1, under the right circumstances hydrogen and oxygen gases chemically combine to form water that has emergent properties. We cannot predict or deduce the presence of these properties of water from knowing the properties of hydrogen and oxygen gases.

Systems theorists reject gross reductionism such as that of molecular biologists, for example, Francis Crick[15] who claims to "explain" life and the human mind in terms of molecular interactions. Mechanistic science claims that physical laws that regulate the interactions among its autonomous parts determine the behavior of any system. In contrast, systems science is holistic. In any system each component must be understood in terms of the whole system; that is, each part must be understood in terms of its interactions with all the other parts of the system. Systems scientists see this pattern of interconnectedness as "traced by evolution in the physical universe, in the living world, and even in the world of [human] history."[16]

Systems science describes "creative evolution" as a sequence of transformations each from a lower level to a higher level of organization. The concept of level of organization implies hierarchy, such as a Chinese-box structural hierarchy of "boxes with boxes." Box B, which contains the smaller box A, is at a higher level of organization. This illustrates the idea of hierarchy in that box B contains box A and box A never can contain box B. Systems science also proposes functional hierarchies. According to this idea, system B not only contains system A but directs the activities of A so that they integrate with and contribute to the overall functioning of system B. The systems science view of evolution is that lower level systems modify their structure and functions so as to form a higher level composite system. For example, sub-atomic particles come together to form atomic nuclei, and these nuclei interact with electrons to form atoms. Atoms interact to form simple molecules. Atoms and different simple molecules interact to form complex molecules, some of which may be called macromolecules. These, in turn, may have interacted under the "right conditions" in the past to form primitive, cellular organisms. From here evolution continues by organizing autonomous cells into multicellular organisms and organizing diverse types of organisms into ecologies.[17]

Reduction to autonomous, potential, new patterns. While rejecting the all-inclusiveness of mechanistic analysis, systems science proposes a subtle form of reductionism. Though the universe consists of interconnected systems, systems theorists reduce any individual system to a particular stable pattern plus several autonomous potentials to transform to new stable patterns

---

[14] V. Csanyi. *Evolutionary Systems and Society* (Durham, N. C.: Duke University Press, 1989), pp. 8-11.
[15] F. Crick, *The Astonishing Hypothesis: the Scientific Search for the Soul* (New York: Simon & Schuster, 1995)
[16] E. Laszlo. *Evolution. The Grand Synthesis* (Boston: Shambhala, 1987), p. 5.
[17] E. Laszlo. *Evolution. The Grand Synthesis* (Boston: Shambhala, 1987), p.24.

(the next section shows how this implies a subtle form of reductionism). Any particular new pattern results from the partial breakup of the old pattern, thus releasing components to reform into a new pattern or, more commonly, to interact with other systems to form a new pattern. An analogous example would be when two people divorce: each is released to find a new mate or join a commune or develop a new life centered on being single.

Laszlo described these transformations to new patterns in the context of forming progressively higher levels of organization. The "right conditions" for this to happen is that the systems must be in a high degree of chaos as a result of being far from equilibrium and of having a high flux of energy passing through them. These so-called "third-state systems" can change into one of many possible new states that have some small degree of stability. When a new semi-stable, steady state emerges, it immediately may breakdown, but in so doing it provides the opportunity for one of many other possible new, semi-stable states to emerge. This new second semi-stable steady state may continue for a long time, trying out, so to speak, many different new possible steady states. Eventually one of the newly emerged steady states survives long enough to achieve a greater stability. The context of its extended survival may be such that this type of steady state that emerges elsewhere in some evolving macro-system also will have a greater stability.

The emergence of this new level of organization is not a mechanistic process. Gross mechanistic theories assume total determinism; that is, they assume that the starting characteristics of a system and the characteristics that describe how the system at its boundaries will interact with its environment totally determine how the system will change. But according to systems science, knowing these starting characteristics and the boundary conditions of a system does not totally determine what new level of organization can emerge when it goes into the chaos of "third-state systems." Nevertheless, systems scientists do construct a pattern of creative evolution. The pattern is that a third-state system leads to the emergence of a new level-1 organization that become stable. The level-1 organization system becomes a third-state system from which emerges a new level-2 organization, which includes a modified level-1 organization. Evolution produces successively higher levels of organization where each higher level includes a modified version of the lower level organization from which it emerged.[18]

Subtle Reductionism of Systems Science

Bohm takes a non-hierarchal view of systems science's description of nature that is utilitarian and a subtle form of mechanistic reductionism.[19]

Systems scientists and in an analogous way, Bohm, describe creative evolution as a sequence of transformations, each from a lower level to a higher level of organization. However, this *systems concept of organization level* reduces our tacit, subjective understanding of qualitatively different levels of organization in nature to a quantitative, objective representation of levels of organization.

For example, we may represent the development of an embryo to a fetus just before birth as the increase in *length* from a few millimeters to 16-20 inches. This is an objective valid representation of development, but it ignores the subjectively obvious, non-quantifiable,

---

[18] E. Laszlo. *Evolution. The Grand Synthesis* (Boston: Shambhala, 1987), p.36.
[19] Ken Wilber, editor. *The Holographic Paradigm and Other Paradoxes* (Boulder, CO: Shambhala Pub. Inc., 1982), p.191.

organizational changes that occur during this development. In like manner, a systems science description of this development would represent it as a large decrease of entropy. This is an objective, valid representation of fetal development, but it also ignores the subjectively obvious, non-quantifiable, organizational changes that occur during this process.

Wilber's critique of systems science brings in the importance of analogical thinking in various types of knowing. He assumes that the world of living systems is fundamentally different than the world of non-living systems. Likewise, he distinguishes the world of the human mind from processes of all other living systems. Various evolutionary systems scientists from Bertalanffy to Laszlo to Jantsch list dynamic patterns in terms of logical, conceptual statements, usually in the form of mathematical relations. This means that the expressed pattern is identical for all three realms. Wilber calls these same patterns "homologue laws," which, in turn, implies that the three realms expressing these patterns are not fundamentally different from one another. Wilber suggests, to the contrary, there are "analogue laws" that are similar to one another that apply to all three realms.[20] For example, while transformations occur in all three realms, these patterns of change are similar because they involve the mutuality of order and chaos rather than involving the same logical, conceptual definition of order and chaos for all of these realms.

Wilber's point is that if these "homologue laws" are not supplemented by other ways of knowing especially as applied to life and to mind, then this in effect is a subtle way of reducing life to non-life and mind to non-mind. However, I argue that science uses "homologue laws" only when this way of knowing is reduced to utilitarian reductionism or utilitarian positivism. At a deeper level of understanding, utilitarian science, that is , normal science involving problem solving, emerges from a "primordial science" that requires metaphorical concepts and analogies for expressing meaning. Thus, at this deeper level of understanding, science does propose "analogical laws." Nevertheless, Wilber is correct in accusing systems scientists of being subtle reductionists because, for the most part, most of them reject or, at least, do not acknowledge the metaphorical foundation of science. To put it differently, they do not understand or recognize that the conceptual, utilitarian models emerge from the *more fundamental narrative understanding*, which uses metaphors and analogies for understanding reality.

Systems Science Pseudo-integration of Knowledge

According to Laszlo, the systems science evolutionary perspective is a positivistic way of knowing that is totally based on empirical observations rather than introspection and the subjectivity of philosophy. Furthermore, it proposes logical, conceptual models that are testable and that describe the same pattern – the evolutionary process – expressed in all realms of the universe. As a result, it is now possible to advance a grand evolutionary synthesis (GES) based on unitary and mutually consistent concepts derived from the empirical sciences. In other words, all we need is scientific constructivism; there is no need for narrative constructivism involving metaphorical, conceptual insights.

The systems science new evolutionary perspective claims to overcome the modern fragmentation of knowledge, but this is a sham and is banal. Laszlo forbids introspection and the subjectivity of philosophy, but he proclaims the evolutionary process is becoming conscious

---

[20] Ken Wilber. *Sex, Ecology, Spirituality (the Spirit of Evolution)* (Boston: Shambhala Pub. Inc., 1995), pp.114-115.

of itself in individual humans. This means that individual humans are conscious of themselves being conscious of the evolutionary process. How can an individual experience this consciousness other than by introspection? Indeed how can any individual be aware that he/she exists other than by introspection? By eliminating all subjectivity, which is the major flaw of scientific constructivism from its first emergence, the all-inclusiveness of the systems synthesis removes the most important aspect of all disciplines. This especially can be appreciated with respect to music. A piece of music can be objectively represented by a set of symbols written on a piece of paper. But can this objective understanding of music really represent music when such "knowledge" does not elicit in a person subjective feelings and pleasure? What systems scientists and scientists in general fail to acknowledge is that subjective insights and the feelings associated with them represent the more important aspect of being human. The systems perspective is an explicit extreme version of reducing flesh and blood individual humans to automatons that can be controlled and manipulated by dictatorial leaders. Such a view not only destroys constitutional, liberal democracy, it readily can lead to the atrocities of Hitler's Nazi Germany, Stalin's Russia, and fundamentalists, Islamic terrorists.

NARRATIVE, CONSTRUCTIVE DEMOCRATIZATION

Metaphorical Aspects of Scientific Paradigms

The unifying paradigm of normal science always involves imagining nature to be structured in a particular way that allows one to solve problems about how components of it interact under some set of known conditions. The how aspect of nature always involves some kind of measurement based on counting and geometry, which depend on metaphorical knowing. For example, the measurement process of counting the number of apples in a basket depends upon imagining that all the apples are identical in at least one respect. Then the "measurement" involves assigning in a one-to-one manner the sequence of positive whole numbers beginning with one to each apple. The number assigned to the last apple in the basket represents the measure of "how many objects" are in the basket. The theory of statistics, based on the theory of probability, prescribes a process of assigning a "measure" called *probability* to a set of outcomes of a random process, which may involve infinitely many possible outcomes (mathematicians say the "how many" of such a process is infinitely non-denumerable, that is, it cannot be "counted," but sometimes each outcome can be represented by a possible frequency of occurrence). The point is that, because of its dependence on measurement, science always involves metaphorical, conceptual thinking where mathematics converts the subjectivity and ambiguity of such thinking into objective, conceptual, non-ambiguous statements. Chapter 17 also illustrates how measurement depends on metaphors and how Newton used metaphorical thinking to construct the universal law of gravity.

As described in chapter 6, pp.60-62, metaphorical, conceptual thinking enables humans to expand their knowledge of reality by "seeing" a similarity between an unknown aspect of nature and some consensually agreed upon empirical knowledge, such as all things tend to fall down. Thus, as described above, quantitative measurement knowledge of nature depends on the more fundamental, qualitative similarities humans are able to see. Scientists since Galileo and Descartes chose to divert acknowledging this dependence on seeing similarities toward focusing on the objective, quantitative "how" aspect of knowing nature. Newton's positivism reinforced this collective scientific choice so that the scientific enlightenment became modern, Manichaean, radical skepticism as, for example, described by Monod, see chapter 18. Paradoxically, for over three centuries (that is, since the 1683 publication of Newton's *Principia*) science has flourished in spite of being embedded in a Western culture dominated by a Judeo-Christian vision that totally opposes this extreme skepticism.

Thomas Kuhn's model, described in chapter 23, eliminated this paradox by distinguishing between scientific ideology and science as process. Kuhn argued that any community of people who study nature is analogous to a community of organisms, that is, a species. Any species on a moment to moment basis and over many generations is totally engaged in "solving problems" related to its survival. In like manner, the way any community of scientists study nature is to solve problems related to the "how" aspect of the structure of nature. For any particular community of researchers it does not matter whether the "how" structure of nature is true or false. All that matters is that each community has a unifying paradigm of *non-ideological pragmatism*, which involves "trial-and-error" independent of ideology. This means that two communities each associated with a "how" structure of nature that contradicts the other nevertheless can co-exist in the same institution. For example, my department consisted of holistic ecologists and molecular biologists committed to mechanistic reductionism. Competition for resources between the two groups led the ecologists to leave and form a new department that also incorporated the geology department. This kind of split has occurred to many biology departments across the US. However, the split did not occur as a result of an ideological conflict that prevented one or both groups from solving problems governed by their respective paradigms. Rather the split occurred because of communal ambition and intolerance.

Non-logical Creativity of Science

Kuhn described the idea of paradigm in a way that focuses on the non-logical creativity of scientific problem solving. A scientist learns his trade by teachers giving him/her problems that metaphorically represent the kind of situation a particular paradigm can deal with. The student by trial and error solves "typical problems" and begins to get a "feel" for solving these problems. This cycle of trial-and-error leading to a greater feel for solving problems that leads to more trial-and-error with respect to more difficult problems continues until the student achieves some level of what Kuhn called *tacit understanding* of one type of problem solving. Kuhn attended Polanyi's lectures[21] who proposed the idea of: "the ineffable domain of skillful knowing [problem solving]… that we may say in general that by acquiring [tacit knowing of] a skill, whether muscular or intellectual, [the Uroboric paradox of learning to ride a bike or learning to solve a type of mathematical problem] we achieve an understanding which we cannot put into words and which is continuous with the inarticulate faculties [of chimps, hominids, mythic intelligence in archaic humans][22] of animals.[23] The point is that one does not solve paradigmatic problems by learning and then following a set of rules. Rather, one transcends the Uroboric, learning paradox by gaining, as described above, a "Polanyi-like" tacit knowledge of this intellectual skill. The practicing scientist is like the pro-golfer who knows how to hit a golf ball straight and far. One can represent the golf swing by a mechanistic model that can help one

---

[21] http://en.wikipedia.org/wiki/Thomas_Kuhn

[22] M. D. Bowds. *The Origins and Structures of Mind, Brain, and Consciousness* (Bethesda, Maryland: Fitzgerald Science Press, 1999), pp. 89-91.

[23] M. Polanyi. *Personal Knowledge* (New York: Harper Torch Book, 1964), p.90.

improve his/her swing. But one does not learn a golf swing by first learning a set of guidelines, and if one focuses on the mechanistic model while trying to hit a golf ball, the swing will get worse rather than better, as I experienced many times when trying to learn to play golf.

Kuhn's Creative Insight about Science as a Narrative

My experience of different kinds of intellectual creativity is that when new ideas first emerge, they are ambiguous, overlapping and also disconnected sometimes even apparently contradictory. After reflection, which may take days, months or years, one may arrange the ideas to form a pattern that eventually may become a perspective, a theory or even a world view. I suspect that Kuhn had intuitions more or less at the same time about paradigm, tacit knowing involved in problem solving, science as similar to biological evolution, and creative learning that involves what systems scientists now understand as transcending a Uroboric paradox. Whether one agrees or disagrees with the biological theory of evolution, everyone agrees that this model can be understood as a story about creativity involving the unpredictable changes of any species and the sequence of emergence of new species as other species become extinct. In proposing that science is like biological evolution, he, in effect, proposed that overall science is the story of humans solving problems about "how nature works." Moreover, *each individual or community solving a problem is a story.* This insight follows from a similarity between machine task and problem solving. Flux of a potential in nature transfers information to an energy coupler whose structure is such that it receives and then transforms the information from nature into a machine task. In like manner, flux of potentials in nature transfer information to humans who, after empirically validating it and by means of a paradigm recognize the information and then transform it into an interpretation guided by a paradigm. Machine collaboration with information in a potential in nature to create a machine task is a story. In like manner, human collaboration with information in nature to construct an interpretation of that information by means of a paradigm is a story.

Sometimes the paradigm of a community does not enable humans to properly receive or interpret the information that nature presents to them. In a process involving trial-and-error linked to creative insights (what St. Augustine called "divine illumination," see chapter 13), humans construct a new paradigm that while including aspects of the old paradigm is radically different from it. The emergent properties of the new paradigm give it the potential to construct interpretations of the new, previously unintelligible, information. As is true of biological evolution, this is a story of humans creating a new paradigm analogous to the emergence of a new species. Thus, *overall science is the narrative about two kinds of creativity: problem solving and paradigm transformation.*

Democratic Scientific Communities

The narrative, scientific constructivism described in the previous section could transcend communal fragmentation without diminishing the effectiveness of specialization. Two communities each with radically different paradigms, such as holistic ecology versus molecular biology, could co-exist side by side where each community continues problem solving guided by its respective paradigm. The order of law based on the vision of narrative, scientific constructivism would see to it that neither of the two communities would impose its paradigm on the other. Nor, by law, would any person receive preferential treatment because of the paradigm he/she commits to. The category of preferential treatment includes availability of resources, working conditions, status, salary, and career advancement. Such a state of affairs is exactly analogous to constitutional, liberal democratization in the US. This narrative, transcendental vision would produce, as it did in American democracy, a complex, diverse, dynamic society that would progressively excel in its quantity and quality of innovations.

Narrative Science Education

Science education, as represented by science textbooks used in high schools and introductory courses in college, demands that students memorize with little understanding abstract ideas presented as pieces of information. Young adults, still influenced by the not totally repressed imagination and creative energy of childhood, on the one hand, are open to experiential understanding abstract ideas and, on the other hand, are intensely stimulated by video games, the internet, and personal high technology gadgets. To these students, science taught as information to be memorized: will be irrelevant to their experiential exploring minds, will be quickly forgotten and therefore not build toward "science literacy," and will eventually lead students, forced to memorize words rather than understand ideas, to shut down the inquiring mind with respect to formal education, especially with respect to science and mathematics – thus *teaching science merely as pieces of information damages the mind.* It reduces humans developing rationality to the lowest common denominator. Youth wanting to understand the "why" of things are infected with the virus of materialistic, non-ideological pragmatism taught with the even greater leveling of merely memorizing the "how," see chapter 23, pp.260-262. This is democratization gone "amuck." Finally, young minds force-fed information not only will not see interrelations among diverse ways of knowing, they will tend to be closed to any spiritual vision, such as the transcendental vision of democracy.

In order for democratic, scientific communities to be stable and intellectually productive, the narrative vision must transform science education. Narrative, science education should produce a college graduate with a major in some discipline and a person who has a metaphorical, conceptual understanding of many of the key ideas in the diverse sciences related to his/her major. The college graduate should have developed the ability to convert newly learned abstract ideas into a metaphorical, conceptual understanding of them and be able to express this understanding orally and in writing. Furthermore, the college graduate may have the beginning of tacit knowledge of problem solving guided by a particular paradigm, but more importantly, he/she should be able to dialogue with scientists representing diverse paradigms. The dialogue, of course, only is possible if all participants have a metaphorical, conceptual understanding of the scientific ideas that are being discussed (science, here, is interpreted to include mathematics). Alternatively, the participants in the dialogue to some extent may have to teach one another the narrative, ordinary language understanding of the abstract concepts of a specialized area of knowledge.

The college graduate then may go on to obtain advanced training (graduate and then post-doctoral work or training by a company) in a particular paradigm, but at the same time, he/she should continue increasing ordinary language understanding of ideas in science. This can be done in many ways such as when the older generation of scientists like myself learned ordinary language understanding of molecular biology at symposia and general lectures designed for this purpose at the Biophysics Society meetings. Many of the most urgent problems confronting postmodern, technological societies require the integration of diverse paradigms for their solution. Narrative, science education not only would stabilize democratic, scientific communities by dialogue across paradigm boundaries, it would prepare scientists from very different paradigms, even contradictory ones, to collaborate to solve modern problems.

Democratization of Intellectual Communities by Narrative Constructivism

The transformation of scientific constructivism to narrative, scientific constructivism may be expanded to non-ideological interpretations of empirical information as guided by a paradigm germane to each community of non-science disciplines. This expansion produces *narrative constructivism*. When narrative constructivism is applied to the systems theory of evolution, each instance of an evolutionary process exhibits three characteristics: (1) order, chaos, hierarchal, new order process (2) that is achieved by trial-and-error selection (3) of possible, stable and most efficient and most cooperative collaborations intrinsic to a particular environment. Evolution defined by these three characteristics applies to self-organization, called *individuation*, of any system and to many types of human creativity including leadership, parenting, teaching, experiential learning, all forms of artistic expression, athletic competition, market place capitalism, personal transformations, and narrative constructivism in all the disciplines of knowledge. Narrative constructivism is a sequential process starting with a vague, everyday language understanding of an empirical process or subjective insight that then is represented by the construction of a logical, conceptual model that can be validated by some set of consensually agreed upon criteria. The validated model then is converted into an objective narrative that can be understood by any layperson and that relates to other narratives constructed by other disciplines. This narrative always can be reduced to the precise, technical language of a particular model for purposes of practical applications.

Paradigms in each non-science discipline, like scientific paradigms, also evolve to diverse variations or transformations of them. Such differentiation would include, for example, narrative, historical constructivism, narrative, psychological constructivism, narrative, philosophical constructivism, and narrative, theological constructivism. Each type of *narrative constructivism* would specify the starting point for creating their constructs. Some method would specify certain types of empirical information or subjective insights that become objective by virtue of a group of thinkers who share the metaphorical, conceptual understanding of the subjective insights.

Each narrative constructivism should specify the criteria by which a theory is judged to be valid or not. Again, of course, each person using a particular narrative constructivism type of knowing should realize that this subjective or group collaboration for creating a new understanding of reality is itself an evolutionary process, which, of course, implies that this way knowing and the models it generates can evolve to higher levels of sophistication. This narrative constructivism of knowing avoids dogmatism such as found in medieval realism philosophy. At the same time, this constructivism transcends the fragmentation of modern science. It allows for diverse points of view, and yet it avoids total relativity such as found in existentialism, avant-garde modernism, and the current humanities postmodern perspective. It does so by providing a method by which a community of thinkers can reach consensus about the validity of a particular model.

If American culture could embrace narrative constructivism and apply it to education, then our society could transcend the fragmentation of knowledge produced by scientific constructivism and postmodernism. It now would be possible to construct educational programs for high schools and colleges that have unifying themes that inculcate values and that define what it means to be educated. The unifying themes include: (1) the evolutionary process as exemplified by all the disciplines and by all artistic creativity; (2) the evolutionary process as exemplified by human, spiritual transformation including but not limited to mystical transformations associated with various religions; and (3) narrative constructivism that comprehends diverse expressions of the evolutionary process and leads to a radically new vision and teaching style called student-teacher constructivism. *Student-teacher constructivism* is *creative learning* in which the learner "constructs" a subjective understanding of abstract ideas in terms of his/her personal experiences. A constructivism teaching style involves describing/presenting abstract ideas in terms of metaphorical concepts, analogies, and stories. A metaphorical concept, analogy, or story represents a pattern that each person can subjectively understand. When this pattern is superimposed onto an unknown, abstract idea, then each person "seeing" this superimposition will begin to understand the unknown abstract idea in terms of his/her personal experiences. My textbook for teaching *Survey of Biology*, designed for non-science majors, exemplifies this approach to student-teacher constructivism.[24]

Narrative constructivism also may point to other unifying themes such as: (1) operation of any machine and machine structure of various systems and institutions; (2) homeostasis including discussion of negative feedback as opposed to positive feedback; (3) mechanistic thinking; (4) mystical, heroic creativity, see chapter 13; (5) epigenesis (developmental process); (6) ecology and the mutuality of ecology and evolution; (7) communication in relation to coupling and collaboration and distinguishing between information and non-information; (8) individuation and hierarchal levels of self-consciousness; (9) creative, mindful dialogue; and (10) various representations of an ultimate SOURCE such as creator God, Gia Goddess, Being (described by Heidegger), Brahman (of Hinduism), Emptiness (of Buddhism) all of which are associated with various religions and/or types of spirituality. The values of constructivism education include a modified description of the values associated with the traditional view of liberal education. The definition of what it means to be educated would include the ideas that one would see interconnections among diverse disciplines, would develop personal creativity and would develop the ability to participate in creative dialogue.

The third enlightenment – the scientific revolution – led to the transcendental vision of tolerance of diversity that became formalized as constitutional, liberal democratization in the American colonies in the late 1700s. The emerging fourth enlightenment is grounded on narrative constructivism that provides the transcendental vision of tolerance of diverse paradigms for interpreting empirical information about reality. The rebirth of US democracy – and indeed, of Western culture – will occur with the institutionalization of this vision by the community of intellectuals in the US, especially those who govern the education institutions. Thus, the fourth enlightenment, like the third, must produce something like constitutional, liberal democratization that legitimizes and encourages diversity of paradigms both in science and in non-science disciplines. However, this drive to democratization of public education probably, only will occur when a ground-swell movement among students compels the intellectual elite to adopt it.

---

[24] Donald Pribor. *Universal Creativity and Individuation of the Biosphere. A Constructivism Epic* (Dubuque, Iowa: Kendall/Hunt Pub. C., 2004)

Chapter 25
SYSTEMS THEORY OF INDIVIDUATION IMPLIES AN ULTIMATE SOURCE

NARRATIVE UNDERSTANDING OF NATURE

Machine Narratives

The emergence of a theory of machines in the early to mid 1800s (1824, Sadi Carnot, ideal heat machine; 1840s, first law of thermodynamics, which is the conservation of energy, proposed independently by James Jouole, Hermann Helmholtz, Julius Mayer, and Baron Justus von Liebig; 1852, William Thompson, machine version of the second law of thermodynamics; 1867, Rudolf Clausius, first mathematical version of the concept of entropy) simultaneously provided a direct attack on total, objective knowing of scientific constructivism expressed as mechanistic analysis. It also laid the foundation for innovations contributing to the industrial revolution. As a result, many famous scientists thought thermodynamics based on the theory of machines only provided principles of engineering rather than laying out fundamental laws underlying all of science. Most scientists still avoid the subjectivity intrinsic to metaphorical, conceptual thinking as a necessary component of some scientific theories. However, metaphorical, conceptual thinking is unavoidable with respect to the entropy principle of thermodynamics[25] and to the overall operation of any machine, see chapters 1 & 2, and 24, especially pp. 262-265.

In particular, any machine collaborates by means of energy couplers with a potential in nature to produce a machine task. For example, a water wheel focuses the energy of falling water into the rotation of one stone over another that maybe used to grind wheat to flower. The order in nature, which is the potential of water to fall, is converted into the new order, which is the machine task to grind wheat. Metaphorically speaking, the old order in nature goes to chaos of energy flux where some of this energy flux is captured by the energy coupler that converts it into a new order, which is the machine task. Thus, a machine "creates" new order out of the chaos of some energy flux in nature. If we focus only on the mathematical description of machine creativity, then machine work (task) is not fundamentally different from the order of a potential in nature; mathematically speaking, this creative process does not generate a hierarchy of orders. However, if we focus on an *interpretation* of what is happening during machine creativity, then it is readily seen that this is an historical process that produces a new, two-level hierarchy of orders.

Any potential in nature does not imply any particular machine task, but according to the first law of thermodynamics (the conservation of energy) machine work (task) always implies an order in nature, which is a potential, that led to the machine work. Furthermore, according to the second law of thermodynamics, the narrative of the machine creativity implies that it is an irreversible historical process producing the hierarchy: machine order equal to machine task that includes by implication order in nature, which is a particular potential that is expressed as an energy flux into the machine. Machine order and the order in nature must be understood metaphorically in that both are patterns that are qualitatively different. The machine work is a

---

[25] D. Pribor, "Transcending subjectivity double binds of science." <u>Proceedings of the Institute for Liberal Studies</u>, vol. 10, pp. 27-37.

higher level of order than the order in nature that by an energy flux and an energy coupler led to the machine work. Overall, any machine creativity involves a narrative interpretation of an historical process, that is, a story interpretation equal to narrative knowing.

Narrative knowing of machine creativity points to a new, higher narrative understanding of science. There are three levels of understanding machine creativity. The first level is a vague insight expressed as metaphorical concepts about how to construct energy couplers to accomplish a desired task. In the second level, the vague insight is converted into a logical, conceptual theory that guides the creation of a machine to accomplish the desired task. If the machine performs according to agreed upon criteria, then the model is valid. The third level understands this model as a narrative involving metaphorical concepts. Thus, the three hierarchal levels are: (1) Subjective, vague understanding of a particular machine creativity; (2) validated objective logical, conceptual understanding that includes the first level from which it emerged; and (3) the metaphorical, conceptual, narrative understanding of the validated objective, logical, conceptual understanding of a machine. A metaphorical, conceptual understanding of machine creativity involves an empathetic, participatory understanding of Nature, which is ambiguous. This understanding may be at the first or the third level. When it is at the first level, it cannot guide accomplishing some desired, machine task, though it is a basis for moving to the second level of understanding. When the understanding is at the third level, it can guide accomplishing some desired machine task by being reduced to the second level of understanding, which is a logical, conceptual model. Most scientists and virtually all science textbooks (as of July 2009) ignore this way of understanding science. However, some scientists are acknowledging the necessity of three levels of knowing for creating scientific theories as a result of studying transformation of systems.

Circular Paradox of Systems Self-Organization

When a system's self degenerates to death and then rebirths to a new system's self, scientists say that the system has "self-organized." This leads to a more fundamental circular, paradox: How can a self transform itself? When a system's self degenerates, the system loses its ability to maintain or re-establish a stable self, and that is all the system's self has, namely an attractor (see chapter 24, p.267) that maintains a particular pattern. System scientists say that the degenerating self dies and is reborn via a strange attractor that is a new intrinsic unity and cohesiveness toward which the system is evolving. Where does this strange attractor come from? According to the Western mentality, "Something cannot come from nothing."

Another way of making this circular paradox more evident is to ask the question: "Which is more primary, the strange attractor or the new self?" This is analogous to the question: What is more primary the chicken or the chicken egg. We know of the existence and the pattern of a strange attractor after the emergence of a new self. In other words, scientists construct the narrative of the emergence of a new system's self and then construct the idea that a particular strange attractor generated the new self. Some systems scientists' position on this is analogous to Bohm's idea of implicate Order in matter-energy that progressively becomes explicate Order over time.[26][27] The position is that there are infinite, non-denumerable strange attractors

---

[26] David Bohm. *Wholeness and the Implicate Order* (London: Routledge & Kegan Paul, 1980)
[27] Renee Weber. *The Holographic Paradigm and Other Paradoxes* Ed. Ken Wilber (Boulder, CO: Shambhala, 1982), pp.191-192.

embedded in the implicate Order of mass-energy. Random fluxes of the universe determine where and when a particular strange attractor will be "uncovered" as a result of the degeneration of a particular system's self. In other words, transformation does not produce a new self or creativity in general is not new Order emerging from chaos but rather is explicate Order emerging from implicate Order. But then a particular set of random fluxes becomes the partial cause of the emergence of an explicate Order from the implicate Order of a strange attractor. As Bohm would argue, this set of fluxes is random at one level of analysis in which it is merely one possibility among an infinitely non-denumerable "number" of possibilities. But at a "higher" or "deeper" level of analysis, this particular set of circumstances is itself the Order of a higher level strange attractor that determines when and where a lower-level strange attractor will be "released" to generate a new system's self. But then, what "causes" this higher level strange attractor order to come into play? The answer, of course, is a still higher level strange attractor order thus indicating the classical Aristotelian infinite regression. Aristotle's answer would be that there must be a really unique strange attractor order that is itself "uncaused." The uncaused strange attractor is Aristotle's idea of God.

Mathematical Description of Evolutionary Self-Organization

Overview of the theory of probability. The first major idea to emerge that later could be used to describe evolutionary processes is the mathematical theory of probability for describing processes that have an infinite, non-denumerable "number" of possible outcomes. For example, suppose one repeats a measuring process to determine the length of a table. It is reasonable to assume that there is a unique number of linear units that represents the length of the table. A particular measuring process may produce results that range from 7.8 to 8.2 units. Modern number theory (theory of real numbers) implies that between and including 7.8 and 8.2 there are an infinite, non-denumerable "number" of numbers (real numbers). Probability theory provides guidelines for constructing one of an infinite number of *probability distribution functions* where probability is like a mass of paint unevenly distributed in the interval bounded by 7.8 and 8.2. The way this probability mass of paint is distributed allows infinite variation, but the total mass of this paint is 0 at the point just before 7.8 and progressively increases as one moves toward 8.2. The total probability mass in the interval bounded by 7.8 and 8.2 is 1. Some of the possible distribution functions are what mathematicians call "well-behaved," and one of the most useful of these distributions is the Gausian distribution. This distribution can be converted into a frequency function that associates a predicted frequency of occurrence of measurement results to every point in the interval bounded by 7.8 and 8.2. The frequency associated with each point in this interval is interpreted as a probability of the point representing one of the possible results of the measurement process. This Gaussian frequency function looks like a bell-shaped curve where the top point of the bell is a unique characteristic of this function called the *mean value*. The mean value is assumed to be the best estimate of the actual length of the table.

Probability interpretation of thermodynamics. The next major idea that contributed to mathematically describing an evolutionary process is representing the laws of thermodynamics by probability functions. Any system is viewed as consisting of many particles where each particle has internal energy called *electro-chemical potential* that does <u>not</u> contribute either to heat or to work events when the system's total internal energy changes. Thus, the first law states that any change of internal energy of a system will equal to the energy that produces a work event plus the energy that produces a heat event plus the sum of changes of electro-chemical potential energy of each particle making up the system. The second law represents the entropy change of a system resulting from the system changing infinitely slow so that the system is very near to equilibrium throughout the change. It further is assumed that (1) each particle can exist in many different energy configurations, such as described by quantum mechanics, and (2) all the possible configurations of a given system are equally likely to occur. This means that the system will be in the statistically most probable configuration.

Then the classical definition of entropy is equivalent to a probability distribution function representing the entropy of the system. This *statistical entropy* represents uncertainty or ignorance about the exact state of the system. Certainty would imply that one knows the energy state of each particle and the one set of relationships among all the particles of the system. In both the classical and the statistical mechanics version of the second law, when a system changes, it produces an irreversible hierarchy of two types of energy: there is *organized energy* producing a work event, which always is accompanied by *disorganized energy* producing a heat event. According to the second law, after the change of a system, its entropy has increased. This implies any one of the following three statements: (1) the system has less energy available to do work (some of the energy it had before the change was converted into disorganized energy of a heat event), (2) the order of the system has decreased corresponding to an increase of its chaos, (3) the uncertainty of the structural state of the system has increased.

In the MaxEnt interpretation of thermodynamics[28] the statistical entropy only has predictive value that is tied to the subjective assumptions one makes about a system. (In fact, this always applies to using probability theory to predict outcomes of any concrete situation.). Statistical mechanics assumes determinism so that there is only one possible outcome of any change to a system. Statistical entropy merely represents the quantity of uncertainty about the state of any system after a change of its internal energy. While there is a net increase of entropy of the environment plus the system that changed, the statistical entropy of a system far from equilibrium may actually decrease, which seems to contradict the second law. However, from the MaxEnt perspective, this discrepancy only means that there was information about the system before the change that only became evident after the change; it really was there all the time. Such a view may be applied to understanding how oxygen and hydrogen gases react to produce water with emergent properties. The MaxEnt perspective would claim that the information necessary for predicting water with emergent properties always was there but only came to light after observing this "evolutionary" event. In any case, the classical and the statistical mechanics interpretations of thermodynamics preclude the possibility of evolution. Any system in nature cannot spontaneously decrease its entropy; that is, it cannot increase its information, which is to say it cannot accumulate information. In his article, "Thermodynamics vs. Evolutionism (Exposing the Myth of Evolution)" Timothy Wallace makes the strong argument that thermodynamics is the most fundamental and universally accepted perspective in modern science, and it refutes the possibility of evolution.[29] However, Wallace is unaware of the formulation of the second law in terms of open systems far from equilibrium, such as described by the Principia Cybernetica Project.

---

[28] (http://en.wilipedia.org/wiki/MaxEnt_thermodynamics)
[29] (http://www.trueorigin.org/steiger.asp)

Principia Cybernetica Project: Narrative, Scientific Constructivism.

The people in the Principia Cybernetica Project proclaim that while there is no absolutely true model of reality, they have constructed a universal model in which a system far from equilibrium may spontaneously internalize information. This model incorporates the idea, developed by Szilard, Gabor, Rothstein, and Brilloun (1940 – 1950), that a decrease in statistical entropy, called *neguentropy* (negative entropy) is equivalent to information. Even Wikipedia's description of MaxEnt thermodynamics states that "Quite possibly it [the decrease of statistical entropy] arises as a reflection of the evident time-asymmetric evolution of the universe on a cosmological scale."[30] The cybernetic model, called *Metasystem Transition Theory (MSTT),*

> Is constructed by the subject or group [and uses] methodology to build … [a] complete philosophical system … [that] is based on a "bootstrapping" principle: the expression of the theory affects its content and meaning, and vice versa … [thus, moving back and forth from a metaphorical knowing to a logical, conceptual model then to a metaphorical understanding of this model that leads to a new, better logical, conceptual model and then finally understanding the logical, conceptual model so as to achieve a philosophical goal.] Our goal is to create, on the basis of cybernetic concepts, an integrated philosophical system, or "world view", proposing answers to the most fundamental questions about the world, ourselves, and our ultimate values.[31]

Thus, in creating this cybernetic model, these scientists have exemplified converting scientific constructivism to *narrative, scientific constructivism* that uses metaphorical concepts. According to the Metasystem Transition Theory,

> [evolutionary] Self-organization is a process where the organization (constraint, redundancy) of a system spontaneously increases, i.e., without this increase being controlled by the environment or an encompassing or otherwise external system.[32]

First of all, there is a hierarchal view of the second law of thermodynamics wherein the cybernetic version of it includes and therefore can be reduced to the classical or statistical mechanics versions of it. The cybernetic version proclaims the universality of the mutuality of Eros-chaos and Eros-order. Eros-chaos, which is the universal drive described by the classical version of the second law, leads to some degree of breakdown of the order of a system. This order may be described in terms of constraints on the interactions among particles making up a system. For example, a system of hydrogen gas is made up of hydrogen molecules each composed of two hydrogen atoms. The two hydrogen atoms in a hydrogen molecule are like members of a married couple. Theoretically each member is not free to date other people, and

---

[30] http://wikipedia.org/wiki/MaxEnt_thermodynamics
[31] http://pespmc.vub.ac.be/MSTT.html
[32] http://pespmc1.vub.ac.be/SELFORG.html

likewise, each hydrogen atom is not free to chemically interact with other atoms. The breakdown of hydrogen molecules releases electro-chemical potential energy stored in the bond between two hydrogen atoms. Some of the released energy may do work, but there always will be some of this energy expressed as a heat event. Thus, the overall potential of the system to do work has decreased; so the entropy, which measures the <u>inability</u> of a system to do work, has increased.

Eros-chaos that breaks down some constraints in a system opens up new possibilities for new particle interactions. This sets in motion a random "trial and error process" of particles "attempting" to form new associations involving new chemical bonds. Many new associations are not stable, and therefore they do not persist. The new associations that are stable do persist. The characteristics of any new stable association provide the totally internal criteria for rejecting many possible new associations and selecting others. The result of this selection process is a new set of constraints on the particles of the system, and correspondingly, the system has taken on a new order. *I designate this selecting process as Eros-order.* Order represents the cohesive, stable unity of the system that metaphorically may be said to represent the *system's self.* Eros-chaos breaks down an old system's self into the chaos of many new possible particle associations. Eros-order is the selection process that converts the chaotic system into a new system's self equal to a new order. Overall this is a creative process involving the *collaboration* of Eros-chaos and Eros-order to produce a new system's self, characterized by new negative entropy equal to new information. Thus, overall this Eros-chaos, Eros-order collaboration is the process of self-organization that in this case may also be called evolution.

Example. Eros-chaos involving high energy flux drives a mixture of hydrogen and oxygen gases far from equilibrium to the higher entropy state of positive hydrogen ions and negative oxygen ions. There are many possible new associations between these ions. Eros-order representing quantum mechanics rules for new stable associations selects water molecules to emerge as the new order defined by new constraints equal to new information. The old order does not contain this new information. Rather, the new information only becomes manifest after Eros-chaos opened up new possibilities. The new order emerged from the chaos of the old order. Water implies hydrogen and oxygen but not vice versa. If we never observed this creative event, we would not be able to predict water as an outcome. After observing the emergence of water as an outcome, we can set up the situation where the same self-organization will be repeated. But this emergence of new information is time asymmetric. The particular Eros-chaos, Eros-order collaboration involving high energy flux at non-equilibrium and quantum mechanics rules insures that stable water molecules will form rather than hydrogen and oxygen molecules. Water at room temperatures is very stable and will not breakdown into hydrogen and oxygen.   In general those types of self-organizing processes that are equal to an evolutionary process produce the hierarchy of new order that implies the chaotic old order from which the new order emerged.

An open system far from equilibrium and exposed to high energy flux through the system will *expose* new possible collaborations between or among structures within the system or between/among structures within the system and structures in the environment of the system. In the process of the system trying out these possible collaborations there is a selection of collaborations that are more stable over collaborations that are less stable. Over time there may occur a gradual emergence of progressively more stable collaborations expressing a new order that also may be called a *new system's self.* According to this cybernetic, thermodynamic model, *there still is no need for a creator God hypothesis.*

Since the mechanisms of self-organizing evolution satisfactorily explain the origin and development of the universe, and our place in it, there is no need to postulate a personal God, in the sense of a conscious entity outside of the universe which created that universe…. It is possible to consider the universe as a whole, or the process of evolution itself, as God-like, in the spirit of pantheism.[33]

The authors of this model claim that it can satisfactorily answer all the eternal philosophical questions, and thus by implication, there is no need for religion that also answers these questions. With regard to fundamentalist religions: "Proclamation of any truth as absolute because of being given through a 'revelation' is sheer self-deception."[34] This cybernetic model seems to explain universal creativity understood as evolutionary self-organization by assuming that collaboration between Eros-chaos (the drive to chaos) and Eros-order (the drive to order) impels the emergence of a new self from a chaotic old self.

But the question remains: *why is there this kind of collaboration*? However much bootstrapping there is in constructing this model, the assumptions imply the model and not the other way around. Each instance of self-organization is a narrative in which there is a series of existential events interpreted by the model to be a collaboration that leads to another existential event interpreted by the model to be a new self. *But why is there any one of these collaborations that participate in producing a new self?*

ULTIMATE SOURCE

The Contradiction of Randomness Being the Source of Creativity

Some scientists choose *randomness* as the "source" of the emergence of new collaborations, but this leads to the following contradiction. Evolution implies ecology, which involves collaborations among all the elementary parts of a system so that there are no autonomous, elementary parts. Random event is a concept in the theory of probability that is applied to a system such as an evolving system described by the cybernetic model described above. Probability theory assumes that any system to which it is applied consists of elementary, autonomous events. Thus, the contradiction: no events participating in the ecology of an evolving system are autonomous, but any probability understanding of an evolving system requires the assumption of autonomous, elementary parts or events.

Transcendental Agnosticism

In order to avoid the contradiction associated with randomness one must make a choice. One choice is agnosticism, which in this case is refusing to be able to know why there are collaborations and rules governing the stability of collaborations. This is exactly analogous to Newton refusing to understand the force of gravity as action at a distance. He embraced the

---

[33] http://pespmc1.vub.ac.be/ETERQUES.html
[34] http://espme1.vub.ac.be/KNOW.html

positivistic perspective of agnosticism with regard to "why" questions such as this. Rather, he proclaimed that all one needs is valid, mathematical representations of empirical patterns in nature. A second choice is what may be called a *transcendental agnosticism* wherein one acknowledges that there is SOURCE that simultaneously is immanent in each episodic self-organization event and is not in any way defined by that self-organization so that SOURCE also transcends each episodic self-organization. While acknowledging that the SOURCE in some unknown manner is in each episode of evolution, IT remains unknown and unknowable. It even cannot be said to be a Being. In agreement with Zen Buddhism, the SOURCE may be said to be both Being and non-Being. What I am calling SOURCE is called in Zen Buddhism Emptiness or absolute Nothingness from which all things-events emerge.

This transcendental agnosticism may be thought of as *transcendental, subjective, narrative constructivism* involving choosing belief in SOURCE. Such a choice has profound consequences that continually reinforce one's belief in SOURCE. When one's life becomes chaotic producing great anguish and despair, one may choose to believe in SOURCE that eventually will manifest new possible collaborations. Then, on a trial and error basis one may select collaborations that are somewhat stable and that lead to a new order emerging out of the chaos in one's life. Thus, believing in SOURCE leads to human creativity. As one experiences this order out of chaos many times, he/she comes to a mode of awareness in which one always is disposed to creativity and individuation.

Democratization of Commitment to Ultimate SOURCE

I propose that the insight to be gleaned from scientific nihilism, avant-garde nihilism, postmodern nihilism, and Heideggerian, Nazi nihilism is the commitment to an ultimate SOURCE that not only is totally unknowable but is not specifically, uniquely associated with any human individual or any human society or institution in preference to other possible human associations. Analogous to the vision of Gautama Buddha (563-483 B.C.), this vision transcends humanism while at the same time celebrating individual human creativity expressed in the context of a spiritual vision of democracy. This is not to say that all religions now should disappear. To the contrary, these diverse religions serve an important function, which is to bind a group of people into a society whose cohesiveness is maintained by a common vision, a genuine care for one another, and a common set of laws that guide ethical behavior and preserve the ideal of justice. SOURCE may be interpreted – rather must be interpreted – to be the unifying principle in any religion or mystical tradition. Thus, SOURCE may be the creator God of Jewish, Christian, or Moslem religions, the TE of Taoism or pure consciousness, the ONE, of Hinduism or Emptiness of Buddhism or Being of Heidegger's mystical philosophy. However, while SOURCE is immanent in all these religions or mystical traditions, it also transcends all of them. This means that no specific societal embodiment of the ultimate SOURCE is the one true vehicle for authentic knowledge producing authentic actions. It is acceptable that participants in a particular, societal embodiment of SOURCE believe that their shared vision is the one true vision. But, as happens in American type democracies, these believers also must submit to two imperatives. One, they must not seek to impose their vision onto anyone else outside their community, and two, they must remain open to create collaborations, by means of creative dialogue, with all people with diverse, different visions.

Chapter 26
HUMAN CREATIVITY

HUMANITY IS AN EVOLUTIONARY PROCESS

The realization that humanity is an evolutionary process is an expansion of the generalization that science is like biological evolution. Scientists manifesting evolution express a type of creativity that is rooted in the third enlightenment. This represents a portion of Western society coming to a new way of seeing the world, hence the use of the word, *enlightenment*. This new way of seeing comes from humans who have evolved, according to Wilber and others,[35] from ego expressing formal rationality to *centaur* expressing *vision-logic*. In relation to the themes discussed in chapters 23 and 24, centaur is the evolutionary stage, which I designate as the scientific, constructivism mind-self, and the first and second stages of vision-logic are narrative, scientific constructivism and narrative constructivism respectively. Thus, modern science creativity has emerged in the course of evolution of the biosphere leading to non-self-conscious hominids that further evolved to various stages of human self-consciousness. This, of course, implies that scientific creativity is a stage in the evolution of human creativity, which, in turn, is a stage in the evolution of the universe. From the perspective of transcendental agnosticism presented in chapter 25, universal creativity described by systems science is <u>not</u> identical to ultimate SOURCE as some thinkers claim. Rather, IT is both immanent in and transcendent of every episodic event of the evolutionary flow of all that is, past-present-future. Wilber's way of saying the same meaning is that the ultimate SOURCE is Spirit, which is non-dual in that IT is non-different from the One descending to produce (maintain in existence) the Many and from the One that draws the Many ascending back to IT. To the individual human going through this cycle the One descending is different from any of the Many including one's individual self ascending back to the One, but with respect to the One, the *descending* is non-different from the *ascending*.

This book, which presumes the above perspective of creativity, traced human creativity in Western cultures beginning with ancient Greeks (1,000 BC) who evolved (from and in association with ancient civilizations, especially the Persians, in the Middle East) to modern cultures expressing scientific creativity. I find it useful for the sake of clarity to conceptualize the evolution of human creativity as occurring in three phases: *pre-personal*, *personal*, and *transpersonal* levels of human consciousness. Each level of consciousness occurs in individual humans. Any particular level of consciousness also may be applied to a particular society/culture with the qualification that any society/culture exemplifying a particular level does so only as a result of a significant subgroup of the society – often only the minority of an "educated elite" – who express that level of self-consciousness. Yet, according to the "perennial philosophy," after humans have evolved to the personal phase of consciousness, there always are a few humans who manifest the highest levels of transpersonal consciousness.

Pre-personal consciousness. Wilber[36] presents the sequence of three sub-phases of this evolution in terms of a general self-sense as <u>first</u>, material self that represents archaic humans; <u>second</u>, bodyego that represents magical humans, and <u>third</u>, late bodyego that represents magic-mythic humans expressing the "representative mind." Chapter 4 describes these first two levels.

---

[35] Ken Wilber. *Integral Psychology* (Boston: Shambhala, 2000).

[36] Ken Wilber. *Integral Psychology* (Boston: Shambhala, 2000), pp. 197-198, charts 1a and 1b.

I describe the first level as archaic humans that evolved from apes to primitive hominids, pp. 33-34, that further evolved to *group ego*, which I designate as *feeling, action self*, pp. 36-37. In agreement with Wilber I designate the second level as *body ego*, pp. 41-43. Chapter 5 describes the third level, which I designate as the *polar mind-self*, pp. 43-45. Wilber designates the transition to a particular level of self-consciousness as a "fulcrum," wherein the individual de-embeds from a lower level and then identifies with a new, higher level and consolidates it. Thus, the first fulcrum, F-1, leads to the physical self; the second fulcrum, F-2, leads to the emotional self [for reasons described in chapter 3, especially pp. 27-29, I would say F-2 leads to a *feeling self*]; the third fulcrum, F-3, leads to self-concept, that is, a polar self. Thus, the polar mind-self successively transcends but includes the physical self and then the feeling self. Chapter 6 describes the transition from participatory, embedded knowing to metaphorical, conceptual knowing, which prepares the way for the emergence of the personal phase of human consciousness.

Personal consciousness. Wilber[37] represents the sequence of three sub-phases of this evolution of personal consciousness in terms of general self-sense as <u>first</u>, the persona that represents civilized humans who develop the mental abilities of adopting roles and following rules in a society and fulcrum F-4 that leads to role self; <u>second</u>, the ego that represents humans who develop formal rationality (formop equal to formal, operational thinking) and fulcrum F-5 that leads to mature ego; <u>third</u>, centaur that represents humans who develop vision-logic and fulcrum F-6 that leads to the centaur, the "existential, integrated self." I find it useful for greater clarity to designate these next three levels of human consciousness as: 4. *persona mind-self*; 5. *rational, autonomous mind-self*, and 6. *scientific constructivism mind-self*. The center of gravity of any level of consciousness is to hover around one basic level where 50% of the time one is at this level, 25% at one level below it and 25% at one level above it: As a result, according to Wilber:

> …each time the self identifies with a particular level of consciousness, it experiences the loss of that level as a death – literally, as a type of death-seizure, because the very life of the self is identified with that level. Letting go of that level is therefore experienced only with great difficulty. In fact, I believe that each of the major milestones [fulcrums] of self-development is marked by a difficult life-death battle, involving the death (or disidentifying with, or the transcendence) of each level, which can often be quite traumatic. The only reason the self eventually accepts the *death* of its given level is that the *life* of the next higher level is even more enticing and ultimately satisfying. The self therefore disidentifies with (or de-embeds from) its present level, "dies" to an exclusive identity with that level, and identifies with (or embraces and embeds in) the life of the next higher level, until its death, too, is accepted.[38]

---

[37] Ken Wilber. *Integral Psychology* (Boston: Shambhala, 2000), pp. 197-198, charts 1a and 1b.
[38] Ken Wilber. *Integral Psychology* (Boston: Shambhala, 2000), p.36.

The above comments by Wilber fit with the way since 1980 I have described some of the fulcrum changes. Chapter 7 describes the persona mind-self and the life, death, rebirth process that led to it, pp.68-71. Chapter 12 describes the life, death, rebirth processes of transforming from childhood to adolescence, pp.117-118 and from adolescence to adolescent adulthood, pp. 120-121. Chapter 13 describes the emergence of full adulthood thus completing the differentiation of rationality to what Wilber calls "static systems/context." Wilber also describes the transition after F-4 but before starting F-5 as *mythic-rational* involving "rationalization of mythic structures" and "demythologizing."[39] Chapter 10 describes this transition in terms of the emergence of the Greek humanistic ideal of heroism that led to pre-Socratic arete and to the classical, not-yet rational democracies of classical Greece, which were the Spartan militaristic city-state, pp.101, and the Athenian direct democracy, pp.101-102. The first emergence of the third enlightenment also is the emergence on a larger social scale of the final stage of the ego (mature ego). Chapter 17, pp. 192, describes it as an order, chaos, hierarchal new order process, which began with Europeans disidentifying and thus de-embedding from a Thomistic analogical philosophy-theology that had degenerated to a sterile, mythic-rational world view. After the publication of Newton's *Principia* in 1683, more and more thinkers accepted the *death* of the medieval world view and embraced the *life* of the new world view, which was Manichaean modernism, see pp.193-194. They were enticed to do so because it enabled some to live the life of the spirit now defined as life of the mind that is totally detached from interactions with the body and the material world. More importantly, it allowed many to break away from religious and state rigid hierarchies of power. This revolution eventually led to constitutional, liberal democracy in the American colonies in the late 1700s. The resulting democratized, scientific individualism collaborated with free market capitalism that even by the mid 1790s began to produce a higher standard of living for many American citizens.

VISION LOGIC OF POSTMODERN SCIENCE

Wilber describes the transition between after F-5 and before F-6 as "pluristic systems." In the first two decades of the 20th century, scientists were confronted with several scientific world views – pluristic systems – that contradicted one another: Newton versus Einstein, relativistic mechanism; mechanism versus thermodynamics, see chapter 24, pp.262-263, quantum mechanics versus theories postulating classical atomism and the idea that measurable matter-energy is continuous, biological theory of evolution versus diverse anti-evolutionary theories such as Newton and Einstein mechanics, quantum mechanics, classical thermodynamics, classical, statistical thermodynamics, and mechanistic, ecosystem theories. By the 1950s (or so) scientists began committing to all of these theories, and introductory textbooks began to describe them. At least some scientists though fully aware of the associated contradictory world views began to commit to two or more of these theories (though no textbooks that I know of noted the contradictions). By the 1970-1980s some scientists from diverse paradigms and contradictory world views began to collaborate to solve complex problems. In this context and in conjunction with the 1960s countercultural revolution "… vision-logic … [gave] rise to the extensively elaborated versions of [dynamic] systems theory [cybernetics, chaos, social autopoiesis, complexity theories] in the natural sciences, and it would stand as well behind the

---

[39] Ken Wilber. *Integral Psychology* (Boston: Shambhala, 2000), pp.197-198, charts 1a and 1b.

postmodernist' recognition [that became explicit in the 1980s] that meaning is context-dependent and contexts are boundless."[40]

Wilber also referred to vision-logic as "constructive postmodernism"[41] that differentiated into a sequence of four sub-stages: 1. dynamic relativity pluralism that involves the capacity to take multiple perspectives; 2. unity, holism, dynamic dialecticism [that is, collaboration of Eros-chaos and Eros-order] wherein there is evidence of some sort of postconventional, universal, panoramic awareness (meta-systematic); 3. systems of systems, "paradigmatic;" and 4. systems of systems of systems, "cross-paradigmatic."[42] The systems view is that the world not only is perceived but more fundamentally is interpreted so that the universe not only manifests consciousness but is co-created by human consciousness. As described in chapter 23, p.258, normal science guided by a single paradigm and revolutionary science that transforms to a new paradigm involves integral-aperspectivism, also see chapter 24, pp.273-274, in contrast to humanities postmodernism that is totally aperspectival, that is, absolutized diversity. While systems science postmodernism avoids this humanities absurdity, it rejects introspection and participatory subjectivity that can provide subjectively true knowledge (see end of this chapter) and valid, qualitative insights.

The narrative, scientific constructivism described in chapter 24, pp.264-265, operationally defines the systems science version of the first phase of vision-logic. On the one hand, a narrative understanding of diverse scientific theories enables one not only to comprehend each of them in terms of one's personal experiences (as a result of metaphorical concepts and analogies) one can give a relativistic commitment to each of them. Alternatively, one can empathize with how someone else could believe in the world view associated with a particular scientific theory. This "capacity to take multiple perspectives" comes from the transcendental vision, see chapter 20, pp.215-216, that no one ideology is absolutely true – some are valid according to a particular set of criteria and others, again in relation to a set of criteria, are invalid or represent degrees of validity. On the other hand, a particular scientific narrative always can be reduced to a logical, conceptual theory that can be empirically tested for validity and be empirically applied to solving particular problems. This reductionism always is possible because any scientific theory is formulated in three stages: vague narrative of empirical patterns, logical conceptual model empirically tested for validity, and scientific model converted into a narrative using metaphorical, conceptual language. As a result of narrative, scientific constructivism, scientists from diverse specialties each using very esoteric language can communicate with one another because of the common, everyday language they share. Furthermore, they can use these narratives to teach their theories to non-scientific communities that may attain various levels of understanding of them. A narrative understanding enables people to interrelate "objective," valid scientific theories to objective social institutions. More importantly, people can relate abstract science and objective social aspects (what Wilber calls the *it* aspect of knowing) to subjective knowing and introspection and to shared subjective understanding via metaphors (what Wilber calls the *I* and *We* respectively of knowing).

The rationality of Socrates, Plato, and Aristotle, for all its benefits, cut humans off from the experience of soul by totally rejecting pre-Socratic arete (as noted by both Nietzsche and Heidegger though using different terms to describe this). In different ways Plotinus, Augustine,

---

[40] Ken Wilber. *Integral Psychology* (Boston: Shambhala, 2000), p.168.

[41] Ken Wilber. *Integral Psychology* (Boston: Shambhala, 2000), p.73.

[42] Ken Wilber. *Integral Psychology* (Boston: Shambhala, 2000), p. 239.

and Aquinas partially reconnected humanity to soul. The Manichaean third enlightenment in totally rejecting participatory subjectivity again cut humanity off from soul. Narrative, scientific constructivism includes the necessity of reductionism and thereby preserves the objectivity of scientific knowledge and its application to society. But by insisting on the *I* and *We* aspect of scientific knowing through narratives, this first phase of vision-logic reconnects humanity with soul. In doing this it saves the scientific community from destroying itself by facilitating personal creativity but more importantly by encouraging collaborative creativity. At the same time, it is the necessary preliminary phase leading to narrative constructivism.

Narrative constructivism, described in chapter 24, "democratization of intellectual communities," pp.276-277, operationally defines the systems science version of the second phase of vision-logic. On the one hand, a narrative understanding of perspectives associated with diverse paradigms enables one to understand a particular perspective and empathize with how someone else could commit to that perspective. This "universal, panoramic awareness involving unity, holism, and dynamic dialecticism (collaboration between Eros-chaos and Eros-order)" comes from the vision of universal individuation and creativity expressing evolutionary processes that exhibit three defining characteristics, see chapter 23, pp.255-256. This vision also includes commitment to integral-aperspectivism. (Actually the narrative understanding of evolution involving three defining characteristics implies integral-aperspectivism.) On the other hand, each particular disciplinary narrative is analogous to any scientific narrative with respect to the way it is constructed (three stages) and its empirical objectivity that enables it to be validated and to solve certain types of problems. Moreover, because of narrative constructivism, people from diverse disciplines each using very esoteric language can communicate with one another because of the everyday language they share. Furthermore, any community with specialized knowledge can use everyday language to teach their theories to non-specialists communities – that most of the time includes each of us – that may attain various levels of understanding of them.

Narrative constructivism promotes sequential dialectical collaboration between subjectivity, Eros-chaos, narratives and objectivity, Eros-order, logical models. The dialectics involves the sequence: vague narratives, logical models, higher level narratives, narratives reduced to logical models. As a result of this dialectics, narrative constructivism integrates the three kinds of knowing – the objective It, subjective I, and shared metaphorical We. This second phase of vision logic could save postmodern societies from destroying themselves, especially the US, by facilitating personal creativity but more especially by enabling collaborative creativity. At the same time, it is the necessary preliminary phase leading to the transpersonal phase of human consciousness. I believe that narrative constructivism also "covers" the third and fourth phase of vision-logic, though I confess I do not understand in detail the meaning of paradigmatic and cross-paradigmatic that characterize these phases. But according to Wilber,[43] these two phases are not necessary to comprehend in order to enter transpersonal consciousness.

---

[43] Ken Wilber. *Integral Psychology* (Boston: Shambhala, 2000), note 4 for chapter 8, pp.238-240.

NARRATIVE CONSTRUCTIVISM GENERALIZATIONS ABOUT HUMAN CREATIVITY AND INDIVIDUATION

With respect to humans, the ultimate SOURCE generates a Feeling Center that generates feeling dialogue in the new born baby. The feeling dialogue is feeling Love that, in turn, generates higher levels of individuation. This feeling Love is the result of many episodic feeling dialogues between two feeling centers, that of the baby and that of the mother. Eventually this dialogue leads to the 1½ to 2 year old child to see that its feelings associated with its body and the willful actions resulting from its feelings are different from the feelings and willful actions of its mother or mother equivalent. As a result, the child begins to episodically "see" its body expressing feelings and willful actions as a "body self" representing the child's human self at this stage of individuation. The body self is ambiguous in that it is episodically "embedded" in mother by participatory, feeling engagement (bonding) and it episodically separates from mother in an attempt to assert its autonomy. This ambiguity leads to episodic union with mother by Eros-chaos, the drive away from the order of an autonomous self, in opposition to episodic separation from mother by Eros-order, the drive to maintain the order of an autonomous self. The body self now becomes the battle field of conflict between Eros-chaos and Eros-order.

Sometime between 4 and 5 years of age most children transcend the conflict of the body self as a result of the emergence of mind that enables the body self to "see itself" participating in many different polar relationships. This body self seeing itself in polar relationships is the first instance of *self-consciousness*. Thus, the body self, as a result of the emergence of mind, is able to "see itself" as a polar-self; that is to say, the newly emerged mind of the body self defines its transformed self as the polar-self. Some developmental psychologists, such as Wilber[44], describe this sequence of human individuations as: a physical body self becoming an "emotional self" (equivalent to "feeling self") that is said to be at the archaic magical stage of development. The archaic magical stage is followed by the magical stage, which I designate as the emergence of mind producing a polar-self with magical-mythic consciousness as described by Gebser.[45]

The emergence of the polar-self is associated with several fundamental characteristics of human individuation. First, the polar-self is transcendental in that the child sees itself as embedded in many different episodic relationships. For example, the polar-self child remains in feeling oneness with mother whether the child is in harmony with her by thinking of her as the great Good Mother or the child is in disharmony with her by thinking of her as the Terrible Mother. So the child is partly embedded in each of these relationships and yet not defined by any one of them. That is, the child polar-self transcends each of these relationships. Second, this transcendental aspect leads to the child seeing itself as relatively autonomous and the unique center of all his/her relationships. Third, this first instance of a self defining itself is the beginning of *illusion*. In this first instance, the child self creates the illusion that its self is real and is the center of the universe of its experience. This is extreme egocentrism wherein the polar-self sees itself as an autonomous center of power. Further individuation will chip away at this illusion until at the highest stage of self-consciousness, the stage that Wilber and others call the causal level, where the self no longer can define itself. One "sees through" the illusions of

---

[44] K. Wilber, *Integral Psychology, Consciousness, Spirit, Psychology, Therapy* (Boston: Shambhala, 2000)

[45] Jean Gebser, *The Ever Present Origin* (Athens, OH: Ohio Univ. Press, 1953, English trans. 1985)

mind and therefore no longer can define a mind-self. This is the first step in loosing a self and "loosing one's mind." The mind does not disappear but rather remains to express a particular self or construct a newly defined self for accomplishing a particular set of tasks. This or any other mind-self that one constructs has no permanence, and it no longer will be the unique center of human consciousness. Rather it will be subordinate to the Feeling Center as the unique center that is not human consciousness but generates human consciousness.

Fourth, the first emergence of a mind-self converts the ambiguity of the body self into a creative tension between two centers of human individuality. The word, self, hereafter only should be used to represent one center of individuality called a *mind-self*. The other center of individuality, the body self, now should be called the *body ego*. There are several aspects to this creative tension. (1) The creative tension is between Eros-chaos and Eros-order. The body ego represents participatory engagement with mother or more usually with some other aspect of reality. This drive toward participatory engagement also is a drive toward chaos. It is called Eros-chaos because it opposes mind-self consciousness that involves separation from any object in order to control aspects of reality. (2) Eros-chaos leading to participatory engagement is the process leading to what may be called *participatory (no-self) consciousness* and the process generated by Eros-order leading to separation and control by mind-self may be called *control self-consciousness*. (3) Eros- chaos may be associated with either *Will-to-egoness* or *Will-to-openness* and Eros-order may be associated with either *Will-to-control* or *Will-to-power*. (4) A mutuality between the two centers of individuality – Feeling Center generating body ego and mind-self – leads to *human individual stability* and involves the collaboration between Eros-chaos, Will-to-egoness and Eros-order, Will-to-control. (5) Creative mutuality between the two centers of individuality leads to creating a new aspect of the mind-self or a transformation to a higher level mind-self and involves the collaboration between Eros-chaos, Will-to-openness and Eros-order, Will-to-power. (6) Often the two centers of individuality (body ego and mind-self) oppose one another in such a way as to lead to a mixture of mind creativity and individual instability or total individual instability. This total instability is the occasion for individual destruction – psychic death – or rebirth to a new level of self-consciousness. (I think only the mind-self evolves; the body ego remains unchanged except that it perhaps becomes progressively "reabsorbed" into the Feeling Center that "sees" that it is non-different than what Eastern mystics call the Inner Witness [that from a Christian point of view is non-different from the ultimate SOURCE – "God the Father" – manifesting Christ].)

Five, the tension between body ego and mind-self represents the tension between the ultimate SOUORCE and the mind-self. The ultimate SOURCE manifests the Feeling Center that usually is designated as *individual soul*. Individual soul is non-different from SOURCE according to the following analogy. When Don runs, his running is embedded in and thus is non-different from Don. If "running" could see it would see itself as non-different from Don, but Don also transcends running in that he is not defined by running. In like manner, individual soul is embedded in SOURCE and may eventually see itself as non-different from SOURCE, but of course, SOURCE is not defined by and thus transcends any individual soul.

Six, the polar self not only sees itself in many different polar relationships; it sees itself as being polar. The polar mind-self emerged from the ambiguity and resulting conflicts producing chaos of the body ego. The new order that emerged from this chaos is the polar mind-self that represses the body ego and regulates to some extent the feeling willfulness of the body ego. However, the body ego along with Eros-chaos conflicting with Eros-order emerged from the Feeling Center that now also must be viewed as Will that produced the "willfulness" of feelings and the Eros drive ambiguity of the body ego. The more usual term for this deep center that initiates human individuation is *soul*. On the one hand, soul is equal to the Feeling Center that is equal to Will that produces two centers of individuality in each human being, body ego and evolving mind-self expressing self-consciousness. On the other hand, the soul is the direct manifestation of an ultimate SOURCE, and in the sense already described, is non-different from the ultimate SOURCE.

Seven, individuation of a human soul is a cycle analogous to the Yin-Yang cycle of Taoism. In some manner analogous to the emergence of new possible collaborations in an evolving system in its chaos phase, the ultimate SOURCE generates a soul that may choose to prepare itself in stages to *passively be returned to ultimate SOURCE*. The Yin half of this cyclical journey consists of the stages: (1) Group-ego with mother that is a feeling, action self; (2) body ego; (3) Polar mind-self; (4) Persona mind-self; (5) Rational, autonomous mind-self; and (6) Scientific, constructivism mind-self. The Yang half of this cycle that brings the individual human soul to its SOURCE consists of six levels of human individuation: (7) *Subjective, narrative, constructivism mind-self*; (8) Empathetic mind-self with respect to one of the three mind centers (action mind center, feeling mind center and knowing mind center)[46]; (9) Empathetic mind-self with respect to two of the three mind centers; (10) Integral mind-self involving empathetic mind-self with respect to all three mind centers; (In Wilber's system: 8) is early subtle, 9) is late subtle where both #8 and #9 are associated with the path of saints and 10) is early causal, F-9, associated with path of sages[47]); (11) No-Mind-self that thereby allows the body ego to be the only center of individuality of the human individual, and (12) body ego becomes the "outer face" of the soul that drives human individuation and thus also is non-different from the Inner Witness that is the "inner face" of the soul that is non-different from the ultimate SOURCE. At this stage of human individuation the "individual" does not see where its humanness ends (or begins) and the presence of ultimate SOURCE. The human is the cycle of the ultimate SOURCE flowing downward into the individual human that is flowing upward into the ultimate SOURCE. I surmise that this flowing upward continues after biological death to greater identity with the ultimate SOURCE (from a Christian point of view this is the unknown and unknowable God the Father manifesting Christ that is *Chaos* leading to the rebirth of *Order* that is *LOVE*).

NARRATIVE CONSTRUCTIVISM PSYCHIC FRAMEWORK FOR HUMAN CREATIVITY

Dialectics of Human Creativity or Non-Creativity

I distinguish two levels of creativity. The first is *vision creativity* that generates the transition from a lower to a higher level of consciousness or is an "enlightenment" such as in science going from one paradigm to another, for example, going from Newtonian mechanics to Einstein's relativity theories. The second is *problem solving creativity*, such as what creative scientists do as guided by a particular scientific paradigm or what creative non-scientists do as guided by a particular non-scientific paradigm. *All creativity is non-ideological* in that vision

---

[46] D. Pribor, *Spiritual Constructivism: Basis for Postmodern Democracy* (Dubuque, Iowa: Kendall/Hunt Pub. Co., 2005), pp. 183-184 in conjunction with pp. 231-232.
[47] Ken Wilber. *Integral Psychology* (Boston: Shambhala, 2000), p.197, chart 1a.

creativity produces an ideology and problem solving creativity is guided by a tacit understanding of a paradigm, see, for example, "Non-logical Creativity of Science," pp.273-274, rather than following a set of rules for solving a problem. Overall there are two types of "psyches" at the personal level of consciousness: personality types disposed to express Eros-order and personality types disposed to express Eros-chaos. In figure 29.1, Eros-order personalities include: 1) sensation volcanic, 2) sensation territorial, 3) thinking territorial, and 4) thinking asthereal. Eros-chaos personalities include: 1) feeling volcanic, 2) feeling oceanic, 3) intuitive oceanic, and 4) intuitive asthereal.

Personalities disposed to express Eros-order. As described earlier, such a person can choose either Will-to-power or Will-to-control. If such a person chooses Will-to-power, then 1) if one chooses Eros-chaos, Will-to-openness, he/she will express vision creativity in a particular context; 2) if one chooses Eros-chaos, Will-to-egoness, he/she will tend to degenerate to Will-to-pleasure or Will-to-no activity and thus not express creativity. If the Eros-order personality chooses Will-to-control, then 1) if one chooses Eros-chaos, Will-to-openness, he/she will express problem solving creativity; 2) if one chooses Eros-chaos, Will-to-egoness, he/she will not express creativity.

Personalities disposed to express Eros-chaos. Such a person can choose either Will-to-openness or Will-to-egoness. If the Eros-chaos person chooses Will-to-openness, then 1) if one chooses Eros-order, Will-to-power, he/she will express vision creativity; 2) if one chooses Eros-order, Will-to-control, he/she will express problem solving creativity If the Eros-chaos person chooses Will-to-egoness, then 1) if one chooses Eros-order, Will-to-power, he/she will express problem solving creativity; 2) if one chooses Eros-order, Will-to-control, he/she will express no creativity.

Some generalizations about human creativity. When one chooses Eros-chaos, Will-to-openness associated with Eros-order, Will-to-control, this may be due to combining fear of failure with the desire to not be limited by other people's control or in a desire for "freedom from structure." In general, when one is overcome by fear of failure, he/she is unable to develop the power necessary for creativity. If a person choosing Will-to-openness also chooses Will-to-power but is unwilling to develop discipline and/or to achieve relevant objective knowledge, then Will-to-openness does not bear fruit but rather degenerates into Will-to-egoness. By not choosing discipline and acquiring appropriate knowledge, the person chooses the easier road, the path of least resistance, where Will-to-egoness becomes Will-to-pleasure.

If a human chooses to be not creative in all major areas of human endeavor, then he/she expresses a severe imbalance with dire consequences. The person who rejects creativity by a Will-to-control eventually may be overwhelmed by the suppressed Eros-chaos associated with a Will-to-egoness, which may show up as psycho-somatic illnesses or uncontrollable compulsions arising from the unconscious. The person who rejects creativity by choosing Will-to-openness in circumstances where simultaneously he/she does not develop the necessary discipline and objective knowledge to embrace Eros-order associated with Will-to-power, then he/she will be overwhelmed by Eros-chaos associated with Will-to-egoness that results in psycho-somatic illnesses or compulsions.

The creative person disposed to express Eros-order will be creative in one type of context but may be sterile in all other types of context. The creative person disposed to express Eros-chaos may be creative in one type of context but in other contexts will express Eros-chaos associated with Will-to-egoness. This leads to self-destruction or destruction of others. The extreme form of this is the artistic person who is aesthetically creative but anti-creative with respect to human individuation. Such a person may become an addict or he/she may become very depressed that sometimes leads to suicide.

Two Kinds of Creativity

In general, creativity in adults occurs in two ways. I designate the first way as *destructive creativity* that is expressed as problem solving creativity or vision creativity. It also may be called *death, rebirth creativity* that may be associated with Christ's Passion, Death, Rebirth story also metaphorically representing a universal phenomenon. The death aspect of this story is the heroic openness that allows the individual to receive Grace that can lead to the creative emergence of something new. I also designate this process as *passion creativity*. During passion, the mind-self is drawn out to some aspect of the world that separates the individual from the stability and, at least tolerance or acceptance of his/her life situation. The attraction – passion for – the new, such as, a new love mate, a new job, a new idea, simultaneously makes one dissatisfied with his/her present situation. That is, passion leads one into chaos, which then becomes the occasion for creativity or for one's intense unhappiness or even one's self-destruction. Consequently, I argue that to try to repress passion during the personal phase of individuation in order to avoid unhappiness and self-destruction is a mistake. This repression simultaneously represses all creativity, and in particular it represses further individuation unless one chooses to retreat from the world to pursue the contemplative life of the mystical path to individuation. The danger of such a retreat is that one fails to individuate to one or more stages of personal consciousness. According to the perennial Wisdom, for successful individuation no stage can be skipped. Passion lived out in the world does, indeed, produce suffering, but it also eventually may lead to the emergence of the empathetic self, which then may transform to the integral self.

Zen creativity is the second type of creativity that also may be called *Life creativity* that may be associated with the Christmas story. The birth of a baby, the Christ who will create new meaning for all humans, metaphorically represents a universal phenomenon. A new idea (Christ) emerges out of one's openness to receiving new ideas; that is, an idea (metaphorically represented by the baby Jesus) is born, that is, emerges in the still of the night (Christmas night). This new idea is a pure gift; the human receiving it did not know of its existence and therefore could not anticipate it or ask for it. This is the spontaneous creativity that occurs in children, which is a pure gift of GRACE that if acted upon leads to a creative process. Life creativity does not occur very often, if at all, during the personal phase of individuation, but it starts to occur regularly in the self expressing vision-logic moving toward trans-personal individuation. It is the predominate type of creativity for the integral self, which is stage 10 in my system. With the consolidation of the integral self in the state of conscious union with SOURCE, one no longer is disturbed by passions, but, at the same time, one may be actively creative in the world and in the process of individuation to higher levels of oneness with SOURCE. With individuation to transpersonal levels of consciousness, Zen creativity begins to displace passion, but on occasions where a passionate response is appropriate, one will express it.

Mystical, Heroic Creativity

This perspective described in chapter 13, pp. 129-131, also provides an overview of human creativity.

SUBJECTIVE, NARRATIVE, CONSTRUCTIVISM MIND-SELF

This is the first stage of the transpersonal phase of human consciousness, which according to many authors, is the *psychic self*, F-7, that is, psychic vision that leads to the path of yogis.[48] It is possible during a crisis that one may experience this level of transpersonal consciousness, but it is a different, usually prolonged phenomenon to convert such an experience into a permanent structure of consciousness. The text below describes such an experience in relation to the current trans-national movements of Alcoholics Anonymous (AA) dealing with addictions and Evangelical religious movements not confined to a particular religion – it may be found in many different fundamentalists or non-fundamentalist religious communities. An important qualification is that not all – in fact perhaps only a minority – of addicts in AA or "born again" Christians have this experience.

Overview

The subjective experience of one's personal evolution has the same three characteristics as described in chapter 23, pp.255-256, but there are five added features. <u>First</u>, the Order, Chaos, hierarchal, New Order process is better represented by a *Life, Death, Rebirth* process. Life metaphorically represents a person's state of mind that gives him/her meaning and/or ability to accomplish many if not all desired goals. When life's meaning and/or goal achievement irreversibly declines, such as loss of a loved one, divorce, loss of a job that cannot be replaced, failure at school, then Life goes to psychological Death that has both negative and positive aspects. The negative aspect, of course, is that this personal chaos is painful producing great anguish. Of course we all prefer order to the anguish of chaos; so the first challenge of personal evolution is to choose to confront our personal reality that our perspective no longer is working.

If one can confront and endure this personal suffering for an extended period with an attitude of openness to new possibilities, then the positive aspect of chaos emerges. One begins to see new possibilities that, when activated, lead to a New Order, that is, a Rebirth of psychological stability, meaning and goal achievement. The <u>Second</u> new feature is that often what enables a person to endure the mental anguish is an existential Faith leading to belief in an ultimate SOURCE identified in various ways such as Creator God or the VOID of Buddhism. The <u>Third</u> new feature is that the person in the midst of psychological Death has the courage and commitment to attempt to activate each new possibility that emerges and/or try over and over again the same possibility. The <u>Fourth</u> new feature is that the individuating person continues to have the Faith-Hope, courage, and commitment to activate a new possibility in spite of all past failures. Thus, the third and fourth new features are the conscious choosing the *trial and error* aspect of evolution. The <u>Fifth</u> new feature overlaps with the fourth feature. One must have patience combined with Faith-Hope to have courage and commitment to activate new possibilities again and again until the changing environment becomes a context in which the activated possibility becomes a success that leads to the emergence of a new order. This is a Resurrection of a person in the extended chaos of repeated failures. This Rebirth process, of course, is the *natural selection* aspect of evolution. The ultimate test of one's Faith-Hope is that

---

[48] Ken Wilber. *Integral Psychology* (Boston: Shambhala, 2000), p.197, chart 1a.

one goes to biological death with the firm belief that his/her heroic efforts will benefit humans after his/her biological death.

Some Examples

The personal evolution of an alcoholic to a recovering alcoholic dramatically exemplifies this process. The first major challenge to this process often is a major barrier to a transformation. It takes some alcoholics many years before they finally acknowledge that they have a problem they cannot control. They fear not being in control; so they delude themselves into thinking they have some control. Alternatively, a drinking pattern is a kind of perverse order that is preferred over the chaos between the old order and the transformation to a new order. Sometimes – perhaps often – it takes a flash personal enlightenment such as: "I could have died because of being drunk;" or "If I continue this drinking, I am going to die!" The next major challenge is to give up personal control or to give up blaming other people for their problem(s) or to expect some significant other to solve their problem(s). This is where alcohol anonymous (AA) is so helpful to many drunks. A group of alcoholics tell their wrenching stories of acknowledging the drinking problem and of turning themselves over to a "higher power," what I call the ultimate SOURCE, even if one has had bad experiences with one or more religions or if one does not *believe* in the existence of any higher power. This second challenge is not about belief, though religious or philosophical belief in a creator God may predispose a person to do what is necessary.

What is necessary is Faith-Hope, which is Trust. *Trust* is not belief, which is a knowledge-based attitude of the mental-self. Trust is the act of the Will to abandon reliance on the mind-self and in so doing give oneself over to that which cannot be "seen" by the mind-self. As a result of this "giving oneself over," the person enters the mystical realm of "no-self" implying "no-knowing," that eventually produces a "no-self awareness" of a *transcendental knowing*. What enables a person to carryout this humanly impossible act? After experiencing this new kind of knowing, one may be able to say that it was Grace from the ultimate SOURCE. Particular circumstances may dispose a person to be open to receive this Grace. The AA meetings for many people provide the supportive environment for one to become temporarily vulnerable and open to receive this Grace. And the wrenching stories may lead one to say: "Hey, this higher power stuff worked for these guys; I'll give it a try." Here is where the third and fourth features come in. Sometimes after only one try, which nevertheless must be repeated over and over again in the context of ongoing attendance of AA meetings, one experiences a preliminary "seeing" associated with Trust. Sometimes one must make the leap of Trust many times before this preliminary "seeing" emerges. As the preliminary "seeing" continually is experienced over time, Trust becomes more permanently established in a person. In some cases this leads to a personal belief system. However, Trust must be continually renewed because it remains radically different from any belief system.

As one continually renews Trust and shares this personal experience with others, comforting them, the experience of Resurrection from personal chaos begins to occur. Now one chooses to be in the AA community and may change his/her participation in other relevant communities. Moreover, as one becomes more enlivened by Grace, one repels some former "friends" and attracts new associations and events that support one's new life of no-drinking. Eventually one sees himself/herself as a transformed person. This is a joyous time, but *the joy poses a great danger*. The transformation to a new order is not the end of one's life story. The new order becomes the necessary possible old order because it breaks down and requires a rebirth to yet another new order. Evolution is on going. No new order is permanent; sooner rather than later it will degenerate. For the "recovered alcoholic" this means a reversion to an out-of-control drinking pattern. A born-again alcoholic is empowered, but if one does not regress to drinking again, the danger is that a person may believe he/she *has power* rather than being continuously empowered by Grace through Trust. The power comes to one through choosing no-mind-self associated with no-mind-knowing. Therefore, a person who believes his/her mind-self has power becomes a born-again fool. Such a person will tend to have an inflated self-confidence in one's abilities and not look at one's life situation in a detached, non-biased way. When the course of events go against one's wished-for expectations, one may remain in a state of denial just as he/she did before acknowledging a drinking problem. As Gdescribed in Greek and Shakespearean tragedies, the hubris (the alcoholic believing he has power) of the hero leads to his tragic downfall. If the hero is in a position of power/leadership, many others suffer the consequences of his foolish hubris. A whole nation plus millions of others may be taken down.

I am not now and I never have been an alcoholic. I am not now and I never have been an Evangelical, born-again Christian. I have experienced many major life transformations including developing hubris that required another terrible transformation. I have learned about the inner workings of evolution from these personal experiences that supplemented my scientific understanding about the external expression of evolution. All of this has embolden me to claim that, while each personal transformation is unique in being tied to an individual life history brought to a crisis in a particular set of circumstances, all such transformations exhibit a metaphorically similar pattern. This insight leads to the ironic insight that many religious people, such as some Evangelical born-again Christians, who have undergone such a transformation, reject all the ideas associated with the scientific theories of evolution. The irony is that while personally experiencing an evolutionary process, they are unable to acknowledge and understand a metaphorical, conceptual description of this process. Another irony parallel to this one is that while intelligent, rational people acknowledge and understand rational descriptions of evolutionary processes, they are unable to apply this understanding to personal crises in their own life stories or to religious people describing their own personal transformations. The key to understanding these parallel ironies is that many, if not most, people today do not embrace the necessary collaboration between metaphorical, conceptual thinking and logical, conceptual thinking, which, of course, further implies a collaboration between subjective and objective thinking.

Chapter 27
ECOLOGY OF HUMAN CREATIVITY AND KNOWLEDGE

UNIVERSAL HARMONY VIA TRANSFORMATIONS

The perspective of this book is that ecology is the expression of SOURCE, which produces harmony among diverse things and diverse potentials. Ecology of the universe includes randomness that produces disharmonies, that is, a random event is equivalent to a disharmonious event. However, the ecology of the universe transcends randomness as a result of implying a harmonious story. Not only does ecology and evolution imply one another, as indicated in chapter 23, pp.255-256, but also the disharmonious events that appear are transcended by emergent creativity that is evolution. Random events are aspects of a mini-story plot. Some changes in the harmony of a portion of the universe produce disharmony, but these random events open up many new, non-harmonious possibilities. By a process analogous to natural selection in the biological theory of evolution, some of the random possibilities come together to form a New Harmony that includes a modified version of the Old Harmony. The network of mini-story plots occurring over time results in a *GRAND STORY* in which the cosmos expresses a continually emerging, harmonious hierarchy of harmonies.

Birth Exemplifies Universal Creativity

Birth exemplifies the most fundamental feature of the unfolding story of the universe. Disharmonies occur and, when taken to be autonomous events, are *evil*, but when these evil events are understood in relation to the unfolding story, they are *transcended* by virtue of being an aspect of a mini-plot, which most fundamentally may be represented as: Harmony leading to Disharmony leading to New Harmony that includes a radically modified Old Harmony. I find it convenient to represent this mini-plot metaphorically as life, death, rebirth. With respect to this particular mini-plot, the New Harmony is a *holarchy* consisting of two levels of organization, the higher level new harmony and the lower level of the modified old harmony. In general, the unfolding story of the universe is a many-leveled holarchy. From the perspective of this book, which is consistent with a similar idea of quantum mechanics, there is no lowest level Harmony or fundamental building blocks of matter. (This paradox of the no lowest level starting point of the unfolding story of the universe is *transcended* though not rationally comprehended, by saying that the emerging cosmos holarchy simultaneously is descent of the One into the Many and the ascent of the Many to the One. These ideas of Plato and Plotinus are described by Wilber.[49]

Life. Before birth the baby is in paradise. No matter how uncomfortable the mother is because of very hot or very cold weather or because of lack of food or water, so long as she is not severely ill, the baby will be in total comfort. It literally is taking a nice warm bath in a superbly controlled environmental room. Thousands of biological servants (mechanisms) take care of all its needs. Food is brought to it and wastes are carried away without the fetus even helping in these processes. The fetus sleeps when it wants to and moves around if it so desires. The womb experience is the ultimate leisure vacation that provides escape from the hard realities of survival. It is Heaven on earth, where all is growth, development, and pleasure without any pain.

Death. Birth is death! After what seems like an eternity in which all is One, True, Good, and Beautiful, in a matter of hours, the baby is literally thrown out of paradise; its life line to mother is torn off and destroyed, and it is given the nonnegotiable command: "You're on your own, kid!" There is only one other life catastrophe that will be so dramatic and so totally nonnegotiable, irreversible, and non-modifiable so as to produce an ultimate disruption of life; that event, of course, is a person's death.

Rebirth. The birth trauma would quickly lead to biological death if an equally dramatic series of events were not immediately set in motion. After the "fall from Heaven," the baby (usually) is *reborn* into a world of dualities. Initially these dualities consist of pleasure-pain and Nature-Society, which complement one another so as to sustain the baby. Pain expressed in

---

[49] K. Wilber, *Sex, Ecology, Spirituality* (Boston: Shambhala Pub. Inc., 1995), pp. 319-344

anger or fear leads to the baby's being fed or changed or comforted by a human "mother" (male or female) who represents the nurturing of human society. Instincts and emotions representing Mother Nature take care of some needs but for the most part, nature "pushes" the infant and a mother toward bonding with one another. Once bonding is established the baby's life is sufficiently stabilized to begin to burst forth, like a bud in the first warm day of spring, into a human psychic flowering. The infant now is an embryonic ego that will spontaneously create the next most central duality of humankind: I versus everything else!

Levels of Creativity in the Human Biosphere

There are four levels of creativity in the human biosphere punctuated by two radical discontinuities. The first level is the continuation of the Big Bang origin of the cosmos that produced a hierarchy of non-life systems on earth, that is, subatomic particles, atoms, molecules, and macromolecular systems. This hierarchy of non-living systems provided the appropriate ecology for the emergence of Life, the first radical discontinuity. The second level results from creativity that produced the diverse ecologies that include non-conscious, living organisms. The third level results from creativity that produced conscious, feeling hierarchical societies, such as, conscious animal pair bonding, family units, and societies organized by feeling communications. The emergence of self-consciousness is the second radical discontinuity that led to the fourth level of creativity, which produced self-conscious, language knowing that in turn, produced human, self-conscious, hierarchical societies with shared vision, such as culture.

SEVEN CHARACTERISTICS DEFINING THE ECOLOGY OF CREATIVITY

First Characteristic

Creativity begins with a particular ecology that expresses disharmonies. A particular ecology refers to a whole system expressing a particular level of organization. Some aspects of this level of organization may be known at least in a vague way. Other aspects may not be known even in a vague way, or one may have incorrect opinions about the level of organization. Experimentation will reveal whether the system is at a higher or a lower level than one thought. If the whole system is oneself, then one may discover that he/she is at a higher or a lower level of consciousness than was initially believed to be the case.

The whole system is a unified "thing-process"; that is, one can point to a thing that is continuously in the process of changing, such as, developing or evolving. The process of developing or evolving may be represented by a story that uses metaphors. In one's continual reevaluation of the story, the metaphors may have to be removed or added to or replaced by new metaphors. Likewise, the plot of the story may have to be modified. The story should be such that it can be represented by (transformed into) a rational model. The transformation to rationality involves converting metaphors (really metaphorical concepts) into pure concepts that can be logically defined, and whenever possible operationally defined. Some concepts such as are used in philosophy or mystical disciplines, may be operationally defined in terms of one's personal experiences. This is one situation where a spiritual guide or teacher may direct a person to create for himself/herself these kinds of operational definitions. The story which uses well constructed operational definitions now can be related to criteria for validity. One may set up his/her own criteria or he/she may accept the criteria given by some community or some spiritual guide or teacher.

A whole system may express disharmonies many different ways. Expected outcomes occur "as expected," but have unexpected distorting effects. Alternatively, expected outcomes do not occur or unexpected outcomes occur along with the expected ones. Over time a seemingly harmonious system expresses internal conflicts that were always there but not observed or the system evolves or develops aspects that produce internal conflicts. For example, puberty is a time when a teenager develops internal conflicts that "push" him/her to create a new self-identity. Sometimes a system is maintained in order as a result of one part of the system dominating other parts, such as a husband dominating his wife or one ethnic group dominating another ethnic group. When the dominated or repressed aspect rebels or one way or another expresses negative symptoms stemming from being dominated or repressed, then the once stable system expresses disharmonies. Sometimes systems express disharmonies as a result of some aspect not being adequately differentiated in some way. For example, a manager or a parent may have a well-differentiated ability to make rational decisions but the individual is unable to be intimate or to be open to mindful creative dialogue, described in chapter 28.

Second Characteristic

There is the paradoxical emergence of a new semi-stable ecology. Paradoxical emergence refers to the emergence of an ecology with a new set of properties that seemingly came from nothing because they were not present in the old ecology. For example, the emergence of life from non-living matter is a circular (uroboros) paradox. We can appreciate that this can happen, but we cannot understand how it happens. Thus, instead of trying to explain how "paradoxical emergence" occurs, we say that this emergence is a creative process that transcends the circular (uroboros) paradox, see chapter 24. The process is non-rational, that is, it cannot be rationally described, and is not predictable – creativity always produces surprises. However, we can describe aspects of the process that then enables us to dispose ourselves to participating in the creative process; we may dispose ourselves to allowing the creative process to occur.

The newly emerged ecology is unstable. If it is not further modified or able to differentiate, it will disappear. The new ecology is like a new-born baby; if the baby is not taken care of while it further differentiates, it will die. A lot of new systems emerge from time to time, but only those systems that are sufficiently stable to further differentiate or are protected long enough for them to further differentiate will survive.

Third Characteristic

In becoming more stable via differentiation the new system produces internal diversity and expresses diversity with respect to its environment. For example, when the young child created for itself a masculine or feminine gender identity, it simultaneously created internal and external diversity. Internally, a male child will begin to differentiate its masculine aspects and leave undifferentiated its feminine aspects, and in so doing the child will progressively differentiate itself from female children. The same dynamics applies to a female child. Thus, *the third characteristic of creativity is differentiation that produces diversity.*

Some new aspects resulting from differentiation may be totally independent of one another. For example, the differentiation of secondary sex characteristics that occur during puberty produce some independent traits such as increased height and production of sex cells (sperm in males, eggs in females). Other traits produced at this time may complement one another, for example, characteristic fat deposit distribution and developing breasts in women. Overall each aspect in any pair interaction may enhance one or both members of the pair, but the stability and effectiveness of neither member is required for the stability of the other.

On the other hand, differentiation may produce aspects that conflict with one another. For example, when a baby who has emotionally bonded with its mother, differentiates some degree of self consciousness, it comes into conflict with its mother in some circumstances. When a hydrogen molecule breaks into two hydrogen atoms and each then loses an electron, the resulting two positive hydrogen ions repel one another. Likewise when an oxygen molecule breaks into two oxygen atoms and each atom picks up two electrons, the resulting two negative oxygen ions repel one another. In general, conflict leads to instability; we say the differentiating new ecology goes into chaos. Sometimes the instability is overcome by one aspect dominating and/or repressing the other aspect. For a time, the mother may dominate the differentiating child. The differentiating masculine ego in boys will dominate and repress their feminine aspects. Alternatively, the unstable system disappears or reverts back to its original structure. The two negative oxygen ions become one neutral oxygen molecule and the two positive hydrogen ions regain electrons and then become one neutral hydrogen molecule.

Fifth Characteristic

Conflict is transcended as a result of the two differentiated aspects "developing" *Cooperative Individuality*. For example, the conflict between the mother and the baby differentiating feeling self consciousness usually leads to their developing a new way of bonding with one another. Likewise the chaos in the population of conflicting positive hydrogen ions and conflicting negative oxygen ions is transcended by two hydrogen ions chemically bonding to one oxygen ion to form a neutral, stable water molecule. We say there is cooperation among all components of the same order of differentiation because a higher-order pattern directs the interactions (orchestrates the interactions) among the components. The new mother-child bonding directs how the baby child and mother relate to one another. Likewise, the chemical bonds between oxygen and two hydrogen ions direct how oxygen and hydrogen relate to one another. We also say that the components of the new higher-order pattern have greater individuality because by being a component of a higher pattern, each has a reduced "degree of freedom." That is to say, each has reduced possibilities, since a potential for a particular higher order has been selected over potentials for other possible higher order patterns. These components are more determined (by the higher order pattern of which they are a member) and therefore are less random. Being less random means they have greater individuality.

Sixth Characteristic

Cooperative Individuality produces a new order via *Hierarchal Incorporation*. Thus, via cooperative individuality, the mother and child have been incorporated into the higher-level pattern of a new mother-child relationship. Likewise, the two hydrogen ions and one oxygen ion have been incorporated into the higher-level pattern of a water molecule. In general, lower-level patterns are modified so as to interact to form a higher-level pattern. Some properties of this new pattern are *emergent properties*. In particular: (1) these properties only can be described in terms of the whole higher-order pattern; (2) these properties cannot be understood in terms of (reduced to) interactions among components; that is, the properties cannot be understood in a gross, mechanistic way; (3) these properties cannot (or could not) be predicted to occur from knowing the components of the new pattern; (4) the properties are the expression of immanent potentials of a set of apparently autonomous entities that interact in a context that actualizes the immanent potentials. One can know the existence of these immanent potentials only as a result of knowing their manifestation, which shows up as emergent properties. One can know the context for the expression of these immanent potentials and thus count on their expression when the appropriate context is set up, but one never can know how-why the context leads to their expression. Knowing how-why implies a mechanistic description, and the expression of emergent properties is a non-mechanistic and indeed, a non-rational process.

Seventh Characteristic

Creativity as narrative. The emerging of a new ecology is a STORY. The narrative perspective of the emergence of life exemplifies this seventh characteristic of creativity. In this perspective the cosmos may be viewed as an unfolding story of an ecology of monologues becoming progressively more complex until the emergence of life dialogues. A life dialogue is a stimulus-response process, where the idea of *stimulus* implies the idea of response and conversely. This complementarity may be represented by the uroboros, a snake eating its own tail, and thus, the emergence of a life dialogue is a uroboros Paradox. We see the same sort of paradox in philosophy with respect to the "thing-process" or "thing-event" complementarity. A thing has "meaning," that is, is knowable, only in terms of a process, that is, an observable event which is an interaction between two things. Likewise, process has "meaning" only in relation to things. In like manner, the stimulus-response uroboros paradox may be stated as follows: A stimulus has occurred only if a response has occurred, but a response has occurred only if a stimulus has occurred.

PARTNERSHIP OF NATURE AND HUMANITY

Before the emergence of humanity, Nature (the universe) was exclusively the individuation process produced by the Eros drives coming from SOURCE. Within this process all individual entities expressing all the levels of individuation were totally subordinated to the overall universal evolutionary process of individuation. With the emergence of humans, SOURCE manifested as Will that initiated human self-consciousness for its own sake rather than be subordinate to universal evolution. In accordance with the trans-systems theory of creativity, this conflict of Wills, that is, Will of nature versus Will in each human individual, led to a creative partnership between Nature and humanity that transcends the conflict. At first this partnership produced participatory knowing in humans at the magic stage. The insights from participatory knowing guided the group-ego of magic humans to differentiate control knowing. This, in turn, led humans to separate more and more from nature and correspondingly to replace some participatory knowing with objective knowing of Nature. That is, the creative partnership between Nature and group ego of magic humans became a dialogue in which Nature stimulates – *talks to* – humans who respond – *talk back* – by creating an objectification of the stimulus by means of language representations leading to actions on nature. This dialogue is analogous to the stimulus-response, organismic dialogue characteristic of any living system. However, the

Nature-Human "knowledge dialogue" is radically different in that it produces a new reality, which we may call the *Knowledge-Universe*. The Knowledge-Universe is identical to interconnecting subject-object polarities. The object of each polarity is some "known" aspect of Nature, wherein Nature now includes an individual or a group of individual humans. The subject of each polarity is a society of self-conscious humans. Thus, each group-ego of magic humans or later each individual self-conscious human lives in two worlds: Nature and Knowledge-Universe. Both worlds are continually changing and evolving, but in this postmodern age, the Knowledge-Universe is changing and evolving much more rapidly than Nature.

EVOLUTION OF THE KNOWLEDGE UNIVERSE

The Knowledge-Universe co-emerged with the first humans who manifested the magic structure of a group-Ego. Individuation of the Knowledge-Universe corresponds to the individuation of self-consciousness, and each of these two evolutionary paths continually, reciprocally influence one another. As partially described in chapter 26, p.293, it is convenient to distinguish 10 of 12 levels of individuation of self: (1) Feeling, Action Self, (2) Body Ego(Self), (3) Polar Self, (4) Persona Self (civilized self), (5) Rational, Autonomous Self, (6) Scientific, Constructivism Self (scientific, autonomous self), (7) Subjective, Narrative, Constructivism Self, (8) and (9) Empathetic Self with respect to one or two of the three mind centers and (10) Integral Self involving all three mind centers.

Each type of self corresponds to a language-mental structure which at first manifests a creative aspect and later manifests a degenerative aspect. The creative aspect leads to a differentiation of the Knowledge-Universe with characteristics corresponding to the type of self that differentiated it. The degenerative aspect becomes manifest when the Knowledge-Universe becomes static and therefore resistant to further transformations. That is, it may continue to differentiate horizontally but resists transformations to a higher level of internal organization and intensity that transcends conflicting knowledge components. Instead of manifesting transformational creativity the Knowledge-Universe becomes the basis for control and unbalanced power over nature and over humans in any society. This degenerate phase may lead some individuals to regress to a lower level of individuation of the self. The degenerate phase also produces great chaos that may lead to the extinction of the social culture living through this degenerative process, for example, the fall of the Roman Empire. However, the chaos may be the occasion for the degenerating culture or for an "embryonic new culture" embedded in the degenerating culture to "self-organize" into a new level of social self and a corresponding new level of the Knowledge-Universe.

Each level of individuation is a *holarchy*. That is, on the one hand, each level emerges from the chaos of the immediate lower level structure. The way this happens is that the chaotic lower level "self-organizes" to produce a higher, level structure that transcends the conflict that produced the chaos in the lower level. On the other hand, each emerging higher level structure will or could go into chaos and then self-organize to be incorporated into a still higher level structure.

The *action self of a group-Ego* produces the magic structure of human consciousness. For individuals immersed in this structure, that is, the first humans and more advanced humans who regress to the magic structure, each human is egoless because he/she participates in a group-Ego that is in union with Nature via participatory consciousness. Each individual is aware of itself having episodic experiences (existential awareness) that totally engage him/her with some thing or event in Nature. These point-like, episodic engagements are spaceless and timeless because the individual has no awareness of a whole persisting over time that locates these experiences in space and in time. The magic human no longer identified with nature, that is, no longer subordinate to the evolution of nature, is driven to survive for himself/herself. This drive for survival leads to individuals participating in magic learning cycles that produce primitive technology, progressive differentiation of control consciousness, progressive separation of the group-Ego from nature, and the progressive separation of individual self-consciousness episodic experiences from the group-Ego. These progressive changes are the creative aspects of magic humans that lead to the emergence of a body ego, see chapter 4, and then to the emergence of individual self-consciousness, the first emergence of the Mental Self as described in chapter 5. The goal of the magic learning cycles is for greater control over and separation from nature. Magic is a "doing without individual self-consciousness," but as individual self-consciousness does emerge, the individual may continue to exert magical power for the sake of greater control for its own sake rather than for the further individuation of self-consciousness. This is the degenerate aspect of magic that tends to cause one to lose the newly won individual self-consciousness. Postmodern science totally devoid of participatory, subjective vision is in one sense a degeneration to utilitarian magic with a loss of subjectivity of self-consciousness.

The *polar self* produces the mythic structure of human consciousness. Each mythic human is aware of itself experiencing Nature and representing it via metaphors organized into stories. The archetypes of the Collective Non-conscious guide individuals to *imagine* his/her polar interactions with nature. The individual as yet without a concept of time nevertheless has a imaginative sense of time that leads him/her to represent events in nature as narratives. These narratives locate humans in polar relationships. First and foremost is the individual's inner world in which the Feeling Center in union with SOURCE silently manifests images that are elicited by and mirror things and events in the individual's outer world. The poetic storytellers objectify this inner silence with verse that in a narrative way locate humans in polar relationships in nature. Thus, besides the primacy of inner self versus outer self, mythical humans describe nature in terms of polarities such as heaven-hell and sky-earth. Also, the individual life span is the mythical journey involving the polarities of life-death and death-rebirth. These polarities, in turn, metaphorically represent the polarities of happiness & joy that breaks down into suffering & misery, and the suffering & misery for those who submit to it and endure it lead to the triumphant joy of a transformation. The Life, Death, and Rebirth cycles celebrated in Nature's cycles that mirror human transformations are the creative aspects of mythical humans. The symbol for mythical consciousness is the circle or the uroboros – snake eating its tail that represents some whole such as any living system that does not indicate that from which it came. The uroboros also symbolizes the degenerative aspect of mythical consciousness in that it is self-contained and thereby resists transformations to linear, conceptual consciousness.

The *persona self* co-emerges with the differentiation of mythical thinking into metaphorical, conceptual thinking. As described in chapter 7, metaphorical, conceptual thinking organized into a grand literal myth unites many closely related but conflicting tribes into a single, hierarchal, patriarchal civilization. The literal myth converts the masculine-feminine polarity into the over-arching persona duality: masculine versus feminine. This duality specifies many other personas as well as rules of conduct that determine how each civilized member should relate to one another, to society as a whole, and to its gods, which later are subordinated to or replaced by a monotheistic male God. The idea of persona enables conflicting individuals to individuate to a level in which they can cooperate with one another to carry out a team effort specified by a team leader. The mass commitment to the grand, literal myth brings about the cohesiveness of the team and the subordination of team members to the leader. Many great things are accomplished, such as, the irrigation systems and the pyramids of Egypt, by the creative vision of great leaders who command the team efforts of large masses of humans. Thus, the creativity of persona self is located in each human in society individuating to a specified persona and to a few visionary individuals who are the leaders of society or who guide the leaders of society to command team effort. This arrangement gives great power to leaders in civilization and to the civilization that manifest team effort guided by individual creativity. A civilization degenerates when its leaders seek power for themselves and for their society rather than individuating to a higher level of creativity. This higher creativity, of course, would include providing for a greater number of citizens to individuate to having a personality that incorporates persona rather than stand in place of personality.

Rational, autonomous self emerges when the *persona self* differentiates metaphorical, conceptual thinking into logical, conceptual thinking. This logical conceptual thinking enables the self-aware individual to see that society forces him/her to adopt a *persona*; that is, the socialized individual is able to stand back and conceptually objectify itself as a member of society who is rewarded for adopting the appropriate persona and punished to various degrees corresponding to the extend one does not adopt the appropriate persona. Likewise the rational self conceptually objectifies the meaning of socially prescribed roles and codes of behavior, internalizes them, and then chooses to abide or not abide by them. Thus, such an individual no longer identifies with a socially prescribed persona but rather identifies with a rational, autonomous self that chooses to what extent it will adopt a persona and abide by the socially prescribed roles and rules of behavior. As a result, the rational, autonomous self manifests the Will aspect of the Feeling Center, that is, it makes choices rather than blindly following the dictates of society. This level of differentiation of self-consciousness provides the possibility of developing an individual *personality* that can individuate from the conventional morality of a persona self to the post-conventional morality of a personal self. Undifferentiated persons choose conventional morality, but individuating persons choose to create knowledge for themselves. This knowledge opens up possibilities for free choice and guides one in making these choices. Thus, creativity of the personal self (= rational, autonomous self) expands and intensifies the Knowledge-Universe that flows into all aspects of a society's culture. A personal self degenerates when it chooses to identify with a persona self; this tends to happen when a particular persona, such as a priest, a high level administrator, or a father as head of a family, gives a person a high status and degree of power in a society. When this happens, social status and personal power displace the intent to be creative and to further individuate.

The emergence of the *scientific autonomous self*, described in chapters 16 and 17, produced a revolutionary change in the knowledge universe – hence this third enlightenment is called a scientific revolution. Even though the "fathers of this revolution," Descartes, Galileo, and Newton, thought they could create absolutely true theories, nevertheless, they realized that they *constructed* knowledge that had to be *empirically* validated involving *consensual criteria*. This is radically different from – a revolt from – believing some "divinely inspired person" or Bible or church hierarchy proclaiming how the universe works. Humans at this stage of individuation are explicit enigmas. These knowledge revolutionaries incorporated the Aristotelian-Thomistic commitment to logical, conceptual thinking; the mathematics of their models epitomized logical thinking. Yet, they chose to commit to the self-evident idea that there are no self-evident ideas other than the ones they constructed as validated mathematical formalisms. They came to this insight by means of introspection and by an intense, participatory, subjective engagement with paradoxes of understanding nature and introspection. Then, having apparently resolved – actually transcended – some of these paradoxes (such as Zeno's paradoxes about motion or the mathematical absurdity of a speed at a point in space), they insist that hereafter introspection and participatory subjectivity are forbidden paths to knowledge. They were like a man who for a long time and through many trials pursues a woman, but then once gaining her favor, he no longer "loves" her but only wants to control her. Descartes epitomizes this absurdity by his intense feeling inspiration, see Chapter 17, pp.191-192, that empirical patterns can be represented by mathematical formalisms. Having obtained a profound insight he rejects all feeling insights even though one such intuition got him the insight in the first place. Until the mid- 1800s scientists *forgot* that metaphorical, conceptual thinking enabled Newton to formulate fundamental ideas of the calculus (finding the area under a curve, that is, calculating a definite integral) and the universal theory of gravity, see chapter 17, pp.189-190. This profound "memory lapse" led many scientists and most science teachers to forgetting that all scientific theories start out as vague intuitions often represented as stories – scientific myths.

I speculate that the seeds of postmodernism were planted when mathematicians in the 1800s developed geometries radically different from that of Euclid.. One geometry assumed that two parallel straight lines continually diverge as they "approach infinity." The other geometry assumed that two parallel straight lines continually converge as they "approach infinity." Then! Einstein's general theory of relativity uses one of these non-Euclidian geometries to describe the universe, and this description is more general and in some sense more valid than Newton's theory (because it solves more problems than Newton's theory can and it can be reduced – simplified – to Newton's theory.). So much for Augustine's or Descartes' immediately evident truths about parallel straight lines!

In the mid to late 20th century, as described in chapter 23, pp.259-262, the increasing explicit relativism of constructed knowing collaborated with free market capitalism and its associated materialism to produce the "materialistic, circular, reinforced pragmatism" that generated a mind-self attitude of dismissing or worse yet not being open to any ideology. As this attitude progressively overtook American college education, it led to the emergence to various degrees, especially in the educated elite, of *radical ego constructivism*, see chapter 22, p.226. With this ultimate, nihilistic individualism a person chooses his/her self-validating ego (mind-self) to replace any higher power such as SOURCE (God) or the US Constitution. It is ultimate nihilism in that no society/culture can survive its persistence. The economic crisis of September, 2008, is but one explicit manifestation of this nihilism that points to the absolute necessity for the emergence on an extensive social basis of a new kind of individualism. The vision-logic of science prepares for the emergence of this new mind-self by a transcendental vision that guides the evolution of normal science first to narrative, scientific constructivism and then to narrative constructivism as described in chapter 24. This "transitional mind-self" individuating toward transpersonal consciousness will manage its own collaboration between participatory subjectivity involving Eros-chaos and control objectivity involving Eros-order.

The *subjective, narrative, constructivism self*, described at the end of chapter 26, is the first level of transpersonal consciousness. This "psychic self" begins to see the universe and one's self in it in new ways – natural mysticism, but its core trait is not any kind of mystical vision. Rather, the core trait is the subordination of the mind-self to the soul now vaguely seen as a direct manifestation of SOURCE. But this "seeing" is associated with the mind-self subordinated by the transcendental Faith-Hope received by the will aspect of soul. One differentiates over time the Faith-Hope that subordinates the mind-self to SOURCE, but one cannot work at making this differentiation happen. One only can dispose one's self to receive this Grace. If, why, and when such Grace is received remains a mystery. However, various kinds of meditation that lead to mindfulness – to be described in the last section of this chapter – have proven to be , not the cause, but the occasion for the Grace to be received and developed. As I experienced this first level of the transpersonal, one's belief systems are ruthlessly deconstructed.

For example, I had a dream during the spring semester of my freshman year in college in 1951. I had this dream only once and it did not involve my seeing anything (in my imagination) or hearing anything. Rather, it was a powerful feeling that I was called to the impossible task of integrating science and Thomistic philosophy. The philosophy aspect did not bother me because I already had a deep understanding of the Thomistic analogy of being and many of the ideas that followed from that insight. But though I studied 4-5 hours many days while in high school (military prep school), I only achieved a C+ average at graduation. I was granted a swimming scholarship from Kenyon College that was revoked because my college entrance exam scores were so low. I never did well in math or science courses; I especially hated biology, and I often became sick to my stomach for psychological reasons in the laboratory period of a science course. Nevertheless, I immediately changed from a pre-medical major to a chemistry major. I somehow managed to get a D- in chemistry, a C- in physics, and C in analytical geometry that enabled me to avoid flunking out of college. I changed my major to philosophy and managed to graduate with this major in 1954.

I taught chemistry discussion sessions for a course in introduction to chemistry at the University of Detroit and then somehow got into medical school. I flunked out after my second year; spent a year getting a teaching certificate at Wayne State University, and taught high school geometry and algebra for one year, which caused me great anguish. I almost flunked out of graduate school two times but managed to get a PhD in cell biology in 1964. I hated laboratory research but managed to be promoted to full professor based on my research in biophysics, grants, and publications and teaching advanced biology courses. And from there the failures and disappointments and small successes go on and on. I have gone through three divorces and am no on my fourth marriage of 21years (July, 2009). The point is that at each failure one has the opportunity to choose despair or mindful commitment to the ultimate SOURCE. The mindful aspect involves being open to new possibilities and then being "reborn" as one activates and integrates some of these possibilities. Eventually over time – in my case, a long time – one evolves to higher levels of transpersonal consciousness and the Inner Witness begins to take over one's life. The Faith-Hope one differentiates at this psychic level enables one to experience openness to new possibilities with respect to each of the mind centers. These experiences eventually coalesce into what I call the *empathetic self*.

Chapters 24 and 26 describe the foundation from which the *empathetic self* can differentiate. This new level of self-consciousness is associated first of all with the awareness of the co-equal, complementary types of personal self, as, for example, represented by Sufism described in chapter 14, p.152 and Table 14.1: control individualism in which rational, control objectivity predominates and participatory individualism in which feeling, participatory subjectivity predominates. Secondly, the new awareness depends upon acknowledging the paradoxes of human individuation. For example, as one individuates to the persona (civilized) self, paradoxically one is both better and worse off. On the one hand, the civilized person is ready to individuate to the next level of self-consciousness. On the other hand, the individual who identifies with personas is cut off from the Inner Self that generates all creativity and individuation. Thirdly, the new level of self depends upon acknowledging diverse personality types, see figure 29.1. And finally, the emergence of the empathetic self requires some overview understanding of the process of human individuation. The core of empathetic self is that each individual develops both types of personal self, that is, control objectivity and participatory subjectivity, in a way that each type complements rather than opposes the other as is the case for the rational or the Manichaean version of the scientific autonomous self. This complementarity leads to the ethic of mutuality of subjective and objective knowing as promoted by narrative constructivism, see chapter 24. As a social phenomenon, the empathetic self understood as an emotional, intelligent self, only recently is beginning to emerge and differentiate to create a post-patriarchal, Knowledge-Universe that will restructure postmodern societies to trans-patriarchal societies. The degenerative aspects of this new self-awareness are not yet apparent.

The *integral self*, as a social phenomena, is described by Jean Gebser[50] and Ken Wilber.[51] The integral self is a phase – according to some writers, the highest phase – in trans-personal and therefore mystical individuation. In the Western, Christian tradition it is described in the writings of St. Teresa of Avila and St. John of the Cross as the unitive state in which one is aware of coming into union with or becoming aware of the already present union with God. All the mystical traditions – Eastern and Western – describe in different ways this spiritual phenomenon. A modern description of it, which I read (2001-2002 and re-read in 2006) and which influenced my description of the Feeling Center in union with the SOURCE, is in three books by Bernadette Roberts.[52] [53] [54] Bernadette Roberts describes her own transition to this state in relation to the writings of St. John of the Cross who describes this journey as the dark night of the soul. I suggest that a postmodern experience of this journey would involve several "dark nights of the soul," that taken together would transcend all the internal conflicts & double binds of human individuation summarized at the end of chapter 7, pp.71-76. Thus, the integral self creates and continually maintains by means of the creative energy of SOURCE the mutuality between the body ego and the mind self, the mutuality between individual and collective social consciousness, the mutuality among the Thinking mind center, the Action mind center and the Expressive mind center, and the mutuality between Eros-chaos and Eros-order. The degenerative aspect would be to take undue pleasure in and therefore rest in the great power of the integral self rather than individuate to the state of no-self and realize or attain identity with God or Brahman or whatever one calls ultimate Reality.

HUMAN COMMUNICATIONS

Communication is a *source* and a *receptor interaction* that generates *meaning* that is specified by intrinsic and extrinsic factors. The intrinsic factors are **LOGOS** that is the potential between the

---

[50] Jean Gebser. *The Ever-Present Origin* (

[51] Ken Wilber. *Integral Psychology* (Boston: Shambhala, 2000)

[52] Bernadette Roberts. *What It Self* (Austin, Texas: Mary Botsform Goens, 1989)

[53] Bernadette Roberts. *The Path to No-Self* (Albany, New York: State Univ. of New York Press, 1991)

[54] Bernadette Roberts. *The Experience of No-Self* (Albany, New York: State Univ. of New York Press, 1993)

source and the receptor and **EROS** that is the drive (or Will) to express meaning. The extrinsic factors are the context of the communication that may vary from indefinitely small such as the context of a contact force communication at a point in space to the totality of finite reality in which the communication is embedded.

The source and the receptor are "things" (substances) which interact (process). Thus, communication involves complementarity in two ways: 1) the source and receptor imply one another; neither has meaning in itself, and 2) the thingness and process aspects of communication imply one another; neither thingness nor process has independent metaphysical, ontological, or epistemological status. The directional process of interaction from the source to the receptor is the *meaning* of a communication, but this meaning is not some single, autonomous form. The meaning varies depending on context, which specifies relevant other interconnected communications (events). Theoretically, each event is interconnected to all other past, present, and future events, but any event understood as a communication has a context specified by some finite (usually small) subset of all possible events. In particular, a communication is "located" in some "place" in the space-time continuum, and context specifies other relevant places. Likewise a communication happens at some "moment" in space-time, and context specifies that the moment has direction in that with respect to our experience of space-time, the moment of this particular communication is a *now* which has a past and a future. Context also specifies relevant moments of the past and possible relevant moments of the future.

The meaning of a communication is further specified as being associated with a nameable event or with a non-nameable event. A *nameable event* is one that can be classified or in some way can be associated with a categorical framework, and, of course, a *non-nameable event* is one that cannot be classified or associated with any categorical framework. For example, a young child who is aware of himself/herself walking perceives this phenomenon as a non-nameable event because he is unable to classify it due to his lack of language. The same individual a few years later after learning to speak may identify an analogous phenomenon as walking. The meaning associated with a non-nameable event involves an intuitive, non-rational perception of it. That is, though the non-nameable event does not fit into any classification scheme of the knower, the event is perceptually understood in a non-rational way. The understanding of a non-information communication involves concrete, non-conceptual knowing called *existential knowing*. For example, the score for a piece of music, that is, the notes written down on a page, is a symbolic representation of the piece that tells anyone familiar with this notation what and how to play the music by means of a particular instrument. However, the individual interpretation of the artist playing the music is an individual, concrete, non-rational understanding of the music. Both the individual interpretation and the playing of an instrument are examples of concrete, non-conceptual knowing referred to as *existential knowing*. Likewise, each individual that listens to this interpretation creates his/her own subjective, concrete, non-rational understanding of the music. Thus, the exchange between the artist and the listener is a non-rational communication; music in this context is an event that is not nameable but nevertheless is perceptually understood in a non-rational (non-conceptual) way. Moreover, this music event like other existential, tacitly understood phenomena such as knowing how to ride a bike, will never become a nameable event. The riding of a bike that any two or more people can observe is a nameable event, but the individual riding the bike is able to do so because of his/her subjective understanding of how to do so. This subjective understanding is not an object that can be observed and named; only the outcome – the manifestation – of the subjective understanding, the riding of the bike, can be observed and named.

Meaning associated with a nameable event is called *information*, and a communication whose meaning is a piece of information is said to transfer that information. Any piece of information does not have independent status; it always is a meaning associated with a particular communication. Moreover, information may be either *operationally definable* or *non-operationally definable*. In science the most usual type of operational definitions involve measurement processes. For example, momentum is defined by the process of measuring mass, the process of measuring velocity, and the mathematical process of multiplying the measured quantity of mass by the measured quantity of velocity, that is, momentum = mass x velocity ($p = mv$). The meaning contained in the idea of a banana cake is qualitative and cannot be represented by a quantity that can be measured. However, we can operationally define this idea by the recipe for making a banana cake. Any two people who correctly follow this recipe will be able to have approximately the same sensations of seeing, tasting, smelling, and touching the cake. Thus, by this operational definition – the recipe – and the sensations they have of the product of the recipe will give each person the meaning of "banana cake." Likewise, the meaning contained in the idea of a tree is qualitative and cannot be defined by a measuring process. But, based on the consensus of humans formulating a language, we can define a tree operationally. We direct a person to sense, for example, see, an object we know to be a tree. After the person has followed our directions, he/she will have some idea of the meaning of "tree." Thus, operationally definable information (operationally definable meaning) is obtained by measurement or a recipe (a procedure) or associating a particular word with some sensation like seeing a tree. Any classification scheme that does not involve operationally definable ideas represents *non-operationally definable information*, as for example, ideas in philosophy such as God, moral good, beauty, etc.

In summary, there are two types of communication. 1) *information communication* in which the meaning is associated with a nameable event, is objective implying that there is a way of achieving a consensus about its meaning, and is rational, that is, the meaning is defined by a classification scheme or a system of categories. 2) *non-information communication* in which the meaning is associated with a non-nameable event, is subjective implying that there is no way of achieving a consensus about its meaning, and is non-rational (not irrational), that is, the meaning is understood by direct insight or intuition such as, inspiration underlying creativity or tacit knowing such as knowing how to walk. Information communication involves ideas whose meaning is either *operationally definable* by a set of directions or by some consensually agreed upon categorical system as was the case when humans created language or non-operationally definable language involving metaphorical concepts.

INNER WITNESS AND MINDFULNESS

The emergence of self-consciousness produces the polarity of a consciousness that is conscious of a self (initially a group-Ego) that defines itself in accordance to the way it interacts with reality. Thus, the polarity may be represented as Consciousness – Self that defines itself; let us call it "self-definition self." The "Consciousness" of this polarity is the Inner Witness that is manifested by SOURCE and the other pole is the self that defines itself. This polarity manifests a hierarchy of levels of individuation of self-consciousness. Only with the emergence of the rational, autonomous self does the self-definition self begin to become aware of the Inner Witness. The emergence of the scientific, autonomous self produces paradoxical results. On the one hand, this self proposes that all knowing interactions with reality are invalid except for those that create valid scientific theories. This proposal could lead one to realize that the self-definition self also only is a relatively valid construction that is energized by the totally subjective Inner Witness that emerges from SOURCE. On the other hand, this self's rejection of all subjectivity not leading to producing scientific theories cuts one off from even acknowledging an Inner Witness, let alone a Feeling Center in union with SOURCE. Postmodernism expands scientific positivism to all types of knowing. Postmodernism proclaims that all knowing leads to a construct, such as a theory or a work of art, that only is valid, never True, in a particular context. In particular, postmodernism undermines the ideas of persona and maturity as defined by traditional world religions or philosophies. This attitude would be even more conducive for the postmodern thinker to deconstruct all definitions of self and then acknowledge the Inner Witness that observes both the defining of self and its deconstruction. However, in absolutizing diversity the postmodern thinker paradoxically absolutizes his/her non-definable, autonomous, subjective self that undermines all knowing and awareness, including the subjective awareness of an Inner Witness.

As noted earlier, the Feeling Center is the source of existential, feeling insights that in some humans congeal into various types of knowledge. Thus, the Feeling center(soul) and these feeling insights are prior to all knowing. Likewise, the Feeling Center as Will also is prior to all knowing, in that Will determines what aspects of self-conscious awareness one chooses to acknowledge. Scientific positivism acknowledges only those subjective insights that lead to the formulation of valid scientific theories. Postmodernism acknowledges one's subjective awareness as an absolute that rejects any absolute source for this awareness as well as relativizing all knowledge frameworks. However, Will is prior to all knowing in such a way that one can acknowledge an Inner Witness that is prior to and more fundamental that any mental construction. The postmodern stress of inner conflict and outer chaos has brought many and can bring many others to this same two-fold act of Will: reject one's subjective awareness as autonomous and self-explanatory and then acknowledge an Inner Witness. When a person is able to do this on a regular basis, he/she is said to be *mindful*. The mindful person will come to acknowledge an absolute source of this acknowledged experience of an Inner Witness. For St. Augustine this was the basis for his absolute commitment to belief in God. This same mindfulness is the basis of my commitment to SOURCE that manifests me and everything else that is. The emergence of the third enlightenment and its evolution to postmodernism has made this habit of mindfulness available to many humans rather than just to a few mystics. All the postmodern person needs to do is deconstruct the autonomy and absolute independence of his/her personal subjectivity. The stress of suffering and failure facilitates this process so that eventually one, out of desperation, "falls" into mindfulness. Such a transformation is the doorway to transpersonal individuation.

With the emergence of the psychic self that leads to the empathetic self one begins to acknowledge or to display behaviors equivalent to acknowledging the Inner Witness. In particular, the empathetic self will have developed mindfulness with respect to one of the objective mind centers, that is, the thinking or action mind center, and to the expressive mind center. Then the mind-self operating in the realm of each of these mind centers, will be displaced and subordinate to the Inner Witness. For example, a mindful person using the thinking mind center will be detached from his/her own points of view and thus be able to attend to, without necessarily accepting, opposing points of view. This detachment from the self clinging to a particular point of view will enable a person to see his/her own ideas more clearly and be open to modifying them so as to incorporate new insights that come from the "outside" or from the Individual, Collective Unconscious. In like manner, the mindful person using the expressive mind center will be detached from his/her own desires to be loved and to express one's feelings, that is, one will be detached from the ethic of self-fulfillment and thus be able to attend to the needs of others without necessarily feeling obliged to satisfy those needs. Because of this detachment, such a person will experience greater oneness with nature and with other humans, because the Inner Witness has displaced mind-self-interest.

However, if a third mind center, which in this case would be the repressed mind center, has not been differentiated and then transformed to mindfulness, then there will be a war between the mindful mind centers and the non-mindful mind center. The whole of self-consciousness still is fragmented by having competing centers of consciousness: the Inner Witness versus a mind center dominated by an autonomous mind-self. This may go on for some time wherein the person lives with the conflict and makes a stable truce with the rebellious autonomous mind center. However, the SOURCE of all may give such a person a terrible gift in which that aspect of the person that identifies with the rebellious mind center is deconstructed. Western mystical traditions, for example, St. John of the Cross, describe this painful deconstruction as the "dark night of the [personal] soul." I suggest that the deconstruction of the third repressed mind center also is a dark night of the soul. As with the other two "dark nights," this deconstruction also leads to the transformation to mindfulness. However, the differentiation and then deconstruction of the repressed mind center is more terrible and painful than the deconstruction of the other two centers.

The dynamics of getting through this dark night is the creative process of individuation as described in chapter 26. To some extent this third dark night requires an even greater *active passivity* in which one gives himself/herself over to the SOURCE of all until one is purified of Egoness so that now all the mind centers are mindful. When this occurs, the empathetic self becomes an integral self in which the Feeling Center equal to the Will center equal to the Collective Non-Conscious in union with SOURCE manifests the Inner Witness that is the center of the individuated self-consciousness. I suggest that other descriptions of the dark night experience enumerated stages of the process of transformation of the personal consciousness as a whole rather than in terms of a sequential transformation of mind centers. These traditional descriptions were suited for persons who chose to separate from the world and devote themselves to mysticism and contemplation. My descriptions may be more suitable for persons who remain in the world and constantly are brought down to be "of the world." Nevertheless, in seeking the highest level of individuation that the "SOURCE" may grant them, each heroic individual may approach or actually reach the integral self or beyond.

With this understanding of transformations to mindfulness it is possible to amplify the description of the process of transforming into the transpersonal realm. In this process one transcends the opposition between participatory subjectivity and control objectivity by incorporating these two types of consciousness into this Inner Witness. The Inner Witness is transpersonal, is both subjective and objective and simultaneously provides participation in and control of nature. I now add that each mind center independent of the other two may move into the transpersonal realm wherein the Inner Witness rather than the mind-self is the center of consciousness that orchestrates the activities of that mind center. The Inner Witness operating through the Thinking mind center is separate from but participates in this center carrying out the thinking, vision, narrative process. That is, the Inner Witness attends to participatory subjectivity that generates existential insights appropriate to the realm of thinking. The Inner Witness also attends to control objectivity of creating theories that can be validated in some way.

In like manner, the Inner Witness operating through the Action mind center is separate from, but participates in this center carrying out the action, vision, narrative process. That is, the Inner Witness attends to participatory subjectivity that generates existential insights appropriate to the realm of action. The Inner Witness also attends to control objectivity of creating action programs or strategies for accomplishing an operationally defined goal so that the program or strategy can be validated. Finally, the Inner Witness operating through the Expressive mind center is separate from, but participates in this center carrying out the aesthetic or inter-personal, vision, narrative process. That is, the Inner Witness attends to participatory subjectivity that generates existential insights appropriate to the realm of aesthetics or interpersonal interactions. The Inner Witness also attends to control objectivity, that is, objectifying an insight in such a way that it can be validated by subjective criteria or objective consensual criteria of validity.

In the person where all three mind centers are mindful, the Inner Witness participates in a dialogue between two or among all three of the centers. This now is possible because in such an integral self there is no conflict between the Inner Witness as a center of consciousness and an mind-self as a center of consciousness for any of the mind centers. Furthermore, each mind center is in harmony with the body self, thus also avoiding conflicts between aspects of the body ego that include instincts, sex drive, and compulsions, and the Inner Witness operating through the mind self. The Inner Witness is both trans-human, that is, is a direct, conscious manifestation of the Feeling Center in union with SOURCE, and is human. The human side is that the Inner Witness only operates through the mind self in dialogue with the body ego. The body aspect of the body ego provides a set of limitations associated with genetic traits, health, aging, and the eventual death of the body. The mind-self provides another set of limitations associated with the social and cultural context in which the mind-self differentiates. Some limitations are associated with the way and the extent each mind center of the mind self has differentiated. All these limitations pose the question of what happens to the Inner Witness and to the integral self after the body dies. Answers to these questions come from traditions that describe trans-personal individuation beyond the fully developed integral self.

Chapter 28
CREATIVE, MINDFUL, PARTICIPATORY DIALOGUE

MINDFUL, PARTICIPATORY DIALOGUE

Individuation to Mindful, Participatory Dialogue

Both non-mindful and mindful collaborative dialogue are core aspects of the patriarchal perspective. As such, both types of dialogue are familiar to many people, and both types tend to overshadow and thus obscure creative, mindful, participatory dialogue. Creative mindful dialogue builds upon an understanding of mindful, participatory dialogue. Thus, it is helpful to discuss in greater detail the nature of this type of dialogue and how it became more prevalent in postmodern U.S. society. In 1982 I wrote an article, never published, that described how humans need to evolve to a higher level of consciousness that facilitates intimate conversation. In rereading this article in the context of writing the book: *Spiritual Constructivism Basis for Postmodern Democracy*, published 2005, I came to three new insights. Firstly, intimate conversations started the countercultural revolution in the 1960s. At that time upper middle-class and wealthy male and female students in the elite colleges and universities began to share their feelings/values with one another. Perhaps influenced by the beatniks of the 1950s, for example, Paul Goodman and Jack Kerouac, existentialists, especially Jean-Paul Sartre, and spokes persons for *avant garde* modernism, especially Nietzsche and influenced by naïve idealism that spontaneously emerged in youth in the post-World War II era, these students began to severely criticize the shallowness, instrumentalism, and crass materialism of their parents and of American society. There were several factors that contributed to the emergence of this countercultural movement. One factor was that many parents pampered and overindulged their children. Another was that the smoking of marijuana became a kind of ritual for facilitating sharing feelings, analogous to passing the peace pipe that was a ritual that facilitated participatory dialogue within American Indian tribes. Still another factor was that college students, especially those from the elite schools, were virtually guaranteed a well-paying and high status job after graduating from college or professional school. When the countercultural movement spread to most of American society in the 1970s, sharing personal feelings became a core value of the ethic of self-fulfillment, see chapter 22 . However, another outgrowth of the 1960s revolution was the feminist's movement of the 1970s, which polarized men and women. The typical adult male socialized to repress his feelings, and under attack by feminists' ideologies – there are many types of feminism – now found communicating with women even more difficult than before the 1970s.

The second insight concerns the power of intimate conversation and its destructive effect when it is separated from any objective validation. In the 1960s the American public generally was either in favor of or indifferent to the U.S. involvement with Vietnam. By the 1970s, at least in part because of the youth movement emerging from intimate conversations, public opinion turned against the Vietnam war. Also, in the 1970s and 1980s getting in touch with and talking about one's personal feelings escalated in many Americans – for some it escalated to an obsession. This led many to leave traditional marriages and begin new relationships based solely on shared feelings and sexual bonding. The new relationships devoid of rational commitments and strategies for solving problems that always come up, often deteriorated to divorce or worse, violence, substance abuse. Thus, intimate conversations proved to be a powerful force for initiating major personal and social changes. However, when they are uncoupled from any rational control and validation, they produce consequences that range from deep disappointment to profound depression and despair.

The third insight is that while participatory dialogue usually is much less intensely personal than intimate conversation, it has the same fundamental characteristics. Hence this section begins with my article which presents a free-flowing, non-didactic description of intimate conversation. This is followed by a rational, didactic analysis of my description of intimate conversation in which its characteristics are expressed at a reduced level in participatory dialogue.

Intimate Conversation

In the May issue of *Ms* magazine, Barbara Ehrenreich argued that male-female pairs don't really talk to one another. All too often a woman's attempt at conversation is met with a grunt, or shrug of the shoulders, or a non-sequitur such as "I guess so," or "could be," "uh-huh," "well, how about that." A pseudo-conversation may emerge from bantering about pleasantries or cynical quips, but this only covers up the lack of genuine, intimate conversation. Men and women don't tell one another how they feel about things; nor do they share their personal convictions with each other. Ms. Ehrenreich concludes her article with a mixture of conviction, wish, and challenge:

> My own intuition is that the conversation crisis will be solved only when women and men – not just women – together realize their common need for both social and personal change. After all, women have discovered each other and the joy of cooperative discourse [participatory dialogue] through a common political project – the *feminist movement* [italics mine]. So struck was I with this possibility that I tried it out loud on a male companion: "Can you imagine women and men together in a movement that demands both social and personal transformation [which is what developing mindful participatory dialogue as a cultural mode of expression would entail]?[55]

I can imagine women and men together in such a movement only if they can enter into intimate conversation with one another. How one defines conversational intimacy suggests the kinds of social and personal transformations that must occur. I suggest that an intimate conversation is a mutual creation of a personal, individualized encounter between two people.

An individualized encounter is a *non-rational event*. Babies have individualized encounters with persons and things. Notice how a baby encounters a rattle. The baby hasn't developed to the point where he can say to himself: "Ah ha, here is a rattle which means that..." Rather, the baby looks at it, feels it, puts it into his mouth, notices the noises it makes and notices its odor if it has one. Furthermore, the baby evaluates the rattle. Of course, he doesn't say this is a good or bad thing. Rather, he shows by his facial expression, focused attention, muscle tone, and so on, that the rattle is interesting, fascinating, and also perhaps that it tastes ... "yuk," at which point he flings it across the room. These body changes plus the subjective awareness designated by the terms *interesting*, *fascinating*, and *yuk* together are called the baby's feelings about the rattle. Thus, in his individualized encounter with the rattle, the baby has made a feeling evaluation of it.

Some degree of individualized encounter is the bottom line for any intimate conversation. You can't be intimate with someone if he/she won't look at you or make some sort of sensual contact with you. In fact, the more senses involved, the better. The hearing, touching, smelling, and seeing each serves as a kind of rope to bind the attention of two individuals to one another. The feeling evaluations are like electrical forces. For all the complex and diverse feelings we humans can experience, overall feelings are either positive or negative. Overall we are either attracted to or repelled by a person or thing. When two people are attracted to one another, an individualized encounter always begins to occur. The part of the nervous system that coordinates and interprets sensations and feelings (the limbic system, see chapter 3), for a moment, anyway, sustains the encounter automatically.

What happens next is crucial. Either the individualized encounter will start to become personal or it will dissolve. In a personal encounter with another, you are aware of yourself looking at, talking to, touching, or whatever, the *other*. Furthermore, you can choose to continue

---

[55] G. Ehrenreich, May, 1982. Ms magazine.

to look at and perhaps expand your interaction with the other, or you can look away. Personal encounters involve self-awareness and free choices. Non-human animals and babies do not have personal encounters; only evolved human persons have them.

The growth of an individualized encounter to a personal, individualized encounter involves a continuous modification of self-awareness and a series of mutual choices. The first crucial choice is whether the person to whom you are speaking will continue to think of you as an individual or start to deal with you rationally. Most potential intimate conversations falter right here. For instance, often a man talking to a woman will think of her as a woman, and of course all women are "such-and-such." Right away, he is not relating to an individual who happens to be a woman; he is relating in a personal but non-individual way to a concrete representation of the rational category, woman.

Sometimes a man and a woman talk to each other with the hidden agenda (sometimes not so hidden) of one trying to get the other into bed. This move toward a sexual encounter also is rational. The one person is thinking of the other as a means to an end rather than as an individual who, no matter what happens next, is enjoyed right now. Incidentally, the bed activity also is more likely to be non-individual and therefore non-intimate and therefore non-satisfying. I'm not suggesting that you can't talk intimately with someone and also want to go to bed with that person. It's just that when talking becomes the means to the sexual encounter, then talking is a technique for reaching some goal rather than an intimate conversation. Ironically, the man or woman with no thought of a sexual encounter may be more inclined to go to bed after a truly intimate conversation than after an attempted seduction.

The infrequency of intimate conversations is partly due to the fact that people think of one another in terms of roles, stereotypical ideas, and other rational categories. It also is destructive of intimacy when one person thinks of the other in terms of the ideals he *wants* that person to exhibit or in terms of the ideals he thinks that person *should* exhibit. The real individual person gets covered up by all this wanting or evaluating. In order to be intimate we must suspend, not deny, but for some moments anyway, suspend our rational mode of understanding the world. Yes, this means we must reach out to one another in the same way that a baby reaches out to touch a smiling face. Mother nature enables the baby to do this spontaneously, but then Mother nature is impersonal; all babies do it and any smiling face will suffice. We, on the other hand, have to make the personal choice of looking for the real person hiding behind his/her social mask. This, of course, also requires that we step out from behind our masks. Each of us has to decide to take off the social wrapping and validations we carry around with us and stand naked before some other person. This can be frightening -- terrifying even. What if that other person doesn't also disrobe? Worse yet, what if that other person takes a look, psychologically speaking, and says yuk?

Under the single idea of suspending rationality, I discussed two kinds of choices. Let me distinguish and interrelate them by means of metaphors. In order for there to be intimacy, each person must undress the other; that is, choose to see the individual who underlies the roles, stereotypical ideas, and uses. Also each person must allow himself to be undressed by the other. As in love making, each person must do something for the other and allow the other to do something for him/her. Each must choose to give and to receive.

How does one muster up the courage to be present as an individual to some selected others? This involves further choices. We must look at ourselves naked in the mirror, that is, reflect on ourselves, and like what we see in spite of the many flaws. If we do not like ourselves, we will expect others not to like us. Then we will hide behind our social respectabilities, and others will hide behind theirs. There will be no intimate encounters. There are many ways we may come to know and like ourselves. This is where religion or its equivalent may make a profound contribution to one's life. Unfortunately, there are many so-called religious people who do precisely the opposite; hate yourself as a sinner in order to be humble before the Lord, that sort of thing. One way or another, we must decide to like ourselves if ever we will be able to expose ourselves to others.

There is yet another important decision each of us must make for the sake of intimacy. We must choose to know and acknowledge our feelings as belonging to us. As we develop self-awareness and rational thinking, a split occurs within us. Our bodies, in conjunction with part of our consciousness, continue to make feeling evaluations, just as when we were babies. Simultaneously, our rational consciousness makes intellectual evaluations that enable us to function within society and reap the benefits of its culture. Very often our intellectual evaluations are contrary to our feeling evaluations. Most of us could not survive outside of society, and, anyway, a person is hardly human totally cut off from social interactions. Understandably enough, intellectual evaluations take precedence. Unfortunately, especially in this scientific age, we tend to ignore our bodies and deny many of its feeling evaluations. One way or another this always causes trouble.

For example, suppose a husband borrows his wife's car and returns it with a virtually empty gas tank. The next day the wife is put out by this because she's afraid she may not make it to work. She stops to get gas and then is late for work. Now she's angry! Some wives will deny this anger because they think (intellectual evaluation) they should not be angry over such a little inconvenience; or they don't want to risk a confrontation with the husband; or ... whatever; people are able to invent many different reasons for denying feelings. What do you suppose will happen? The repressed anger kindled by stresses at work will smolder all day, and the woman will come home feeling irritable, depressed or cold and unresponsive to her husband without the foggiest notion why. This situation has the makings of a big blowup, particularly if the husband also has had a stressful day. On the other hand, the woman could acknowledge her anger and talk it over with her husband in the manner of: "This is how I feel," rather than, "Damn you,..." Such an approach at least makes it possible to defuse a potentially explosive situation.

Two people cannot develop intimacy if they do not work through the inevitable negative feelings they will have toward one another from time to time. Nor can they even begin to be intimate if they meet each other as split personalities with the mind and heart operating independent of one another. The two people may be close for awhile on a purely intellectual level or on a purely feeling level. Eventually, however, the neglected human dimension will rebel and take its revenge.

This last example points to another essential prerequisite for intimacy: self-assertion in conjunction with caring. We can see this with respect to the latter part of the example. The woman cared enough about her relationship with her husband to risk confronting him with her feelings of anger. After all she is not merely a wife or a woman who can grant sexual favors. She is a person with needs, one of which is not to be taken for granted and another of which is not to be walked over. She counts for something; her husband matters, her relationship matters. Therefore, she's willing to struggle to work through all the individual conflicts. If a person CARES, then he/she won't just sit around and let things happen. He'll do something about his life and about his relationships.

With respect to intimate conversations, each person must care enough to dare to assert himself. Each must demand something from the other and demand to be heard by the other. At some level of consciousness, each must say to himself: "Never mind what I thought of you in the past or what I will think of you in the future. Right now you are my significant other. I want to give you something of myself and receive from you something of yourself, if only for a short time."

Communication between two people only begins to grow when yet another mutual decision is made. Each must bracket, that is, temporarily suspend, his/her own point of view in order to take in and seriously consider an-*other* point of view. For the moment, anyway, this is like abandoning any particular rational framework. But to lose our individual, rational framework is to fall into the abyss of reality having no structure or meaning. This is like the gap experience Nietzsche described of *avant-garde* artists. The resulting chaos is death to individual consciousness. However, if each perseveres in this living death, then both may experience the ecstasy of birth of a new unit of life: two people in intimate, non-rational communion with one another. Such a unit contains both people, but is much more than the sum of two. The unit has a life of its own that unites two irreducibly different, sometimes even contradictory, points of view. The resulting community emphasizes individuality while transcending separation. This, by the way, illustrates one aspect of creativity that was described in chapter 27; it is *cooperative individuality.*

We now are at the moment where the two people can create an intimate conversation. Each must have a point of view which can be simultaneously bracketed and presented to the other. Each must have sufficient self-awareness to know what his beliefs, convictions, and opinions are. This used to be one of the major goals of formal education, and, indeed, formal or self-directed education can be a great help here. Frankly, however, with today's emphasis on specialization and rationality for the sake of material goals, education tends to be a major block to intimate conversation. As often as not, the more education a person has, the less likely he/she will be willing or able to enter intimate discussions. Furthermore, each person must have developed the skill to put his thoughts into words. This skill comes easier to some than others, but everyone needs practice. If a person only rarely has deep discussions or only rarely tries to put his ideas on paper, he/she won't have this skill. When faced with a personal, individualized encounter, he/she won't be able to do his part in creating an intimate conversation. On the other hand, a person with this skill will also be able to formulate his ideas in a manner particularly suited to the person to whom he is talking. You can't just say what you mean; you must say it in a way that the other person can understand. One also should be willing and able to reformulate his thoughts on the basis of his understanding of the other point of view.

From time to time many people are given moments when they experience an intimate conversation. You fall into such intimacy the way one falls in love; you just go with the flow and let it happen. Everything seems to click as each person is drawn out by the *ambiance* of the situation to unexpected insights and facility with words. At such moments, it all seems so spontaneous and easy that any awareness of what it takes to participate in an intimate conversation is irrelevant.

Inspiration and love, and to some extent intimacy, always are gifts. However, there are social circumstances and personal traits that greatly increase the chances of intimate conversation and its reduced level of intimacy, participatory dialogue, to occur.

Empathetic listening. As indicated in my article, an intimate conversation always is a personal, individualized encounter, which is a non-rational event. The non-rationality of the event implies that in order for it to occur, one must bracket/suspend roles, status, stereotypical ideas along with any other of the rational categories these ideas exemplify. One must suspend the rational mode of understanding the world and correspondingly stop evaluating the other according to some set of rationalized values and stop any utilitarian wanting something from the other. Each person must take off his/her "social wrapping" and validations – each must take off his/her persona, one's social mask. Likewise, each must bracket/suspend one's own point of view in order to take in and seriously consider another point of view; this is what empathetic listening refers to.

In order for empathetic listening in participatory dialogue to occur, one must suspend judging others' points of view based on one's own point of view. Likewise, one must avoid judging by bracketing/suspending social roles and status. Overall one must suspend control individualism in order to express empathy competencies associated with participatory individualism. Yankelovich describes the equivalent of these ideas in what he designates as two of three essential characteristics of participatory dialogue. One is that all people who take part in what I call participatory dialogue reduce all hierarchies such as parent-child, teacher-student, employer-employee, to egalitarian relationships. "There [must be] … no hint of sanctions for holding politically incorrect attitudes, no coercive, influences of any sort, whether overt or indirect."[56] The second one is that the participants in dialogue must already have or develop the ability "to respond with unreserved empathy to the views of other. [That is, each participant must be able] to think someone else's thoughts and feel someone else's feelings."[57]

Self-awareness. As indicated in my article, intimate conversation involves self-awareness associated with free choices. The growth of an individualized encounter to a personal, individualized encounter involves a continuous modification of self-awareness and a series of mutual choices. We must choose to know and acknowledge our feelings as belonging to us. Moreover, each of us must have a point of view, which can be simultaneously bracketed and presented to the other. Each must have sufficient self-awareness to know what his/her beliefs, convictions, and opinions are.

The self-awareness of participatory dialogue is that one must know or during the course of the dialogue come to understand his/her assumptions on which a point of view is grounded. This idea is Yankelovich's third of three basic characteristics of mutual dialogue. "Dialogue must be concerned with bringing forth people's most deep-rooted assumptions."[58] All knowing expressed in conceptual language is based on assumptions. Very often, even well-educated people, are not aware of the assumptions underlying their perspectives. Sometimes these assumptions are linked to a person's sense of self. Bohm stresses that when one's perspectives or opinions are criticized, the person, without being totally aware of this, reacts to this criticism as an attack of his/her self. Thus, unexamined assumptions often lead to misunderstandings and errors of judgment. The empathetic aspect of participatory dialogue provides the safe place where people can be uninhibited in consciously acknowledging their own and other participants'

---

[56] D. Yankelovich, *The Magic of Dialogue* (New York: Simon & Schuster, 1999), pp. 41-43.

[57] D. Yankelovich, *The Magic of Dialogue* (New York: Simon & Schuster, 1999), pp. 41-43.

[58] D. Yankelovich, *The Magic of Dialogue* (New York: Simon & Schuster, 1999), pp. 44-46.

assumptions. "Arguably, the most striking difference between discussion and dialogue is this process of bringing assumptions into the open while simultaneously suspending judgment [about the truth or even the validity of these assumptions.]"[59]

Self-confidence and ethic of care. Again, quoting from my article on intimate conversation, as in love making, each person must do something for the other and allow the other to do something for him. Each must choose to give and to receive. Also self-assertion in conjunction with caring is necessary. Each person must care enough to dare to assert himself/herself.

Yankelovich describes the requirements of self-confidence and care in terms of trust and I-Thou relationships. Participatory dialogue most readily occurs in a group of people, who, as a result of an ethic of care for one another, form a community. If this community also is not dominated by a single, dogmatic worldview, then community members can feel free to share with one another their deep-seated assumptions. They can trust that no matter how outrageous one's assumptions or opinions may appear to other members of the community, he/she will not be ridiculed or condemned by them. Deep down all of us realize that "the most profound forms of communication take place at the deeper level of personal encounter – the level of I-Thou relationships. Given the slightest chance, people gravitate in this direction."[60] But both industrial culture and postmodernism bring people with very different perspectives to communicate with one another at the level of impersonal transactions or discussions. If these people, according to Yankelovich, can be guided to take part in participatory dialogue, they form temporary caring communities. These communities, in turn, have the cohesion and communal integrity to be creative and cooperative to accomplish complex tasks.

Senge and Bohm refer to this idea in terms of the second of three conditions necessary for dialogue: *All participants must regard one another as colleagues.* [The first condition, discussed earlier is: *All participants must "suspend" their assumptions, literally to hold them "as if suspended before us."* The third condition is: *There must be a "facilitator" who "holds the context" of dialogue.*] As indicated earlier, the break down of cohesive communities committed to a dogmatic version of traditional values, is followed by no cohesive communities or communities that only discuss ideas or carryout impersonal transactions. Participatory dialogue definitely does not occur in these traditional communities or in these non-traditional discussion groups. Postmodernism provides the context for a uroboric paradox analogous to the emergence of life from non-life, see chapter 24, pp.267-268. Participatory dialogue (analogous to the egg that becomes a chicken) converts people, who, as a general rule, communicate at a superficial level, into a community of colleagues (analogous to a chicken that develops from an egg). At the same time, a community of colleagues is an absolute requirement for participatory dialogue to occur (analogous to the chicken producing an egg). Thus, the uroboric paradox: in order to engage in participatory dialogue one must create a "community of colleagues." But one only can create a community of colleagues by means of participatory dialogue. According to Senge, the "willingness to consider each other as colleagues"[61] provides a precondition for people to share with one another their deep-seated assumptions and thus transcend this cycle.. This is especially true when people with significantly different viewpoints are forced to come together to resolve

---

[59] D. Yankelovich, *The Magic of Dialogue* (New York: Simon & Schuster,1999), pp.106-108.

[60] D. Yankelovich, *The Magic of Dialogue* (New York: Simon & Schuster,1999), pp.106-108.

[61] P. Senge. *The Fifth Discipline (the Art and Practice of the Learning Organization)* (New York: Currency Doubleday, 1990), p.245.

their differences. Also, colleagues don't feel the need to compete with one another, but rather "feel as if they are building something, a new deeper understanding."[62] Senge hints at a basis for transcending the dialogue uroboric paradox. According to him, all humans participate with one another at a deep unconscious level, something like Carl Jung's collective unconscious. That is, human "thought is participative,"[63] which is to say that all humans already are potential colleagues. Ideas in individuals emerge from this unconscious common base. All that needs to happen is for people to realize their deep connection with one another. Then the choice to commit to being colleagues will enable them to enter into participatory dialogoue.

Participatory chaos. In order to have an intimate conversation with another, one must break away from all rational structures and participate in the individual life sphere of another person. I designate this unstructured participation as *participatory chaos.* My description of intimate conversation indicates various aspects of this chaos. One is that two people cannot develop long standing intimacy if they do not work through the inevitable negative feelings they will have toward one another from time to time. Another aspect is that the two people involved in the conversation must relate to each other at both an intellectual level where they formulate their rational viewpoints and an emotional level where they empathetically go out to one another. For a very rational person, embracing the feeling dimension will tend to bring them into chaos. Likewise, for a very "feeling person," objectively stating one's rational viewpoint or becoming aware of the assumptions of one's viewpoint will put them into chaos.

Bracketing one's viewpoint in order to take in and seriously consider an-other point of view is like abandoning any particular rational framework. But to lose our individual, rational framework is to fall into the abyss of reality having no structure or meaning. This is like the *gap experience* Nietzsche described of *avant-garde* artists. The resulting chaos is death to individual consciousness, or rather death to the mind-self as the center of consciousness. However, if each perseveres in this living death, then both may experience the ecstasy of birth of a new life: two people in intimate, non-rational communion with one another. Such a unit contains both people but is much more than the sum of two. The unit has a life of its own, which unites two irreducibly different, sometimes even contradictory, points of view. The resulting community emphasizes individuality while transcending separation.

Another aspect of this participatory chaos is, first of all, one must develop the skill of orally formulating one's point of view. Then, he/she must be able to formulate the ideas in a manner particularly suited to the person to whom he/she is talking. You can't just say what you mean; you must say it in a way that the other person can understand. One also should be willing and able to reformulate his/her thoughts on the basis of one's understanding of the other point of view.

Potential conflict is a necessary precondition for dialogue. Actual conflict usually leads to one person or point of view winning and dominating all other contending persons or contrary points of view. In those situations where the conflict cannot be resolved in this way, for example, nuclear or economic war among world military and/or economic superpowers, current war between humans and the non-human biosphere, and the current conflict implicit in multiculturalism, the only alternative to someone winning is everyone losing. Postmodernism is

---

[62] P. Senge. *The Fifth Discipline (the Art and Practice of the Learning Organization)* (New York: Currency Doubleday, 1990), p.245.

[63] P. Senge. *The Fifth Discipline (the Art and Practice of the Learning Organization)* (New York: Currency Doubleday, 1990), p.245.

that social condition of multi-factional conflict in which the only possible, rational outcome is everyone loses. Postmodern diversity sooner or later leads to total collapse of society. It is no longer possible for any one religion or "ism," such as, Christianity, Buddhism, scientific humanism, to conquer all others. Hitler may have come close but his near win definitely was the last apparent possibility of any one nation dominating all other nations. The new world order is an emerging postmodern world society. In some manner participatory dialogue converts a win-lose situation or a conflict-everyone-lose situation into a win-win semi-stable harmonious community. This emergence of a win-win situation is analogous to the paradoxical emergence of life from non-life.

What makes this emergence of a win-win situation even more unexpected is that most people are not disposed to participatory dialogue. In fact, the education system, which is the secular new religious institution of postmodern societies, teaches some mixture of radical utilitarian and expressive individualism both of which are opposed to participatory dialogue. Postmodern students are taught to repress participatory mutuality, and modern culture, especially in the United States, reinforces this repression.

How will people learn to engage in participatory dialogue? They will learn it the same way many of them learned to ride a bike. If one's body is sufficiently strong and coordinated, then he/she can "solve" the circular (uroboric) paradox. In order to ride a bike one must have a feel for how to ride a bike, but in order to have a feel for riding a bike, one must first ride a bike. The creative, suitably prepared person intends to ride a bike. He keeps intending until he begins to ride a bike a little and very imperfectly, but then he/she gets some feel for riding and begins to improve his/her bike riding, and each improvement leads to a greater feel for bike riding. In like manner, each person who has some rational discipline and emotional intelligence is suitably prepared to *Intend* participatory dialogue. If he/she continually intends to do it, eventually he/she will participate in this kind of dialogue to some extent. Just as in the bike riding example, he/she will enter the positive feedback cycle of: a little mutual dialogue leads to a little feel for participating in mutual dialogue that leads to more proficient mutual dialogue that leads to a greater feel for mutual dialogue, and so on.

The new human ecology that emerges from the participatory chaos of participatory dialogue is what Senge refers to as the *learning organization*. This book describes creativity as a Life, Death, Rebirth process dealing with psychological developmental transformations, and other types of creativity. Death that is equivalent to Chaos is the necessary prelude to Rebirth to a new vision. In Senge's terms Death is Conflict which is necessary for creating a new vision. He claims that great learning communities are not characterized by an absence of conflict but by the visible conflict of ideas. This is because

> The free flow of conflicting ideas is critical for creative thinking, for discovering new solutions no one individual would have come to on his own. Conflict becomes, in effect, part of the ongoing dialogue.[64]

---

[64] P. Senge, *The Fifth Discipline (the Art and Practice of the Learning Organization)* (New York: Currency Doubleday, 1990), p. 249.

In general, people who have individuated to post-conventional morality (see chapter 4) have some potential to participate in mindful, participatory dialogue. They also have the potential to individuate creative, mindful, participatory dialogue discussed in the next section. The degree to which one can be intimate with others in this way – varying from lowest to highest – is as follows: (1) post-conventional control individualism, with no or only a low degree of incorporation of emotional intelligent competencies associated with participatory individualism, precludes intimate dialogue; (2) Post-conventional participatory individualism that focuses on the self-fulfillment ethic as described in chapter 22 has very little potential to incorporate intimate participatory dialogue; however, such an individual, at least, may be able to express to others how he/she feels, but this person will not be empathetic with or willing to receive the feeling expressions of others; (3) post-conventional control individualism with the capacity to be intellectually open to new ideas has a potential for participatory dialogue; this intellectual openness will spill over into making one be open to feeling intimacy; (4) a person who has gone through and reflected upon personal life transformation may become mindful of the possibility of other transformations to be explored in mindful, participatory dialogue.

CREATIVE MINDFUL PARTICIPATORY DIALOGUE

Definition in Terms of Outcomes

Creative, mindful, participatory dialogue creates: (1) greater mutuality that transcends differences or even contradictions of the perspectives of the participants in the dialogue group; (2) within individuals participating in the dialogue a greater self-awareness including a greater awareness of their points of view and the assumptions upon which these points of view are based; (3) a shared, expanded vision that includes for each person a modified version of the perspective he/she had at the start of the dialogue.

Requirements for Creative, Mindful Participatory Dialogue

First requirement. One must intend to enter into creative, mindful, participatory dialogue. This choice depends of the second requirement to be described next.

Second requirement. One must trust, that is, have experiential Faith-Hope that universal creativity will manifest itself in this particular group engaging in participatory dialogue. Human individuation is a special case of creativity in the human biosphere, which always is an Order, Chaos, New Order process. Entering mindful, creative dialogue is an aspect of human individuation. The chaos, which always produces the anguish of stress of individuation, often is necessary for the emergence of a new order. One aspect of the *ethic of individuation* is that one must have a Faith-Hope commitment to the prospect that personal evolution leads to an integrated personality capable of creative mindful dialogue. Until one has had some experience of personal transformation, he/she cannot know even of its possibility. Unfortunately, something like a profound existential malaise can overwhelm postmodern citizens beginning as early as late childhood. In today's society children and adults alike – though, perhaps in different ways – are disillusioned with all institutions, family, marriage, education, religions, politics, business. What used to hide us from stark reality and buttress us now disappoints us and worse yet, confirms our worst fears: there is no meaning; living – it's just all bullshit! Our only refuge, it seems to me, is faith-hope in individuation. How do we obtain such a faith-hope? We grow into it; we tentatively believe others who have lived through crises and have come out of them happier and more powerful, more effective people. For some, this is the sort of inspired tentative belief that occurs at AA meetings. Then, we try it out; we embrace a challenge and experience a rebirth from that chaos. This gives us some little courage - based on faith-hope - to embrace another challenge and another, and then still another. Gradually we build up faith-hope and courage to enter into ever greater chaos with corresponding greater anguish until we become warriors ready to face the *dread* that the adult person must face en route to reaching vision-logic personhood, see chapter 26, pp.288-290, and then the transpersonal level of consciousness.

In like manner, potential dialogue participants must give themselves the opportunity to grow into faith-hope in creative dialogue. They can do this by tentatively believing others who have lived through the participatory chaos of creative dialogues and have come out of them with new visions and/or solutions to apparently unsolvable problems. The potential participants must attempt the process of creative dialogue, embrace the challenge and experience a rebirth from that chaos. This will give them some little courage based on *experiential faith-hope* to embrace other instances of creative dialogue. As they do this over and over again, they will develop greater experiential faith-hope in the process and become proficient at creative dialogue.

Third requirement. There must be the potential for conflict among participants in the creative dialogue group. If people choose creative mindful dialogue as a basis for forming a discussion community to talk about some topic or problem of mutual interest, then paradoxically, creative dialogue only is possible when there is a potential conflict among strongly held, well thought out diverse perspectives and theories. People may choose participatory mutuality out of love for another (or others) or they may have a kind of "mental participation mystique" as a result of being a committed member of some dogmatic religion, or cult, or academic discipline or profession. People in these kinds of groups may *love one another*, that is, engage in non-information dialogue, or they may discuss and reinforce their beliefs or clarify and refine them, but they cannot engage in creative mindful dialogue. Again, paradoxically, it is more possible to have creative dialogue with an enemy than with someone in your own group, in which everyone is committed to the same perspectives. Here again we see that radical utilitarian and expressive individualism of postmodernism that has fragmented society also has provided a necessary, but not sufficient, condition for creative dialogue. Cultural evolution to the current postmodern, fragmented societies has brought the human biosphere to the brink of collapse, but simultaneously to the brink of creating a new kind of society and culture via creative dialogues. In the past, creative dialogues occurred among a few creative people; now, many are called to engage in it as the only way of avoiding totalitarian states or total social collapse.

The new human ecology that emerges from creative dialogue is what Senge refers to as the learning organization. Creativity is a Life, Death, Rebirth process. Death equal to Chaos is the necessary prelude to Rebirth to a new vision. In Senge's terms Death is Conflict, which is necessary for creating a new vision as described in the section on participatory chaos..

Fourth requirement. Senge and Bohm propose a third (of three) basic condition necessary for creative dialogue: there must be a facilitator who holds the context of dialogue.[65]

---

[65] P. Senge, *The Fifth Discipline (the Art and Practice of the Learning Organization)* (New York: Currency Doubleday, 1990), p.243.

The most important function the facilitator serves is to continually remind people attempting dialogue that they are participating in a larger whole, and they must avoid attitudes that spill over into words that keep them and others from being incorporated into this larger whole.

Senge maintains that humans' habits of thinking, especially in an era of celebration of the competition of free market capitalism, pull one towards discussion where there is a true or a most valid perspective. The truth or the greatest validity is decided by competitive debate. Also, in the early stages in developing a dialogue team, the participants tend to take thoughts that occur to them as literal – with only one meaning – rather than as metaphorical concepts composing an assumed narrative. In this situation, humans tend to believe in their own views and want them to prevail. This instinctive attitude prevents them from suspending their "deep-seated assumptions" and thus bars them from participatory dialogue. The facilitator of a dialogue session reminds members that each of their views is only one of many possible ways of seeing a situation. At the same time, the facilitator does not assume the role of "expert" that would shift attention away from the team members' ownership of their own ideas and responsibility. Helping people to maintain ownership of the dialogue process keeps the dialogue going. At the same time, the facilitator's understanding of dialogue enables him/her to influence the flow of ideas just by participating in the dialogue group. This participation demonstrates dialogue to others who then begin to learn how to do it.

Fifth requirement. There must be the *Intent* to enter and embrace *participatory chaos* of creative, mindful, participatory dialogue. This fifth requirement has four aspects (as opposed to the seven aspects of participatory chaos of creative, mindful, collaborative dialogue, see chapter 29):

1. The commitment to participatory chaos includes acknowledging diverse perspectives where some or all oppose one another and there is no perspective that can resolve these oppositions. Participants engaging in creative dialogue agree to disagree. This lack of consensus leads to unstructured communications that, of course, produces chaos. The only consensus each participant can count on is that each member has made a personal commitment to all the requirements for creative dialogue; or at least, each member will be submissive to the directions of the facilitator.

2. Each participant has the emotional competencies[66][67] associated with empathetic listening. If any individual does not have this emotional intelligent competency, then with the aid of the facilitator he/she
   a. suspends social roles & status
   b. suspends critical thinking applied to judging others or judging a particular perspective.

3. Each participant chooses to be mindful of his/her perspective and mindful of the assumptions underlying this perspective. Again, the aid of a facilitator may be necessary.

4. Each participant chooses to be vulnerable to one another with the help of a facilitator if necessary. This involves the necessity of Trust. Senge and Bohm

---

[66] Daniel Goleman. *Emotional Intelligence* (New York: Bantam Books, 1995).
[67] Daniel Goleman. *Working with Emotional Intelligence* (New York: Bantam Books, 1998).

refer to this idea in terms of the second of three conditions necessary for dialogue; namely, all participants must regard one another as colleagues.

Chapter 29
CREATIVE, MINDFUL, COLLABORATIVE DIALOGUE

RELALTION TO CREATIVE, MINDFUL, PARTICIPATORY DIALOGUE

If potential participants in the dialogue group already share a vision, then they can pursue the possibility of collaborative dialogue. For example, Watson and Crick shared three fundamental aspects of a scientific vision. Firstly, they both were committed to a gross mechanistic perspective for doing creative science. Second, they both were scientific humanists. Third, they both had read and were inspired by Schrodinger's book, *What is Life?*, and were convinced that the "secret of life" would be revealed by understanding the structure of chromosomal DNA. Thus, even though James Watson as an American biochemical mechanist had a very different cultural and philosophical approach to laboratory research from Francis Crick, an English theoretical physicist, the two scientists were drawn to collaborate with one another because of their shared vision. The American generals in various wars the United States has been in and had many personal and philosophical differences, but they could collaborate because of the shared vision of democracy. Collaborative dialogue is possible for any group of people in any particular organization that provides all members with a shared vision.

Senge alludes to this creative aspect of dialogue in relation to science, though he does not distinguish between mindful and non-mindful dialogue; nor does he distinguish between participatory and collaborative dialogue.

> In a remarkable book, *Physics and Beyond: Encounters and Conversations*, Werner Heisenberg ... argues that Science is rooted in conversations [creative collaborative dialogues]. The cooperation of different people may culminate in scientific results of the utmost importance. Heisenberg then recalls a lifetime of conversations [creative collaborative dialogues] with Pauli, Einstein, Bohr, and the other great figures who uprooted and reshaped traditional physics in the first half of this century. These conversions, which Heisenberg says "had a lasting effect on my thinking," literally gave birth to many of the theories for which these men eventually became famous. Heisenberg's conversations, recalled in vivid detail and emotion, illustrate the staggering potential of collaborative learning -- that collectively, we can be more insightful, more intelligent than we can possibly be individually.[68]

---

[68] P. Senge, *The Fifth Discipline (the Art and Practice of the Learning Organization)* (New York: Currency Doubleday, 1990), pp 238-239.

However, if potential participants in a dialogue group do not share a vision, then collaborative dialogue will have limited, if any success. For example, one group of individuals in the Biology department at The University of Toledo (UT) was committed to a gross mechanistic, molecular perspective. A second group in this department was committed to a holistic, systems, ecological-environmental perspective. These two perspectives, the one atomistic and the other holistic, were so radically different that the two groups could not engage in collaborative dialogue. The only way of coordinating the creative energies of these two groups was for one group – in this case, the ecology-environmental group – to be subordinate to the dominance of the other group. This dominance relation already had been going on for some years. With the encouragement of top administrators of UT that included the Arts and Science dean, the natural science associate dean, and the provost, the two groups split, forming two new departments. Creative collaborative dialogue between the two groups could have benefited both of them, and I believe both groups now are less effective and less creative as a result of this split. If appropriate leadership would have guided the two groups to engage in creative, participatory dialogue, all persons involved could have generated a shared vision without destroying each group's characteristic identity. Then, based on this shared vision, the two groups could have begun creative, collaborative dialogue that would benefit everyone.

Teachers at UT and at many other educational institutions have a different vision and definition of mission than research scholars, especially those that receive large sums of money from outside agencies, such as, NSF, NIH, large corporations. These teachers are like women in a patriarchal society or wives in a patriarchal marriage. Their lack of a shared vision and their lack of mutual respect prevent any collaborative dialogues between the dedicated teachers and the "grant getters." As was true of women in the 19th century, it is not feasible for dedicated teachers to divorce or rebel against their dominators. The absence of collaborative dialogue in universities today goes hand-in-hand with these institutions progressively becoming more and more dysfunctional. Perhaps when any of these institutions suffer severe financial and status difficulties, their faculty will choose to attempt creative, participatory dialogue to generate a shared vision and definition of mission and then pursue creative, collaborative dialogue.

PARTICIPATORY CHAOS AND OBJECTIVE KNOWLEDGE OF COLLABORATIVE DIALOGUE

Creative, collaborative dialogue starts when participants share a perspective that governs their activities and shared commitment to relevant rational models, but some aspects of this perspective or rational models break down or contradict one another. Then the dialogue individuals choose a limited objective, participatory chaos (defined in a later section in this chapter) and engage in a creative process that leads to the emergence of a new perspective or new objective knowledge that solves a particular problem. The dialogue group first generates a modified shared vision that guides the activities of the participants. Then, creative collaboration produces new objective knowledge that transcends but includes modified versions of diverse rational models that are contrary or contradictory to one another. This outcome is not the result of a consensus, though it leads to a consensus that is the shared commitment to the new objective knowledge. Likewise, the outcome is not a compromise though it may appear as such to one who has not participated in the dialogue.

For example, the shared vision of Watson and Crick had to overcome oppositions each inherited from their respective scientific cultures. From 1930 to 1950 several lines of research by people who fundamentally disagreed with one another began to form an outline that, once appreciated, could help produce an understanding of the secret of life. One group worked out techniques (applying x-ray diffraction analysis) for understanding the three-dimensional structure of proteins. Not only do proteins consist of very specific linear sequences of amino acids, but these macromolecules can also have a three-dimensional size and shape. In fact, as we know, it is a protein's characteristic shape that gives it specific biological properties. This is especially true of enzymes. Proteins are like miniature crystals except they lack the pattern that monotonously repeats itself as is true of inorganic crystals; so proteins are called aperiodic crystals (nonrepeating crystals). Some scientists speculated that these unique aperiodic crystals held the secret to life. One of the pioneers in quantum mechanics, Niels Bohr, argued against there being the same sort of biological unpredictability for proteins (his term was "indeterminacy") as was found for subatomic particles. Bohr reasoned that life processes are too well coordinated to be explained on the basis of statistical quantifications of indeterminate chemical interactions. Bohr's student, Max Delbruck, was influenced to look elsewhere than classical and quantum mechanics for the physical basis of life. Based on his studies of x-ray-induced mutations in phage (a virus that invades bacteria), Delbruck proposed that chromosomes are aperiodic crystals. The collection of genes in these aperiodic crystals somehow contain and implement the master plan for life processes.

Meanwhile, many biologists firmly held that life could only be understood in terms of biochemical interactions. Archibald Garrod was able to show that the disease phenylketonuria (PKU) is a result of a gene mutation. Furthermore, he showed that this disorder is due to the absence of a particular enzyme. The connection is obvious: genes determine which enzyme will be produced and enzymes coordinate the life processes. Genes are the executives with master plans, enzymes are the junior executives, and all the other chemicals are the community of molecules keeping the organism successful, that is, alive. These results were published in 1914, so, of course, no one paid any attention to them -- Garrod was too far ahead of his time. But in the 1940s, George Beadle and others showed that, indeed, a change at a single site in Neurospora (a type of fungus) caused an enzyme not to be produced. In the 1950s, a sequence of studies demonstrated that the sickle cell trait results from a mutation of one gene that changes only one amino acid in a long-chain, hemoglobin molecule. Garrod's suggestion was correct after all.

By 1952 several lines of evidence had indicated that the chromosomal material containing the genetic master plans is none other than DNA. However, the biochemical mechanists, rightly proud of their accomplishments, would have nothing to do with the wild speculations of physicists, with their new laws of nature. It all sounded so mystical, or at least metaphysical, which is almost as bad. The sophisticated physicists, in turn, thought the biochemists were too caught up with simplistic explanations. Life is much more complex than just chemistry, but then what can you expect from people who mess around with test tubes and such instead of thinking about the deep realities of life?

REQUIREMENTS FOR CREATIVE, MINDFUL, COLLABORATIVE DIALOGUE

The first, second, third, and fourth requirements are the same as for creative ,mindful dialogue.

Fifth requirement. One must Intend to enter and embrace, Objective Participatory Chaos, which for this type of dialogue has seven aspects.
1. This commitment acknowledges diverse perspectives where some or all oppose one another and there is no perspective that can resolve these oppositions. Participants engaging in creative collaborative dialogue agree to disagree. This lack of consensus leads to unstructured communications that produce chaos. The only consensus that each participant can count on is that each member has made a personal commitment to all the requirements for creative collaborative dialogue. At the very least, each member will be submissive to the directions of the facilitator and the referee for Socratic dialogue (see below) and receptive to the input of *philosophical translators* (see below).
2. Each participant, at least for the duration of the particular collaborative dialogue, is committed to the three core ideas of postmodernism , which are:
    a. Constructivism.
    b. Contextualism.
    c. Integral-aperspectivism.
3. As a result of the first and second aspects of objective, participatory chaos, each participant is receptive to understanding different perspectives and open to the possibility that one's own perspective could be modified and/or expanded.
4. As a result of the third aspect of objective, participatory chaos, the participants must be in the same or closely related areas of objective knowledge; or philosophical translators must be available.
    a. According to the narrative constructivism perspective (see end of chapter 24), any valid rational model is a more precise representation of a narrative using metaphorical concepts and analogies.
    b. A philosophical translator converts a rational model into a narrative that a non-specialist – a lay-person – can understand.
5. Each participant must have developed the ability of critical thinking and have specialized knowledge in some area(s) relevant to the subject of collaborative dialogue.
6. Each participant must be aware of the assumptions of his/her rational model or be receptive to understanding these assumptions as made clear by a philosophical translator.
7. Each participant need not like any or all of the other participants, but he/she must respect the legitimacy of each of the other participant's having a particular point of view.
    a. Each participant must have intellectual integrity and <u>trust</u> the intellectual integrity of all the other participants.
    b. Participants' respect for one another must be based totally on intellectual integrity rather than on social roles or status.
    c. As a result of this respect, all participants must regard one another as colleagues.

Sixth requirement. Participants must be able to play the game of Socratic dialogue in which players reach consensus on definition of terms, ways of expressing ideas, rules of argumentation & debate, and the legitimacy of: (1) digressions, (2) factual information, and (3) conceptual interpretations relative to the shared vision that creative collaborative dialogue has produced. There should be a referee for the Socratic dialogue.

CREATIVE DIALOGUE OF SCIENTIFIC CONSTRUCTIVISM

In human creativity, Eros-chaos associated with Will-to-openness allows new insights to emerge during the chaos from what I and other thinkers call *the Inner Self.* Then Eros-order, Will-to-power converts this existential insight into an objective representation. In reflecting on my experiences of creativity over many years in conjunction with ideas about creativity presented in chapters 24 and 26, I propose that scientific constructivism occurs in four steps, where each step has a chaos phase and an order phase. Exemplifying the first step, the fathers of modern science broke away from the order defined by the dogmatism of the medieval Catholic Church. The vision that emerged from the chaos of the intellectual diversity problem is that humans could create valid, rather than absolutely true descriptions of nature in terms of mathematics. Exemplifying the second step, thinkers like Galileo and Newton converted this scientific Enlightenment vision into mathematical equations, which are logical, conceptual models. Then (the third step) various Enlightenment thinkers created strategic plans, that is, they designed experiments that could validate or invalidate these mathematical descriptions of nature. Finally, (the fourth step) some people carried out these experiments and determined to what extent, if at all, the mathematical descriptions are valid.

Scientific constructivism may be generalized to cyclic constructivism present in all types of creative learning. This cyclic constructivism consists of five steps involving four types of creative learning. The cyclic sequence is: (1) create a vision; (2) convert the vision into a logical, conceptual model; (3) convert the logical, conceptual model into a strategic plan; (4) convert the strategic plan into actions that accomplish the goals of the logically conceived vision, and (5) begin a new cycle. If the goals are achieved, then fine-tune the activity. If the goals are not achieved, begin a new cycle starting with a modified strategic plan, or with a modified logically conceived vision, or with a new vision.

FOUR STYLES OF CREATIVE DIALOGUE INVOLVING CREATIVE LEARNING

I have taken a modified version of the idea of Paul MacLean's theory of a triune brain, described in chapter 3, pp.27-29, as the basis to propose that after the emergence of Mind, it, as described in chapter 5, pp.53-54, also differentiates into three mind centers: Expressive Mind Center, (feeling perceptions and intuitions), Thinking Mind Center, and Doing Mind Center (action). I further propose in agreement with other thinkers that gene segments interacting with the "psycho-social environment" of the 5 year-old child leads to one of the mind centers becoming the "preferred mind center." Further psychological development leads to the preferred mind center becoming either the dominant mind center or the repressed mind center.[69] [70] These ideas imply that there are twelve personality types wherein each type consists of (1) a dominant

---

[69] K. Hurley and T. Dobson, *What's My Type (Use the Enneagram System of Nine Personality Types to Discover Your Best Self)* (New York: Harper Collins Pub., 1991)

[70] K. Hurley and T. Dobson, *My Best Self (Using the Enneagram to Free the Soul)* (New York: Harper Collins Pub., 1993)

mind center, (2) a support mind center, and (3) a repressed mind center. I have integrated my model of twelve personality types based on my modification of the Enneagram as developed by Don Riso[71] [72] with some neo-Jungian theories[73] and with David Kolb's classic work on experiential learning[74] to propose that there are four classes of mature individualism corresponding to four styles of experiential learning. Maturity refers to the idea that an individual has the beginnings of a well-developed dominant mind center and some development of the support mind center.

With the second law of thermodynamics integrated with the systems science theory of universal evolution as a base and as described in chapters 23, 24, and 26, I propose that creativity at all levels of organization in the universe including human creativity is an Order, Chaos, New Order process. Humans exemplify this process when an individual or institution goes into irreversible chaos, that is, the chaos cannot be overcome by negative feedback mechanisms. The loss of structure to a system, that is, the Chaos, always eventually shows possibilities that were not available when the intact structure was present. If the individual/institution mindfully endures the pain of "not knowing what to do," eventually he/she/they will see new possibilities and actuate some of them to produce a New Order that incorporates a modified Old Order. Based on the theories of David Bohm that were adopted by Peter Senge as described earlier in s chapters 28 & 29, I propose that when appropriate *requirements* are satisfied, dialogue becomes a creative process. Moreover, as described earlier, creative dialogue differentiates into a sequence of two types: (1) Vision dialogue (participatory dialogue), and 2) Collaborative dialogue among people who share the same vision.

The ideas about personality types associated with experiential learning combined with the nexus of ideas associated with creative dialogue lead to the perspective that experiential learning is better described as creative learning. In this perspective there are four styles of creative learning corresponding to the four styles of experiential learning each of which corresponds to one of the four classes of mature individualism.

Eros-chaos versus Eros-order Individualism

With regard to human creativity discussed in chapter 26, some personality types are disposed to express Eros-order whereas other personality types are disposed to express Eros-chaos. This being the case, we may designate these two categories as: *Eros-order psychetypes* and *Eros-chaos psychetypes*. Eros-order is associated with control individualism and Eros-chaos is associated with empathetic participatory individualism. For the most part, control individualism depends upon objectified conceptual knowing whereas empathetic participatory individualism depends upon objectified non-conceptual, *existential* knowing. However, all humans that reach conventional maturity start with feeling awareness with intuitions that generate metaphorical-conceptual or conceptual knowing. Differentiation of personality types, see Figure 29.1, partly results from some humans who focus on conceptual knowing while other types focus on non-

[71] D. Riso, *Personality Types (Using the Enneagram for Self-Discovery)* (Boston: Houghton Mifflin Co., 1987)

[72] D. Riso, with F. Hudson, *Personality Types* (Boston: Houghton Mifflin Co., 1996)

[73] M. Malone, *Psychotypes* (New York: Pocket Books, 1977)

[74] D. Kolb, *Experiential Learning (Experience As the Source of Learning and Development)* (Englewood Cliffs, New Jersey: Prentice-Hall, Inc., 1984

conceptual knowing. This variation in focus allows for some personality types to be intermediate between a conceptual focus and a non-conceptual, existential focus. The two personality types, 3-T & 7-T focus on empirical information which is represented by concepts, but these individuals shy away from abstract theories that are based on and "explain" empirical information. As a result, the 3-T & 7-T individuals are not cut off from concrete direct experiences of the world and therefore they tend to have an existential as well as a conceptual orientation to knowing the world. The 1-D = (ISFJ) personality type is both "conceptual-controlling" and "feeling-non-conceptual-empathetic." That is, their primary orientation is accomplishing tasks consistent with action, control individualism, but while accomplishing tasks with people, they also are able to be empathetic which is consistent with empathetic, participatory individualism. The 6-F = (ENFJ) personality type also is both "conceptual-controlling" and "feeling-non-conceptual-empathetic." However, the primary orientation of a 6-F individual is establishing and maintaining feeling relationships, but this occurs in the context of some structured community that requires its members to be practical and subordinate to order.

Eros-order Psychetypes. These personality types place a high degree of focus on conceptual knowing and therefore are totally orientated to some type of control individualism. The emphasis on conceptual knowing requires that we distinguish two types of Intellectual-Vision Control Individualism which are: 1) Conceptual-Vision Control Individualism associated with Eros-order psychetypes and 2) Existential-Vision Control Individualism associated with Eros-chaos psychetypes. With this distinction in mind the following is a list of personality types orientated to Eros-order.

1. Action Control Individualism
   a. 1-D = (ISTJ)
   b. 6-D = (ESFJ)
   c. 8-D = (ESTJ) & (ENTJ)
   d. 3-D = (ISTP)
   e. 5-T = (INTP)

2. Conceptual-Vision Control Individualism
   a. 5-T = (INTJ)
   b. 5-T = (ENTP)
   c. 9-T = (INFJ)

Eros-chaos Psychetypes. These personality types either place a high degree of focus on non-conceptual existential knowing or have some degree of both conceptual and non-conceptual, existential knowing. The individuals with a high degree of non-conceptual, existential knowing are orientated to empathetic, participatory individualism. The individuals with some focus on conceptual empirical knowing (in contrast to conceptual, theoretical knowing) are orientated to Existential-Vision Control Individualism. With this distinction in mind the following is a list of personality types orientated to Eros-chaos.

1. Empathetic, Participatory Individualism
   a. 2-F = (ENFP)
   b. 4-F = (INFP)
   c. 9-F = (ISFP)

2. Existential-Vision Control Individualism
    a. 3-T = (ESFP)
    b. 7-T = (ESTP)
    c. 6-F = (ENFJ)
    d. 1-D = (ISFJ)

Four Styles of Creative Learning

The four styles of creative learning, in turn, may be described in terms of four types of creative dialogue, which are Vision dialogue (participatory dialogue) and the sequence of three subtypes of dialogue that makeup Collaborative dialogue. The "so-called" scientific method is a special case of this sequence of four types of dialogue. In this dialogue perspective, creative learning has at least two defining characteristics:

1. It involves an Order, Chaos, New Order process, that is, Life, Death to one perspective, Endure the anguish of not knowing what to do in order to see new possibilities, and Rebirth to a New Life that incorporates the old life.
2. It involves Creative dialogue between two mind centers within an individual or between two or sometimes among three sets of personality types in an institution where the diverse personality types in each set represent the dominance of the same mind center.

I was able to correlate each of the twelve personality types in my model with the sixteen personality types in the Myers-Briggs Type Indicator (MBTI)[75] [76] Many corporations, businesses, and other institutions since 1980 have used the MBTI. Therefore, there is a great deal of data from which to infer the percent in the general population in the U.S. that represents each of the sixteen personality types of the MBTI, as shown in Figure 29.1.[77]

THE FOUR STYLES OF CREATIVE DIALOGUE

I.   *CREATIVE VISION DIALOGUE*: Create Metaphorical, Conceptual Vision.

   A.   Chaos Phase: Eros-chaos psychetypes expressing Empathetic, Participatory Individualism produce chaos by
   1   Expressing what Kolb calls *divergent thinking*.
   2   Allowing ideas from the Inner Self that includes the Individual, Collective Unconscious to be acknowledged and looked at by mind-self consciousness.
   B.   Order Phase: Eros-order psychetypes expressing Conceptual-Vision, Control Individualism
   1   Formulate vague ideas into metaphorical concepts.

---

[75] D. Keirsey, and M. Bates, *Please Understand Me (Character and Temperament Types)* (Del Mar, CA: Prometheus Nemesis Book Co., 1978)

[76] I. Briggs Myers with P. Myers, *Gifts Differing* (Palo Alto, CA: Consulting Psychologists, Inc., 1980)

[77] R. Baron, *What Type Am I (Discover Who You Really Are)* (New York: Penguin Books, 1998)

2   Organize the metaphorical concepts into a narrative or vision.

II.   *CREATIVE, COLLABORATIVE, CONCEPTUAL DIALOGUE*: Convert Metaphorical, Conceptual Vision into a Logical, Conceptual Vision

    A.   Chaos Phase: Eros-chaos psychetypes expressing Empathetic, Participatory Individualism
       1   Explore diverse implications of metaphorical concepts and the metaphorical vision.

    B.   Order Phase: Eros-order psychetypes expressing Conceptual-Vision, Control Individualism
       1   Select from the diverse metaphorical concepts those that will be represented by unambiguous, logical concepts.
       2   Convert the metaphorical vision into a logical, conceptual vision.

III.   *CREATIVE, COLLABORATIVE, STRATEGIC DIALOGUE* (first phase of *HYPOTHESIS TESTING*): Convert Logical, Conceptual, Vision into a Logical, Conceptual, Strategic Plan

    A.   Chaos Phase: Eros-order psychetypes expressing Conceptual, Vision, Control Individualism
       1   "See" diverse implications, possibilities, and problems associated with the logical, Conceptual Vision.

    B.   Order Phase: Eros-order psychetypes expressing Action, Control Individualism
       1   Select those implications and possibilities that can be operationally defined in relation to the capabilities of the individual or institution in a particular context, that is, the vision must be doable.
       2   Select those solutions to problems that can be operationally defined in relation to the capabilities of the individual or institution in a particular context.
       3   Create a hierarchy of implications, possibilities, and solutions to problems, such as, what implications or possibilities should be explored first and what solutions should be tried first; what are the backup solutions.
       4   Combine operational definitions and hierarchy into a strategic plan.

IV.   *CREATIVE COLLABORATIVE TASK DIALOGUE*: Convert Logical, Conceptual, Strategic Plan into the Concrete Events of a Task that Accomplishes the Goals of the Strategic Plan as Judged by Operationally Defined Criteria

    A.   Chaos Phase: Eros-order psychetypes expressing Action, Control Individualism
       1   "See" diverse implications and possibilities and possible problems associated with the strategic plan.

    B.   Order Phase: Eros-order psychetypes expressing Action, Control Individualism
       1   Create individual skills, functional teams with team skills, and dialogue teams that solve unforeseen, existential problems as they arise, and use this integrated creativity

to convert the logical, conceptual, strategic plan into the non-conceptual, existential set of events that achieve the goals of the strategic plan.

2   Hypothesis Testing: determine via objective criteria

    a   Whether the existential events accomplish the goals of the strategic plan.

    b   To what extent accomplishing the goals of the strategic plan is less than desired

    c   Whether the individual/institution needs a new

        (1)   Strategic plan.

        (2)   Logical, Conceptual Vision.

        (3)   Metaphorical, Conceptual Vision.

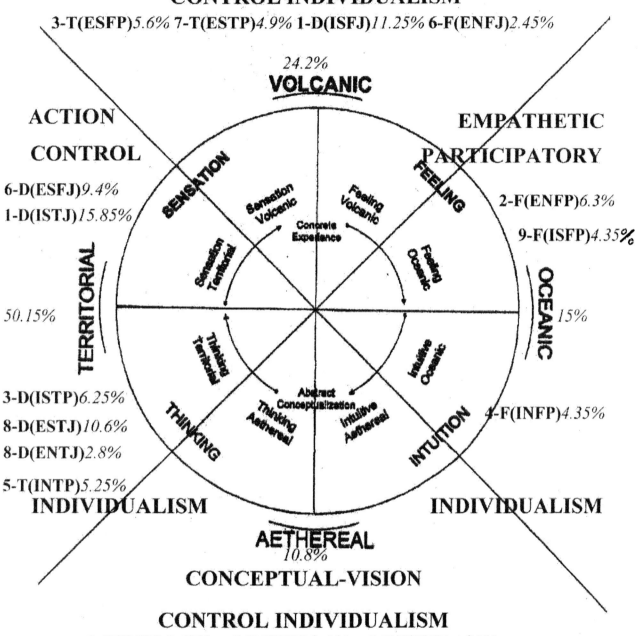

EXISTENTIAL-VISION

CONTROL INDIVIDUALISM

3-T(ESFP)*5.6%* 7-T(ESTP)*4.9%* 1-D(ISFJ)*11.25%* 6-F(ENFJ)*2.45%*

24.2%
VOLCANIC

ACTION

CONTROL

6-D(ESFJ)*9.4%*

1-D(ISTJ)*15.85%*

50.15%

3-D(ISTP)*6.25%*

8-D(ESTJ)*10.6%*

8-D(ENTJ)*2.8%*

5-T(INTP)*5.25%*

INDIVIDUALISM

EMPATHETIC

PARTICIPATORY

2-F(ENFP)*6.3%*

9-F(ISFP)*4.35%*

15%

4-F(INFP)*4.35%*

INDIVIDUALISM

AETHEREAL
10.8%

CONCEPTUAL-VISION

CONTROL INDIVIDUALISM

9-T(INFJ)*2.55%*   5-T(INTJ)*3.5%*   5-T(ENTP)*4.75%*

FIGURE 29.1 Four Types of Individualism Related to Mental Functions, Thinking versus Feeling and Intuition versus Sensation, and Related to Sixteen Personality types of the Myers-Briggs Type Indicator (MBTI)

Chapter 30
SPIRITUAL CONSTRUCTIVISM REFORMATION OF THE DECLARATION OF INDEPENDENCE

THE MIDDLE WAY BETWEEN DOGMATIC HIERARCHIES AND RADICAL EGO CONSTRUCTIVISM

The 17th and early 18th century American democracy, that is, the democratic ideal in America from 1740 to the Civil War, exemplifies an approach to the "middle way." Virtually all citizens were committed to a creator God. Based on this commitment all citizens are equal under the law, later defined by the Constitution, and have God-given Rights. One of these Rights is that no citizen will be forced to join or in any way commit to any particular religion or philosophy. This Right also implied that one may choose radical, ego constructivism. Up until the Civil War this right was limited by socialization to commitment to traditional values; so if one did choose ego constructivism, it became a conservative version of it. In this version one did not reject traditional values outright but rather adopted a modified version of them that best fit one's life situation. For example, in the 1950s many Catholics remained committed to Catholic teachings except to the one that forbids contraceptive birth control. In the 1970s some Catholics chose "right to choice" in spite of the Church's teachings that abortion is a sin equivalent to murder. Scientific constructivism from its emergence in the 1600s was independent of traditional values and by the mid-1800s market place values also became totally secular.

In the 20th century collaboration between scientific constructivism and market place values began to convert the cultural trend of conservative ego constructivism toward *radical, ego constructivism*, see chapter 20, p.226. Then radical self-fulfillment constructivism that arose in the 1960s to oppose "the military-business-science-technology complex" (see introduction) influenced a greater number of Americans to adopt and further degenerate toward radical ego constructivism. Various cultural institutions such as the communications media and education in a non-self-conscious way collaborated to intensify in the 1990s the drive toward what may be called radical, postmodern democratization. In everyday language *radical democratization* is that people do whatever they want to do without any external constraints. As a result, many trends indicate that American society is degenerating toward self-extinction or toward a bifurcation point. One of the major trends is the continued deterioration of the education system which incorporates the worst aspects of two opposing perspectives. On the one hand, a hierarchal, dogmatic education system converts the many aspects of postmodern democratization that lead to radical ego constructivism into "truths" that, on the other hand, students are required to memorize rather than construct their own understanding of them.

Postmodern, democratic middle way is the bifurcation point toward which American society is degenerating. This degeneration "covers" a Zen strange attractor. That is, our situation is "such as it is," and humans can create a new vision – rather than discover one – that transforms American society to a new "postmodern, democratic self." I propose that this new vision is a valid middle way analogous to an idea proposed by Nagarjuna. This middle way results from a collaborative joining of transcendental skepticism with narrative, scientific constructivism. In this view of the "middle way," one chooses to progressively subordinate one's mind-self to the ultimate SOURCE, which I call SOURCE. One may remain committed to a particular religion that would designate this ultimate SOURCE as a creator God or as Brahman,

VOID, or Te (of Taoism). At the same time, all types knowing, even religious theological or religious-philosophical knowing, are interpretations of reality rather than absolutely true representations of reality. This constructivism perspective allows people with contradictory views to co-exist with one another. In this new vision, on the one hand, there are no absolutely true, dogmatic hierarchies, and, on the other hand, each individual mind-self progressively subordinates itself to a "higher power" (analogous to this idea in AA). Moreover, individual and social collectives can create *valid* hierarchies. This middle way absolutely opposes radical ego constructivism without giving up one's choice to create his/her own understanding of reality. The postmodern, democratic middle way is what I call *spiritual constructiviwm*. If this vision were incorporated into colleges and universities as a transformation of radical ego constructivism, it would lead to a "revisioning of science" and general education as summarized in chapters 23 and 24.

OVERVIEW

The three defining principles of Jefferson's document imply (according to one's theistic orientation) a Reformation Christian or a Deistic, that is, non-Christian, theistic, rational democracy. Each of these forms of democracy has built in inconsistencies – they are ambiguous – that would destroy democracy in postmodern societies.[78] I propose that the wisdom of the people of the American colonies that led to the American Revolution transcends the language used by Jefferson to represent this great enlightenment. By all means, not the final word but a more adequate representation of this enlightenment is one that incorporates the ideas of spiritual constructivism and of human creativity involving the mutuality of Eros-chaos and Eros-order. Accordingly, I restate the three definitive propositions as follows:

First proposition: *All humans are born equal because each is SOURCE manifesting individual soul that generates progressively higher levels of self-consciousness associated with corresponding higher levels of free will.*

Importance of looking inward. But w*hy believe in SOURCE or individual soul?* I ask my students each to ask himself/herself: *"Do I know that I exist?"* or *"Am I conscious of myself knowing or doing something?"* that is, *"Do I have self-consciousness?"* In my large classes ranging from 100 to 300 people, all students indicated a Yes to these questions; there were no students who volunteered to say "No." Likewise most people would answer "Yes," though I know some academic philosophers who would say they don't understand the question; or they don't know what consciousness is and therefore cannot answer these questions. This type of No answer indicates an important point to those who answer "Yes." Namely, even though one cannot say what existence is or what consciousness is, one can "see" that he/she exists and has self-consciousness. This acknowledgement can lead to an important realization, namely: my present self-awareness does not explain itself; it must come from a transcendental source. Both Eastern and Western thinkers down through the ages call this source, individual soul. Thus, acknowledging one's self-awareness leads one to be able to choose to realize that he/she has an individual soul that comes into existence at conception or some time after conception but before birth. But again individual soul does not explain itself, that is, it does not explain how or why it is. Acknowledging this existential fact enables one to choose to realize that individual soul that

---

[78] Donald Pribor. *Spiritual Constructivism: Basis for Postmodern Democracy* (Dubuque, Iowa: Kendall/Hunt Pub. Col, 2005), pp.366-372.

produces the Body Ego and then the mind-self-consciousness is a manifestation of some transcendental reality that does explain itself. Judaism and Western world religions call this Reality, God. Hindus call It Brahman, and Zen Buddhism calls It Buddha-nature or absolute Emptiness. I call It SOURCE, with the intent of remaining open to diverse theistic or non-theistic interpretations of this ultimate Reality. In relation to acknowledging subjective experiences each of us may say that each human is SOURCE manifesting individual soul that manifests two centers of human individuality, body Ego and mind-self, which expresses self-consciousness.

Three aspects of soul. The soul as generator of progressively higher levels of human individuation has three aspects, which I call *Feeling center*, *Will*, and *Collective Non-Conscious*. The soul as the Feeling center brings animal feeling consciousness to the first level of human individuation, which is expressed as a "group ego" action self. Later, the action self becomes a Body Ego expressing individual ego feeling awareness. Because soul first manifests awareness in the realm of feelings, which, in turn, are incorporated into all higher levels of human awareness, I refer to the soul as the Feeling Center.[79] Individuation of the body Ego to mind self-awareness involves the mutuality of Eros-chaos and Eros-order. With respect to manifesting these drives, I refer to soul as Will. Finally, various patterns called, in agreement with Carl Jung, *archetypes*, that guide human individuation are "contained" in the individual soul. I refer to soul as something like what Jung called the Collective Unconscious, which contains these archetypes. However, in contrast to Jung I hold that soul is not a kind of consciousness in polar relation to ego; nor is it a Self that guides human individuation. Rather, soul is a fundamentally mysterious manifestation of SOURCE and as such is not a system's self with consciousness, but is that which generates the human experience of having a "self" as a result of having a body Ego and a mind-self-consciousness. Therefore, with respect to containing archetypes that guide human individuation, I refer to soul as the Collective Non-Conscious.

Evolution to civilized self. The emergence of the first four levels of human individuation is guided by soul in union with SOURCE. The individuation from one level to another is the hierarchal mutuality of chaos and creativity. Eros-chaos drives the human individual or society to *chaotic Logos*, which leads to the emergence of new possibilities. Eros-order actuates some of these possibilities so as to incorporate the modified individual consciousness into a new, more powerful level of consciousness. As one moves to higher levels of human consciousness, he/she achieves greater knowledge representing individual experiences and correspondingly expresses greater self-determination, that is, free will. Individuation to the persona self, which is the level at which one is civilized, produces a relatively autonomous mind-self-consciousness consisting of two opposing aspects: the *objective mind-self* and the *subjective mind-self*. The objective mind-self consists of control consciousness associated with Eros-order that uses metaphorical conceptual or conceptual knowledge to accomplish tasks and control one's behavior. The subjective mind-self consists of participatory consciousness associated with Eros-chaos that uses the "seeing" of subjective experiences to be engaged with, that is, participate in aspects of the external reality and of oneself. The persona mind-self also creates the "psychic space" called *Individual Collective Unconscious* that is in polar opposition to mind-self-consciousness. The

---

[79] Actually I got the idea of calling soul the Feeling Center after reading three books by Bernadette Roberts, *What is Self* (Austin, Texas: Mary Botsform Goens, 1989); *Path to No-Self* (Albany, New York: State Univ. of New York Press, 1991); and *The Experience of No-self* (Albany, New York: State Univ. of New York Press, 1993)

objective mind-self focuses on knowledge and represses the expression of the other three lower levels of human consciousness. The subjective mind-self looks inward to access all that is stored there. As a result, the persona mind-self consists of the duality: mind-self-consciousness versus Individual, Collective Unconscious. The persona mind-self also consists of the more fundamental duality: *mind-self-consciousness versus Inner Self.* The *Inner Self* consists of all the contents of the Individual, Collective Unconscious plus soul in union with ultimate Reality represented as SOURCE.

Collaboration between mind-self and Inner self. Individuation beyond the persona mind-self results from a collaboration between mind-self-consciousness and the Inner Self. Eros-chaos brings an individual (or a society) into chaotic Logos (partial chaotic order). Sometimes this happens without one's choosing it; or a creative person, such as a creative scientist, or a person desiring to individuate, will choose by means of the subjective mind-self to be in partial chaos. Chaotic Logos always breaks open new possibilities. If one via the subjective mind-self chooses to endure the chaos, eventually he/she will see some of these new possibilities. Eros-order via the objective mind-self expresses some of these new possibilities, which then are incorporated into a new Logos, that is, a new order that includes a modified old order. Another way of saying this is that participatory subjectivity brings Life representing Logos (order) to Death representing chaotic Logos. Then control objectivity actuates some new possibilities and incorporates them into a Rebirth representing new Logos (order) that includes a modified old Logos. All individuation (and all human creativity) requires that the subjective mind-self look inward to the Inner Self, which enables one to see new possibilities made available by chaotic Logos. Then the objective mind-self actuates some of these possibilities. Thus, individuation involves collaboration between the subjective and objective mind-self. More fundamentally, this mutuality expresses a collaborative dialogue between the Inner Self and mind-self-consciousness.

Possibility of a civilized self becoming a hero and eventually returning to SOURCE. Individuation involves the individual soul seeing possibilities. But before any of these human possibilities can be actuated, soul must generate a self-consciousness that actuates some of these possibilities by defining a mind-self in a certain way, that is, one has a consciousness of an objectified self. The emergence of the persona mind-self generates the fundamental paradox of human individuation. The more individuated the mind-self is, the more power it has to express free will and the more possibilities it has to individuate to a still higher level. At the same time, the individuating mind-self becomes progressively more of a barrier to the light of SOURCE to break through mind-self-consciousness to produce a seeing of new possibilities. This is why individuation beyond the persona self is a hero battle. The potential hero must die to mind-self by enduring chaos so as to let the light of SOURCE break through. SOURCE always is there trying to break through mind-self, but the free will of the mind-self that becomes ever greater can block IT from doing so. Eventually individuation leads to the empathetic self, in which mind-self is partially subordinated to soul. Since from the very beginning soul is totally receptive to the light from SOURCE, now the human person more readily receives the "seeing" generated by SOURCE. After the full differentiation of the integral self, the mind-self is totally subordinated to soul, but still there is self-consciousness that separates the person from SOURCE. Further individuation brings the falling away of self-consciousness so that now all that remains is a Body Ego consciousness that sees that it is non-different from SOURCE.

Second proposition: *When human individuation reaches the logical, conceptual stage of mind-self-consciousness, then redeemed humans have the innate disposition, which Jefferson*

341

referred to as "inalienable rights," to participate in the creation of their individual and collective social individuation.

   Redeemed humans with respect to a secular understanding of the Christ-event. Redemption of humans comes from the three secular core spiritual ideas of the Christ-event. The Christ-event is an historical moment related to a man called Jesus. At the age of thirty he began preaching that not only was he the fulfillment of the Old Testament's promise of a redeemer for all humans; he modified the teaching of the Old Testament to be incorporated into a new vision for all humans. This enraged the patriarchal leaders of the Jews who demanded and succeeded in having the Romans crucify him. During his preaching in the last three years of his life, Jesus attracted hundreds of followers and twelve disciples. Three days after the death of Jesus, eleven of the twelve disciples and several of his followers proclaimed that Jesus had resurrected as the Christ who in some manner provided for the salvation of all humans. Jesus did not come back from the dead, but rather he transformed to a new form that lives in the Inner Self of all humans and yet, for a time until "his ascension into Heaven" was experienced by some of his followers as present in the external world. Some time after his death the followers of Jesus became energized to form a Jesus cult that evolved to the world religion known as Christianity. Then and now there are diverse interpretations of the sermons of Jesus, but the core "truth" of this religion is that the life, death, and resurrection of Jesus as the Christ provides for the salvation of all humans. This central theme of the Christ-event transcends for each and all humans the alienation between the external world and the world of the Inner Self. Then and now there are diverse interpretations of the way "salvation" and corresponding transformation occurs.

   Three secular core ideas of the Christ-event. More than his teachings, the life and death of Jesus is the way for personal salvation. I propose that there are *three, secular, interrelated ideas* of this way of salvation. The ideas are secular in that they are not exclusively grounded in any particular religion. That is, these ideas may be associated with many diverse religions, including some forms of Hinduism and Buddhism, which are explicitly atheistic or agnostic. The *first idea* is that Jesus lived in a way that may be described as *participatory control individualism.* For the most part, Jesus expressed the characteristics of participatory subjectivity. However, in situations that required aggression and control, Jesus expressed some of the characteristics of control objectivity. The *second idea* is that Jesus expressed what may be called narrative, rational knowing. This way of knowing is similar to narrative, constructivism. That is, all knowing is a metaphorical, conceptual interpretation of reality that sometimes may give meaning and purpose to one's life. Each person must be open to inspiration and insights that come from the ultimate SOURCE of all things. Each person must "see" for himself the transcendental, that is, non-ego and non-mental self meaning and purpose in each moment or at least in each situation. The mystic embraces the discipline (Yoga) that prepares him/her to be "open to the spirit" in this way. Jesus usually spoke in everyday language, often using parables. The parables required each listener to strain to understand the intended message in terms of one's personal experiences. This, of course, is very different from being told according to some literal myth or dogma what and how one should think about life. Jesus was a mystic always maintaining that all his insights and power to accomplish good works, such as the miracles he is reputed to have performed, come from "his Father in Heaven." At the same time, Jesus acknowledged the validity of practical knowing. He is reputed to have said: "Give to Caesar what is Caesar's and to God what is God's." But one's subjective metaphorical, conceptual interpretation of reality in terms of personal salvation, that is, one's ultimate meaning, is the more important kind of knowing. Jesus is reputed to have said something like: "Seek ye first the

Kingdom of Heaven"; that is, seek first the inspiration that comes from the Inner Self. This kind of knowing is the more important type; it is the knowing that is essential for one's ultimate happiness. However, it does not preclude utilitarian knowing. The analogy to this is that narrative, constructivism always can be reduced to logical, conceptual constructivism "upon appropriate demand by Caesar."

The *third idea* is that Jesus showed many times throughout his life, and with his attitude toward his physical death, that one's death can be the occasion for personal rebirth to a new higher level of life. However, there is a transcendental, existential requirement for attaining a higher life. The only way that a human, personal death can lead to rebirth to a higher level of life is that each individual must choose personal, existential faith in a higher power, which Jesus referred to as the Father in Heaven. The existential faith commitment refers to one looking inward to the Inner Self, which is equivalent to "the Father in Heaven," rather than a mental choice to believe in some dogma or to obey some hierarchal authority such as a dogmatic, dictatorial political system or church. Each human, including Jesus, only physically dies once, but one may have many "dress rehearsals" for physical death. That is, one's life situation may go into chaos, which, metaphorically speaking, is a kind of death. Then one has the opportunity to practice "how to die." The person must choose existential faith in a higher power and thereby endure the chaos. In effect by "embracing and enduring the chaos" one acknowledges that the situation is "such that it is" rather than "what it is." One does not see a whatness or meaning in the situation. One only experiences darkness, nothingness, the void, and correspondingly one experiences spiritual anguish. One is sorely tempted to despair, that is, not remain steadfast in one's existential faith in a higher power. The choice to give in to despair is a choice of the mind-self to want to be in control or at least to know: "Why is this chaos happening to me?" Alternatively, the person in despair may construct an explanation for the chaos or deny that there is a lasting chaos – "things will get better" or "this is not really happening to me." The commitment to existential faith is a choice of the "middle way" of spiritual constructivism. On the one hand, one chooses transcendental skepticism. The person says to himself: "There is no explanation that I can know for my personal chaos." A Zen person may say: "I choose no knowledge." On the other hand, one chooses to subordinate the mind-self to the Inner Self, which is non-different from SOURCE . The person says to himself/herself: "I give myself over to the suchness that the ultimate Reality presents to me." A Zen person may say: "I choose no-self."

Continued commitment to existential faith leads to existential hope. Anyone who has endured intense, personal chaos realizes that over time the insight or the occasion to create a new, higher level self eventually emerges. My interpretation of this is that one comes into relative harmony with what once was a chaotic situation. The harmony is manifested as what Eastern thinkers call Dharma. *Dharma* is the suchness behavior complementary to the suchness of any situation. Sometimes Dharma is the same as what is prescribed by a moral code. In general, Dharma is subjective and existential, just as transcendental faith, hope, and harmony are subjective and existential. The harmony of the person once in chaos leads to a transformation that, in turn, manifests Dharma. The person is reborn to a new higher level of consciousness. The rebirth may be metaphorically visualized as a self-organization involving the creation of a "spiritual energy coupler." Each human always is immersed in the flux of spiritual energy. The person who has undergone a self-transformation now is able to focus some of this energy into accomplishing Dharma. In my interpretation, existential harmony is non-different from transcendental love. Thus, every personal life, death, rebirth process is analogous to the life, death, resurrection of Jesus of the Christ-event. What drives the personal death to a rebirth is Gexistential faith that leads to existential hope that produces existential love, which manifests as Dharma. Every act of personal, heroic creativity, every personal heroic individuation is a life, death, rebirth process that is analogous to or even that participates in life, death, resurrection of Jesus in the Christ-event. Personal heroic creativity and individuation each is a coming into relative harmony with ultimate Reality.

Third proposition: *Civilized, rational humans who reject any and all representations of ultimate Reality are unredeemed but retain the possibility of redemption.*

Redeemed persons with developed spiritual constructivism have the spiritual energy derived from ultimate Reality – what I call SOURCE – to collectively decide how they are to be governed. The government resulting from this creative dialogue has power only by virtue of the creativity of dialogue among redeemed people, and that creativity is a direct manifestation of ultimate Reality. Such a dialogue group may disavow and/or break from any government that claims its authority from any source other than ultimate Reality manifesting creativity of spiritual constructivism dialogue.

Though Calvin Coolidge's speech was oriented primarily to a Reformed Christian or Deistic democracy, his concluding remarks are consistent with my reinterpretation of Jefferson's document. He states that the pioneers and those who followed them of the early American colonies:

> …were a people who came under the influence of a great spiritual development and acquired a great moral power.

> No other theory is adequate to explain or comprehend the Declaration of Independence. It is the product of the spiritual insight of the people. We live in an age [1920s] of science and of abounding accumulation of material things. These did not create our Declaration. Our Declaration created them. The things of the spirit come first. Unless we cling to that, all our material prosperity, overwhelming though it may appear, will turn to a barren scepter in our grasp. If we are to maintain the great heritage which has been bequeathed to us, we must be like-minded as the fathers who created it. We must not sink into a pagan materialism. We must cultivate the reverence which they had for the things that are holy. We must follow the spiritual and moral leadership which they showed. We must keep replenished, that they may glow with a more compelling flame, the alter fires before which they worshiped.[80]

---

[80] http://www.declaration.net/z-art-coolige-inspiration-declaration.asp